Key works to the
fauna and flora of the
British Isles and north-western Europe

The Systematics Association
Special Volume No 33

Key works to the fauna and flora of the British Isles and north-western Europe

FIFTH EDITION

Edited by

Reginald W. Sims
Formerly, Department of Zoology, British Museum (Natural History), London

Paul Freeman
Formerly, Department of Entomology, British Museum (Natural History), London

David L. Hawksworth
CAB International Mycological Institute, Kew, Surrey

Published for the SYSTEMATICS ASSOCIATION by the
CLARENDON PRESS · OXFORD
1988

Oxford University Press, Walton Street, Oxford OX2 6DP

Oxford New York Toronto
Delhi Bombay Calcutta Madras Karachi
Petaling Jaya Singapore Hong Kong Tokyo
Nairobi Dar es Salaam Cape Town
Melbourne Auckland

and associated companies in
Beirut Berlin Ibadan Nicosia

Oxford is a trade mark of Oxford University Press

Published in the United States
by Oxford University Press, New York

British Library Cataloguing in Publication Data
Key works to the fauna and flora of the
British Isles and North-western Europe.—
5th ed.—(The Systematics Association
special volume; no. 33).
1. Natural history—Europe—Bibliography
I. Sims, R.W. (Reginald William)
II. Freeman, Paul   III. Hawksworth, D.L.
IV. Systematics Association   V. Series
016.57494   Z7408.E/
ISBN 0-19-857706-0

Library of Congress Cataloging in Publication Data
Key works to the fauna and flora of the British Isles and
northwestern Europe.
(The Systematics Association special volume; no. 33)
1. Zoology—Great Britain—Bibliography.   2. Zoology—
Europe, Northern—Bibliography.   3. Animals—Identifi-
cation—Bibliography.   4. Botany—Great Britain—Biblio-
graphy.   5. Plants—Identification—Bibliography.
6. Botany—Europe, Northern—Bibliography.   I. Sims,
R. W. (Reginald William), 1926–   .   II. Freeman, Paul.
III. Hawksworth, D. L.   IV. Systematics Association.
V. Series.
Z7998.G7K49   1988   [QL255]   016.574941   87-7979
ISBN 0-19-857706-0

Typeset by Wyvern Typesetting Limited, Bristol
Printed in Great Britain
at St Edmundsbury Press, Suffolk

Preface

The primary purpose of this work is to provide references to scientific books and contributions in serial publications which can be used to identify living organisms of the British Isles and north-western Europe and the shelf seas.

This book is the fifth in a series produced by The Systematics Association, originally titled the *Bibliography of key works for the identification of the British fauna and flora*, which ran to three editions. The first, published by the Association in 1942 under the editorship of Dr John Smart, was of 500 copies and was exhausted within 10 years. The second edition, for which Dr Smart was joined by Sir George Taylor for the botanical part, was of 1000 copies published in 1953, and remained in print for the next decade or so. For the third, the editorship was placed in the hands of a team of three: Dr G. J. Kerrich for the insects and other mainly non-marine Arthropod groups and to act as senior editor, Dr N. Tebble for the other animals, and Mr R. D. Meikle for the 'botanical' part; this edition of 1500 paperback and 300 hardback copies was published in 1967. The fourth edition published in 1978, found Dr G. J. Kerrich willing to continue as senior editor and arranger of the Insecta and Chelicerata but he was joined by Mr R. W. Sims for other animal groups and by Professor D. L. Hawksworth for all botanical groups, fungi, bacteria, and viruses. This last edition marked two important changes. The first was the broadening of the scope of the contents by extending the geographical area covered to include the whole of north-western Europe and its seas. The second was that, whereas the Systematics Association itself had published the earlier editions, publication of the fourth edition was by Academic Press (London). One effect of these changes was the modification of the title to reflect the wider application of the book (an innovation continued in this present edition).

By 1985 it became apparent that a replacement work was due not only because of diminishing stock but also to take into account more recent literature. Council decided to invite one of the officers, Mr R. W. Sims, to be the senior editor. He continued as previously with editing the zoological entries and was joined by Dr P. Freeman for the Insecta while Professor D. L. Hawksworth resumed his task of marshalling the entries for all botanical groups, fungi, bacteria, and viruses.

This book incorporates the ideas of the four previous editions but reflects the progress in different directions made in the intervening years. A great deal of additional systematic work has been published: for some groups this has greatly increased the number of items cited, though in most others more comprehensive publications have become available causing the total number to be reduced. The language in which the work cited is written is normally indicated only when the language concerned is not readily evident from the title or when the title has been translated or transliterated in the case of those printed originally in a non-Roman script. Brief comments are again appended in appropriate cases. However, a major change has been introduced in the presentation of each reference to assist the user in the pursuit of a particular topic. Since newcomers to a group are seldom familiar with the names of its specialists but more interested in their works, the title of each reference is now cited first to facilitate rapid scanning for subject. The author name(s) now terminate(s) the reference instead of, as formerly, distracting the eye of a searcher when it preceded the title. Nevertheless, by employing a bolder type-face and moving the author name(s) to the right, this important item of information can be readily seen by anyone needing to search the references by this method.

Like the fourth edition, the geographical area covered by the works listed within this volume extends eastwards from the seas off the west coast of Ireland to the Franco-Swiss frontier and northwards from this line to within the Arctic Circle, i.e. the shelf seas of the north-eastern Atlantic, the British Isles, France north of 49° N, Belgium including Luxemburg, the Netherlands, the Federal German Republic, Denmark including Iceland, Sweden, and Norway to Spitzbergen.

In the preparation of this book, thought has been given to the advisory work of the librarian. Thus, a number of general guides to the fauna and flora have been included, as also have the faunal lists of marine biological laboratories. Comprehensive works should be available in the main national and municipal libraries or on loan through them, while many are quoted in the catalogues of the most prominent scientific booksellers. The shorter works, mostly published in journals, may be found in the libraries of the larger natural history societies or more specialized organizations, the larger universities, and national museums. Photocopies of most listed items can be provided through the Photocopy Service, General Library, British Museum (Natural History), Cromwell Road, London SW7 5BD.

Journal titles have been abbreviated, so far as has been found possible, in accordance with the *List of serial publications in the British Museum (Natural History) Library* (2nd edn. 1975; 3 vols).

Many persons generously contributed to the preparation of this book in addition to the compilers of the sections. For example, members of the Department of Library Services in the British Museum (Natural History) commented on and suggested titles in the preliminary General section. To all of them we are most grateful, in particular to Miss P. Gilbert, Miss J. Jeffries and Mrs N. Round. The Zoological Editor wishes especially to express his gratitude to the compilers of the sections for their contributions, namely: Mr A. Warren (Protozoa); Miss S. M. K. Stone (Porifera); Dr P. F. S. Cornelius (Cnidaria, Ctenophora); Drs R. A. Bray, D. I. Gibson, J. D. Lambshead, and H. M. Platt (Mesozoa, Acanthocephala, Gastrotricha, Kynorhyncha, Nematoda, Nematomorpha, Nemertinea, Platyhelminthes, Priapulida); Mr C. G. Hussey (Rotifera); Dr J. D. George and Mr A. I. Muir (Pogonophora, Polychaeta); Mr E. G. Easton (Oligochaeta, Hirudinea); Miss A. S. Baker, Mr P. D. Hillyard, Mr K. H. Hyatt, and Mr F. R. Wanless (Chelicerata, Myriapoda); Dr G. A. Boxshall, Miss J. P. Ellis, Drs A. A. Fincham, R. W. Ingle, and R. J. Lincoln (Crustacea); Mr P. J. Chimonides and Miss P. L. Cook (Entoprocta, Bryozoa), Dr C. H. C. Brunton (Brachiopoda); Miss A. M. Clark (Echinodermata, Phoronidea, 'Protochordata'); Mr A. C. Wheeler (Pisces); Mr A. F. Stimson (Amphibia, Reptilia); Dr C. J. O. Harrison (Aves); Miss P. Jenkins (Mammalia)—all members of the British Museum (Natural History). Gratitude for help is also due to Dr D. J. Hooper (Soil and Freshwater Nematoda) of Rothamsted Experimental Station, Harpenden; Dr P. E. Gibbs (Echiura, Sipuncula) of the Marine Biological Association of the United Kingdom, Plymouth; Mr D. Hepple (Mollusca) of the Royal Scottish Museum, Edinburgh; and, finally, Dr A. C. Pierrot-Bults (Chaetognatha) of the Zoölogisch Museum, Universiteit van Amsterdam.

In the preparation of the lists of references to the Insecta, the Entomological Editor has concentrated on those works that are strictly of use in identification, removing a number that are of a more general nature. With the agreement of the Keeper of Entomology, he has been given much assistance and advice from the staff of the Department of Entomology of the British Museum (Natural History), to whom he wishes to express his thanks for their co-operation. In particular, help in individual insect orders has been given by the following, aided where necessary by their Sectional colleagues: Mrs J. A. Marshall (Orthoptera, Phasmida, Dermaptera, Dictyoptera, Isoptera); Mr D. Hollis (Apterygota, Psocoptera, Phthiraptera, Hemiptera, Thysanoptera); Dr P. C. Barnard (Ephemeroptera, Odonata, Plecoptera, Neuroptera, Mecoptera, Trichoptera); Mr P. M. Hammond (Coleoptera); Mr K. G. V. Smith (Diptera, Siphonaptera); Mr D. J. Carter (Lepidoptera); Dr M. G. Fitton (Hymenoptera Symphyta and Parasitica); Mr B. Bolton (Hymenoptera Aculeata).

The arrangement of the Botanical Sections largely follows that of the 1978 edition of *Key works*, except that the lichen-forming fungi have now been thoroughly integrated with other nutritional groups of fungi in accordance with the seventh edition of *Ainsworth & Bisby's dictionary of the fungi* (1983).

The Botanical editor wishes to thank many advisers and contributors, particularly: Dr B. Ing (Myxomycota); Drs D. A. Reid and D. N. Pegler (Hymenomycetes and Gasteromycetes); his colleagues at the CAB International Mycological Institute (all other fungal groups); Dr D. M. John and Mr J. H. Price (Algae); Dr A. J. E. Smith (Bryophyta); Professor D. M. Moore,

Drs S. L. Jury and J. R. Akeroyd (Pteridophyta and Spermatophyta); Dr I. K. Ferguson (pollen and spore identification); Drs L. R. Hill and J. F. Bradbury (Bacteria); Dr M. Goodfellow (Actinomycetes); and Dr M. F. Murant and Professor R. E. Spier (Viruses).

R. W. S.   P. F.   D. L. H.

Contents

CONTENTS

xi

Introduction

Nomenclature

GENERAL

Biological nomenclature, 2nd edn. Arnold, London. viii + 72 pp.

(1977) **Jeffrey, C.**

An introduction to the various codes of nomenclature.

BACTERIA

International code of nomenclature of bacteria. Bacteriological code. 1976 revision. American Society for Microbiology, Washington, DC. xxxv + 180 pp.

(1976) **Lapage, S. P.** *et al.*, eds.

CULTIVATED PLANTS

International code of nomenclature of cultivated plants. *Regnum Vegetabile.* International Bureau for Plant Taxonomy and Nomenclature. Vol. 64, 32 pp.

(1969) **Gilmour, J. S. L.** *et al.*, eds.

NON-CULTIVATED PLANTS

(Fungi, Lichens, Algae, Bryophyta, Pteridophyta, and Spermatophyta)

International code of botanical nomenclature adopted by the Thirteenth International Botanical Congress, Sydney, August 1981. *Regnum Vegetabile.* International Association for Plant Taxonomy and Nomenclature, Utrecht. Vol. III, 472 pp.

(1983) **Voss, E. G.** *et al.*, eds.

VIRUSES

Classification and nomenclature of viruses. *Monographs in Virology*, No. 5. Karger, Basel. 81 pp.

(1971) **Wildy, P.**

ZOOLOGY

International code of zoological nomenclature, 3rd edn. International Trust for
  Zoological Nomenclature *with* British Museum (Natural History), London
  and California Press, Berkeley and Los Angeles. 338 pp.
                                      (1985) **Ride, W. D. L.** *et al.*, eds.

Source Books and Abstracting Services

Most abstracting services listed are available on-line.

List of abstracting and indexing services in pure and applied biology. *Biol. J.
  Linn. Soc.* 3:277–86.   (1971) **Edwards, P. I.**

Biological Abstracts. Biological Abstracts, Philadelphia.   (1926 on)
  Published every two weeks with an annual author and subject index; abstracts.

BioResearch Index. Biosciences Information Service, Philadelphia.   (1965 on)
  Complements *Biological Abstracts*.

Bulletin signalétique. Centre de Documentation du CNRS, Paris. Part 14, Fungi;
  Part 16, Animals; Part 17, Vegetable biology and physiology.   (1940 on)
  Published monthly; abstracts.

International Abstracts of Biological Sciences. Pergamon Press, Oxford. (1954 on)

Current Contents. Institute of Scientific Information, Philadelphia.   (1970 on)
  Agriculture, biology and earth sciences.

Excerpta Botanica. G. Fischer, Stuttgart. Section **A** 'Taxonomica et chorologica'
  (Systematic botany); Section **B** 'Sociologica' (Plant geography/ecology).
                                                                      (1959 on)
  Published at approximately monthly intervals; covers all plant and fungal groups;
  abstracts.

Progress in Botany. Springer, Heidelberg and New York. Formerly (before Vol.
  36) *Fortschritte der Botanik*.   (1931 on)
  Morphology, physiology, genetics, taxonomy, geobotany; annual series of reviews.

Guide to the standard floras of the world. Cambridge University Press, Cambridge.
                                      (1984) **Frodin, D. G.**
  An annotated, geographically arranged, systematic bibliography of the principal
  floras; enumerations, check-lists, and chorological atlases of different areas.

The Zoological Record. BioSciences Information Service, Philadelphia and The
Zoological Society of London. (1864 on)
> Published annually; references are listed under authors' names with subject and
> systematic indexes; titles only (no abstracts).

Entomology Abstracts. Cambridge Scientific Abstracts, Bethesda, Maryland.
(1969 on)
> Monthly abstracts; annual author and subject indexes.

Annual Review of Entomology. Annual Reviews Inc., Alto, Pennsylvania.
(1956 on)
> Commissioned reviews/surveys with full bibliographies.

CAB International (CAB International Centre, Wallingford, Oxon., UK)
issues monthly abstracting serials on: agriculture, agricultural entomology,
horticulture, forestry, veterinary science, and medical and veterinary
entomology.

Classification

Synopsis and classification of living organisms. McGraw-Hill, New York. Vol. 1,
1166 pp, 87 pls; Vol. 2, 1232 pp, 141 pls. (1982) **Parker, S. P.**, ed.
> A comprehensive illustrated account of all higher groups down to (sub)family level
> of viruses, Monera, Plantae, and Animalia.

Animalia

General

Grzimek's animal life encyclopaedia, English edn. Van Nostrand Reinhold, New York. **1**, Lower animals, 599 pp; **2**, Insects, 541 pp; **3**, Mollusks and Echinoderms, 541 pp; **4**, Fishes I, 531 pp; **5**, Fishes II/Amphibians, 555 pp; **6**, Reptiles, 589 pp; **7**, Birds I, 579 pp; **8**, Birds II, 620 pp; **9**, Birds III, 648 pp; **10**, Mammals I, 626 pp; **11**, Mammals II, 634 pp; **12**, Mammals III, 657 pp; **13**, Mammals IV, 566 pp. (1972–75) **Grzimek, B.**, ed. An illustrated introductory series to animal classification and diversity throughout the world.

Animal life in Europe. The naturalist's reference book, English edn (P. and M. Michael). Warne, London and New York. 595 pp. (1958) **Graf, J.** Diagnoses of the main animal groups, no keys; well illustrated with colour plates.

Fauna von Deutschland. Ein Bestimmungsbuch unserer heimischen Tierwelt, 12th edn. Quelle and Meyer, Heidelberg. viii + 580 pp, 1836 figs, 10 pls. (1974) **Brohmer, P.**

Biologie des eaux souterraines, littorales et continentales. *Vie Milieu* (Suppl.) **9**, 740 pp, 254 figs and pls. (1960) **Delamare Deboutteville, C.** Especially useful as an introduction to interstitial animals.

The changing flora and fauna of Britain. Academic Press, London and New York. xiii + 461 pp. (1974) **Hawksworth, D. L.**, ed. Deals with the better known British groups of plants and animals.

The naturalised animals of the British Isles. Hutchinson, London. 600 pp. (1977) **Lever, C.** An illustrated account with maps; deals mainly with vertebrates.

Methuen handbook of colour, 3rd edn. Eyre Methuen, London. 252 pp, 30 col. pls. (1978) **Kornerup, A. and Wanscher, J. H.**

A dictionary of ecology, evolution, and systematics. Cambridge University Press, Cambridge. viii + 298 pp. (1982) **Lincoln, R. J., Boxshall, G. A. and Clark, P. F.**

MARINE

Marine life: an illustrated encyclopaedia of invertebrates in the sea. Harrap, London. 288 pp, 49 figs, 128 pls, 1300 colour photographs.
(1979) **George, J. D. and George, J. J.**
A systematic account with colour photographs of marine invertebrates.

The illustrated guide to marine life, expanded English edn. Starke, London. 356 pp. (1982) **De Haas, W. and Knorr, F.**

The Hamlyn guide to the seashore and shallow seas of Britain and Europe. Hamlyn, London. 320 pp, 49 figs, 132 pls. (1976) **Campbell, A. C.**
Includes all main groups of animals and plants.

Collins pocket guide to the sea shore. Collins, London. 272 pp, 40 pls (20 in colour), over 200 figs. (1958) **Barrett, J. H. and Yonge, C. M.**
Approaches identification more from an ecological than systematic viewpoint (Protozoa, Mesozoa, and Mammalia omitted).

The young specialists' look at marine life. Burke, London. 356 pp.
(1966) **Haas, W. de and Knorr, F.**
A useful identification guide regardless of the age of the specialist.

The seashore and shallow seas of Britain and Europe. Country Life Books, London. 320 pp. (1976) **Campbell, A. C.**
A well-illustrated identification guide with colour drawings.

The encyclopedia of underwater life. Allen and Unwin, London. 287 pp, 350 colour photographs. (1985) **Banister, K. and Campbell, A.**, eds.
Covers many common British species with useful ecological notes.

Secrets of the seashore. Reader's Digest, London. 192 pp. (1985) **Anon.**
Deceiving title—an informative, well-illustrated work on the British seashore with contributions from 25 experts.

The littoral fauna of Great Britain: a handbook for collectors, 4th edn. Cambridge University Press, Cambridge. xviii + 306 pp, 32 pls. (1967) **Eales, N. B.**
Contains comprehensive introductions and keys for the specific identification of the members of most groups.

Marine plankton. A practical guide, 5th edn. Hutchinson Educational, London. 244 pp, 56 pls. (1977) **Newell, G. E. and Newell, R. C.**
An introduction to the study of plant and animal marine plankton with descriptions of major groups and common species.

Fiches d'identification du zooplankton. Conseil Permanent International pour l'Exploration de la Mer, Copenhagen. (1949 on)
A series of sheets each with keys and diagrams for the identification of a group of marine zooplankton.

An introduction to the marine fauna of the British Isles. Oxford University Press, Oxford. (In press) **Ryland, J. S. and Hayward, P. J.**, eds.
A sound, well-illustrated guide to all British benthic coastal species and fish.

Photographic identification guides to the marine flora and fauna of the British Isles. Marine Conservation Society, Ross-on-Wye. (1979 on)

A series of unique identification guides to many groups of organisms for both specialist and student, comprising colour photographs of each species with text.

LAND AND FRESHWATER

Freshwater biology, 2nd edn (revised Edmunson, W. T.). Wiley, New York. 1248 pp.   (1969) **Ward, M. B. and Whipple, G. E.**

Biologische und morphologische Untersuchungen über Wasser-under Sumpfgewachse. Fischer, Jena. 4 vols.   (1905–24) **Glueck, H.**

A guide to freshwater invertebrate animals. Longmans Green, London. x + 118 pp.
                                                                 (1959) **Macan, T. T.**

Freshwater life. Warne, London. 288 pp.   (1974) **Clegg, J.**

Animal life in freshwater. A guide to British freshwater invertebrates, 6th edn. Methuen, London. viii + 308 pp, 211 figs.   (1963) **Mellanby, H.**

A key to the major groups of British free-living terrestrial invertebrates. Blackwell Scientific Publications, Oxford and Edinburgh. 124 pp, 4 figs.
                                                  (1967) **Smith, K. P. and Whittaker, J. B.**

The pocket encyclopaedia of plant galls in colour, revised edn. Blandford Press, Poole. 191 pp, 10 figs, 292 pls.   (1975) **Darlington, A.**

Faunal Lists

Marine fauna of the Isle of Man. Liverpool University Press, Liverpool. 307 pp.
                                         (1963) **Bruce, J. R., Colman, J. S., and Jones, N. S.**, eds.

Dale Fort marine fauna, 2nd edn. *Fld Stud.* **2** suppl., xxiv + 169, 6 maps.
                                                              (1966) **Crothers, J. H.**, ed.

Invertebrate fauna of the Severn Estuary. *Fld Stud.* **4**: 477–554.
                           (1977) **Boyden, C. R., Crothers, J. H., Little, C., and Mettam, C.**

The marine fauna of Lundy. *Rep. Lundy Fld Soc.* **20** on.   (1974 on)

Seashore life in Cornwall and the Isles of Scilly. Bradford Barton, Truro. 124 pp.
                                                                 (1971) **Turk, S. M.**

Plymouth marine fauna, 3rd edn. The Marine Biological Association of the UK, Plymouth. 457 pp.   (1957) **Russell, F. S.** *et al.*

The ecology of Chichester harbour, S. England, with special reference to some fouling species. *Int. Revue ges. Hydrobiol.* **49**: 233–79.
                                          (1964) **Stubbings, H. G. and Houghton, D. R.**
Contains comprehensive lists of the fauna and flora.

The macrobenthos of Chalk and Greensand shores of southeastern England. (A report for the Nature Conservancy Council). British Museum (Natural History), London. 117 pp + 37 pp (Appendices).
                                          (1986) **Fincham, A. A. and George, J. D.** (Fauna)

Tittley, I. and Price, J. H. (Flora)
Comprehensive lists of animals and plants living on the rocky shores of Kent and
Sussex with extensive ecological notes.

The marine fauna of Whitstable. *Ann. Mag. nat. Hist.* **7** (Ser. 12): 321–50.
(1954) **Newell, G. E.**

Additions to the marine fauna of Whitstable. *Ann. Mag. nat. Hist.* **9** (Ser. 12):
481–96.   (1966) **Maghraby, H. M. and Perkins, E. J.**

The marine fauna of the Blackwater estuary and adjacent waters, Essex. *Essex
Nat.* **32**: 1–61.   (1967) **Davis, D. S.**

The Victoria history of the counties of England. Archibald Constable, London.
(1901–1926) **Page, W.**, ed.
Multivolume series from the early part of the twentieth century providing valuable
reference sources for records of marine fauna and flora.

Fauna and flora of St. Andrews Bay. Scottish Academic Press, Edinburgh and
London. xii + 310 pp.   (1974) **Laverack, M. S. and Blackler, M.**

Inventaire de la faune marine de Roscoff. Station Biologique, Roscoff.   (1951 on)
Numerous volumes, each dealing with a major group of marine animals occurring on
the coast of Brittany.

PROTOZOA

General

How to know the Protozoa, 2nd edn. Brown, Iowa. 234 pp, 394 figs.
(1979) **Jahn, T. L., Bovee, E. C., and Jahn, F. F.**

Traité de zoologie. Masson et Cie, Paris. **1**(1), 1071 pp; **1**(2), 1160 pp, 833 figs.
(1952) **Grassé, P. P.**, ed.
General textbook, systematic description of Protozoa except ciliates.

An introduction to the study of Protozoa. Clarendon Press, Oxford. 506 pp, 180 figs..
(1961) **Mackinnon, D. L. and Hawes, R. S. J.**

Protozoology, 5th edn. Thomas, Springfield, Illinois. 1174 pp, 388 figs.
(1966) **Kudo, R. R.**
A general textbook with keys to families and descriptions of many species.

Protozoology. Springer, Berlin. 554 pp, 443 figs.   (1973) **Grell, K. G.**

Protozoa. Blackie and Son, Glasgow. 325 pp, 157 figs.   (1976) **Westphal, A.**
A general textbook with systematic descriptions to orders.

A newly revised classification of the protozoa. *J. Protozool.* **27**: 37–58.
(1980) **Levine, N. D.** *et al.*
Classification to suborder with taxonomic characters.

The Protozoa. Introduction to protozoology. Mosby, St. Louis. 732 pp, 600 figs.
(1980) **Farmer, J. N.**

A general textbook with keys to genera and higher categories and descriptions of many species.

An illustrated guide to the protozoa. Society of Protozoologists, Lawrence, Kansas. 629 pp, 3000 figs. (1985) **Lee, J. J., Hunter, S. H., and Bovee, E. C.**, eds. Descriptions and keys to orders and genera.

Protozoology. A manual for medical men, veterinarians, and zoologists. Baillière, Tindall, and Cox, London. 2 vols, 1563 pp, 565 figs, 20 col. pls.
(1926) **Weynon, C. M.**

Dofleins Lehrbuch der Protozoen-Kunde, 6th edn. G. Fischer, Jena. 2 vols, 776 pp, 762 figs. (1949–52) **Reichenow, E.**

Fauna von Deutschland: ein Bestimmungsbuch unserer heimischen Tierwelt. Quelle and Meyer, Heidelberg. 582 pp. (1982) **Brohmer, P.** Keys to genera and species.

Protozoa of underground waters in caves. *Annls Spéléol.* **25**: 91–106, 5 figs.
(1970) **Gittleson, S. and Hoover, R. L.**

Free-living

SARCOMASTIGOPHORA

MASTIGOPHORA

For further references, see under Algae on p. 271.

Die Süsswasser-Flora Deutschlands, Österreichs und der Schweiz. G. Fischer, Jena.
(1913–27) **Pascher, A.**, ed.

Containing:

Flagellatae I. Pantosotomatinae, Protomastiginae, Distomatinae. 138 pp, 252 figs. (1914) **Pascher, A. and Lemmermann, E.**

Flagellatae II. Chrysomonadinae, Cryptomonadinae, Eugleninae, Chloromonadinae. 192 pp, 398 figs. (1913) **Pascher, A. and Lemmermann, E.**

Dinoflagellatae (Peridineae). 66 pp, 69 figs. (1913) **Schilling, A. J.**

Volvocales = Phytomonadinae. Flagellatae IV = Chlorophyceae I. 506 pp, 451 figs.
(1927) **Pascher, A.**

Morphologie in der Teil in Bildern. Teil I Flagellaten. Berlin 106 pp, figs.
(1921) **Kühn, A.**
Keys to families and genera.

Les algues d'eau douce. Initiation à la systématique. III Les algues bleues et rouges. Les Eugleniens, Peridiniens et Cryptomonadines. Bourbée et Cie, Paris. 512 pp, 137 pls. (1970) **Bourrelly, P.**

A guide to algal keys (excluding seaweeds). *Br. phycol. J.* **11**: 49–55.
(1976) **George, E. A.**

Check-list of British marine algae—third revision. *J. mar. biol. Ass. UK* **56**: 527–94.   (1976) **Parke, M. and Dixon, P. S.**

A note and key to the orders of Phytomastigophorea Calkins 1909. *Marathwada Univ. J. Sci. (Biol. Sci.)* **16**: 87–8.   (1977) **Kalyankar, S. D.**

The palaeobiology of plant protists. Freeman, San Francisco. 1028 pp.
(1980) **Tappan, H.**

The free-living unarmoured Dinoflagellata. *Mem. Univ. Calif.* **5**: 1–562, 388 figs.   (1921) **Kofoid, C. A. and Swezy, O.**

The dinoflagellates of northern seas. Marine Biological Association, Plymouth. vi + 250 pp, 53 figs, 35 pls.   (1925) **Lebour, M. V.**

Dinoflagellatae. *Rabenh. Krypt.-Fl.* **10**(2,i) 617 pp; **10**(2,ii) 590 pp.
(1931–33, 1935–37) **Schiller, J.**

Taxonomie des phytoplanktons einiger Seen in Uppland, Schweden. *Symb. biol. upsal.* **9**(3): 399 pp, 39 figs.   (1948) **Skuja, H.**

Taxonomische und Biologische Studien über das Phytoplankton schwedischer Binnengewässer. *Nova Acta R. Soc. Scient. upsal.* (IV) **16**, 404 pp, 63 pls.
(1956) **Skuja, H.**

Phytoplankton of the Gulf of Mexico. Taxonomy of calcareous nannoplankton. *Geosci. Man* **20**: 1–62.   (1979) **Pierce, R. W. and Hart, G. F.**

Systematik und Ökologie der farblosen Flagellaten des Abwassers. *Arch. Protistenk.* **121**: 73–137.   (1979) **Hänel, K.**
Keys to species.

A revision of the *Diplopsalis* group of dinoflagellates (Dinophyceae) based on material from the British Isles. *Bot. J. Linn. Soc.* **82**: 15–26.
(1981) **Dodge, J. D. and Hermes, H.**

Marine dinoflagellates of the British Isles. HMSO, London. 303 pp, 35 figs, 8 pls.
(1982) **Dodge, J. D.**
Keys to genera and species.

Dinoflagellate taxonomy. In *Dinoflagellates* (ed. D. L. Spector), pp. 17–42, 12 figs. Academic Press, Orlando, Florida.   (1984) **Dodge, J. D.**

Monographie du genre *Trachelomonas* Ehr. *Rev. gén. Bot.* **38**: 358–380, 449–469, 518–528, 580–592, 646–658, 687–706; **39**: 26–51, 73–98, 8 figs, 15 pls, 1 graph.   (1926–27) **Deflandre, M. G.**

A treatise on the British freshwater algae. Cambridge University Press, Cambridge. xviii + 534 pp, 207 figs.   (1927) **West, G. S. and Fritsch, F. E.**
Keys to families and genera with notes on the commoner British species of all but the exclusively holozoic classes.

Étude systématique du genre *Lepocinclis* Perty. *Mém. Mus. r. Hist. nat. Belg.* **1**(2): 1–84, 84 figs.   (1935) **Conrad, W.**
Key to species.

Synopsis der gattung *Phacus*. *Arch. Protistenk.* **95**: 81–252, figs.
(1942) **Pochmann, A.**
Key to species.

Contributions to our knowledge of saprophytic algae and flagellata. *New Phytol.* **41**: 171–205, figs. (1942) **Pringsheim, E. G.**
Astasia key to species.

A comparative study of the species of *Volvox. Trans. Am. microsc. Soc.* **63**: 265–310. (1944) **Smith, G. M.**
Descriptions with keys to species.

Waste treatment Protozoa: Flagellata. Florida Engineering Ser. No. 3, Gainsville, Florida. 140 pp, 149 figs. (1962) **Calaway, W. T. and Lackey, J. B.**
Keys to species.

Les choanoflagellés des côtes de la Manche. 1 Systématique. *Bull. Soc. linn. Normandie* **7**: 191–209, 28 figs. (1966) **Boucad-Camon, E.**

Euglenoid flagellates. Prentice-Hall, Englewood Cliffs, New Jersey. 242 pp, 176 figs. (1967) **Leedale, G. F.**
Descriptions and keys to orders, genera, and species.

Die Gattung *Chloromonas* Gobi emend. Wille (*Chlamydomonas* und die nachstrerwandten Gattungen 1). *Beh. Nova Hedwigia* **34**: 1–283, 223 figs, 25 pls.
(1970) **Ettl, H.**

[Cryptogamic plants of the USSR.] Vol. 7 *Silicoflagellatophyceae.* [Russian]. Translated by the Israel Program for Scientific Translations, Jerusalem. **5698**, 363 pp, 58 figs. (1970) **Glezer, Z. I.**

Coccolithen Kakige Plankton seit Jahrmillionen. Ziemsen, Wittenburg Lutherstadt. 99 pp, 183 figs. (1972) **Reinhardt, P.**

SARCODINA (RHIZOPODA)

The Protozoa Sarcodina. Oliver and Boyd, Edinburgh and London. vi + 183 pp, 80 figs. (1956) **Jeeps, M. W.**

Fresh-water rhizopods of North America. *Rep. US geol. geogr. Surv. Territ.* **12**: 324 pp, 48 pls. (1879) **Leidy, J.**

Faune Rhizopodique du Bassin de Leman. Kündig, Geneva. 714 pp, figs and pls.
(1902) **Penard, E.**

The British freshwater Rhizopoda and Heliozoa I–V. Ray Society, London. xiv + 674 pp, 176 figs, 74 pls.
(1905–21) **Cash, J., Wailes, C. H. and Hopkinson, J.**

Rhizopoden. In *Morphologie der Tiere in Bilden.* Berlin. Vol. 2, pp. 107–272, 206 figs. (1926) **Kühn, A.**

Taxonomy of the amoebas, with descriptions of thirty-nine new marine and freshwater species. *Pap. Dep. mar. Biol. Carnegie Instn Wash.* **24**: 1–116, 35 figs, 12 pls. (1926) **Schaeffer, A. A.**

Urtiere, Protozoa Wurzelfüssler, Rhizopoda Sonnenterchen, Heliozoa. *Tierwelt Dtl.* **56**: 1–176, 86 figs.   (1968) **Rainer, H.**

The Mycetozoa: a revised classification. *Bot. Rev. Lond.* **36**: 59–89, 6 figs.
(1969) **Olive, L. S.**

The lobose amoebas. 1 A key to the suborder Conopodina Bovee and Jahn, 1966 and descriptions of thirteen new or little known *Mayorella* species. *Arch. Protistenk.* **112**: 178–227, 13 figs.   (1970) **Bovee, E. C.**

Studies on pathogenic and non-pathogenic small free-living amoebae and the bearing of nuclear division on the classification of the order Amoebida. *Phil. Trans. R. Soc.* **B259**: 435–476, 12 pls.   (1970) **Singh, B. N. and Das, S. R.**

Einführung in die Kleinlebewelt. Wechseltierchen (Rhizopoden). Keller, Stuttgart. 80 pp, 73 figs.   (1972) **Grospietsch, T. G.**

Taxonomy and phylogeny. In *The biology of Amoeba* (ed. K. W. Joon), pp. 37–82. Academic Press, New York and London.
(1973) **Bovee, E. C. and Jahn, T. L.**

An illustrated key to freshwater and soil amoebae. *Scient. Publs Freshwat. biol. Ass.* **34**, 155 pp, 64 figs.   (1976) **Page, F. C.**

The mycetozoans. Academic Press, New York. 293 pp.   (1978) **Olive, L. S.**
Keys to families and genera.

Marine flora and fauna of the northeastern United States. Protozoa: Sarcodina: Amoebae. *NOAA Technical Report NMFS Circ-419.* 56 pp.
(1979) **Bovee, E. C. and Sawyer, T. K.**
Keys to families, genera, and species.

The genera of Myxomycetes. University of Iowa, Iowa. 102 pp.
(1983) **Martin, G. W., Alexopoulos, C. J. and Far, M. L.**

Marine Gymnamoebae. Institute of Terrestrial Ecology, Cambridge. 54 pp, 172 figs.   (1983) **Page, F. C.**
Descriptions and keys to genera and species.

Le genre *Arcella* Ehrenberg. Morphologie-Biologie. Essai phylogénétique et systématique. *Arch. Protistenk.* **64**: 152–287, 403 figs. (1928) **Deflandre, G.**

Étude monographique sur le genre *Nebela* Leidy (Rhizopoda-Testacea). *Annls Protist.* **5**: 201–86, x–xxvii, 161 figs and pls.   (1936) **Deflandre, G.**

Beiträge zu Rhizopoden fauna Deutschlands. 1 Thekamoeben des Rhoen. *Hydrobiologia* **10**: 305–22.   (1958) **Grospietsch, T. G.**

Faune terrestre et d'eau douce des Pyrénées-Orientales. Fasc. 5 *Thécamoebiens du Sol.* Hermann, Paris. 103 pp, 182 figs.   (1960) **Bonnet, L. and Thomas, R.**

The distribution of testate amoebae in some fens and bogs in northern England. *J. Linn. Soc. Zool.* **44**: 369–82.   (1961) **Heal, O. W.**

Amoebida testacea (Rhizopoda). *Zoology Iceland* **2**(1): 1–58, 40 figs.
(1965) **Decloitre, L.**

An illustrated introduction to the testate rhizopods in *Sphagnum*, with special reference to the area around Malham Tarn, Yorkshire. *Fld Stud.* **3**: 801–38, 11 figs, 2 pls.   (1973) **Corbet, S. A.**

An atlas of freshwater testate amoebae. Oxford University Press, Oxford. 222 pp, 95 pls.   (1980) **Ogden, C. G. and Hedley, R. H.**
Descriptions and scanning electron micrographs of many common British species.

Report on the Foraminifera dredged by HMS *Challenger* during the years 1873–76. *Rep. scient. Results Voy. Challenger Zoology* **9** (in 2 vols.), 814 pp, 115 pls.   (1884) **Brady, H. B.**

On the recent and fossil Foraminifera of the shore-sands of Selsey Bill, Sussex. *Jl R. microsc. Soc.* **1908**: 529–43, 1 pl; **1909**: 306–36, 422–46, 677–98, 6 pls; **1910**: 401–26, 693–5, 6 pls; **1911**: 298–343, 5 pls, 436–48.
                                              (1908–11) **Heron-Allen, E. and Earland, A.**

Foraminifera. Clare Island Survey, no 64. *Proc. R. Ir. Acad.* **31**: 1–188, 13 pls.
                                              (1913) **Heron-Allen, E. and Earland, A.**
Contains a full list of references up to 1912.

Foraminifera of the shore-sands and shallow water zone of the south coast of Cornwall. *Jl R. microsc. Soc.* **1916**: 29–55, 5 pls.
                                              (1916) **Heron-Allen, E. and Earland, A.**

The Foraminifera of the west of Scotland. *Trans. Linn. Soc. Lond., Zool.* **11**: 197–299, 5 pls.   (1916) **Heron-Allen, E. and Earland, A.**

The Foraminifera of the Atlantic Ocean. *Bull. US natn. Mus.* **104** (1–8), 866 pp, 199 pls.   (1918–31) **Cushman, J. A.**

The Foraminifera of the Plymouth region. *Jl R. microsc. Soc.* **50**: 6–84, 3 pls.
                                              (1930) **Heron-Allen, E. and Earland, A.**

Catalogue of Foraminifera. American Museum of Natural History, New York. 59 vols, index, and bibliography.
                                              (1940 *et seq.*) **Ellis, B. F. and Messina, A. R.**
This publication includes the original descriptions and figures of all new species of Foraminifera.

Principles of micropalaeontology. Melbourne and Oxford University Presses. xvi + 296 pp, 64 figs, 14 pls, tables.   (1945) **Glaessner, M. F.**

Foraminifera. *Zoology Iceland* **2**(2): 1–79, 14 figs.   (1945) **Norvang, A.**

Foraminifera of the Gullmar Fjord and the Skaggerak. *Zool. Bidr. Upps.* **26**: 1–328, 312 figs, 32 pls.   (1947) **Hoglund, H.**
Maps, tables, diagrams, figures, and descriptions of many species likely to be found along the British coasts.

Taxonomic notes on the species figured by H. B. Brady in his report on the Foraminifera dredged by HMS *Challenger* during the years 1873–1876. Accompanied by a reproduction of Brady's plates. *Am. Ass. Petrol. Geol. Publ.* **9**: 1–238, 115 pls.   (1960) **Barker, R. W.**

Ecology and distribution of recent Foraminifera. Johns Hopkins University, Baltimore, Maryland. viii + 297 pp, 82 figs, 6 pls.   (1960) **Phleger, F. B.**

An index to the genera and species of the Foraminifera, 1890–1950. Stanford University Press. 393 pp.   (1961) **George Vanderbilt Foundation**

Principles of zoological micropalaeontology. Pergamon, Oxford. Vol. 1, 652 pp, 548 figs. [Translated from the German by Allen, K. A.] (ed. Neale, J. W.).

(1963) **Pokorny, V.**

Treatise on invertebrate palaeontology. Part C. Protistsa 2. The Geological Society of America, New York and University of Kansas Press, Lawrence, Kansas. **1**, xxxi + 1–50 pp. *Sarcodina. Chiefly Thecamoebans*; **2**, 511–900 pp. *Foraminiferida*.

(1964) **Loeblich, A. R. and Tappan, H.**

On the Foraminiferida of the Plymouth region. *J. mar. biol. Ass. UK* **45**: 481–505.   (1965) **Murray, J. W.**

Foraminifera. Families Globigerinidae and Globorotalidae. *Fich. Ident. Zooplancton* **108**: 2–3, figs.   (1976) **Be, A. W. H.**

Atlas de foraminifères planctoniques du Crétacé moyen (Mer Boréale et Téthys). *Cah. Micropaleont.* **1979**: 7–185.

(1979) **Anon. (European Working Group on Planktonic Foraminifera)**

British nearshore foraminiferids. *Synopses Br. Fauna* (NS) **16**: 68 pp.

(1979) **Murray, J. W.**

Keys to genera.

Foraminifera. Macmillan, London. 433 pp.   (1981) **Haynes, J. R.**

Systematic index of recent and Pleistocene planktonic Foraminifera. University of Tokyo, Tokyo. 190 pp. (1981) **Saito, T., Thompson, P. R. and Breger, D.**
Keys to genera and species.

SARCODINA (ACTINOPODA)

Les héliozoaires d'eau douce. Kündig, Geneva. 341 pp.   (1904) **Penard, E.**
Descriptions and keys to orders, genera, and species.

Zoetwaterrhizopoden en Heliozoën. *Fauna Ned.* **9**: 303 pp. [Dutch].

(1940) **Hoogenraad, H. R. and Groot, A. A. de**

Comparative morphological study of radiolarians: foundations of a new taxonomy. *Zool. Zh.* **55**: 485–96. [Russian].

(1976) **Petrushevskaya, M. G., Kashon, Z., and Kashon, M.**

Acanthariea, Protozoa of the world oceans. *Fauna SSSR* **123**: 1–223. [Russian].

(1981) **Reshetnyak, V. V.**

Keys to species and genera.

Revision of the taxonomy of the Heliozoa with attention to electron microscopical criteria. *Annls Inst. oceanogr., Paris* **58** (Suppl.): 173–8.

(1982) **Febvre-Chevalier, C.**

Keys to genera.

Radiolaria. Springer, New York. 355 pp.   (1983) **Anderson, O. R.**
Taxonomic descriptions of genera.

Nordisches Plankton. Kiel and Leipzig.   (1901–29) **Brandt, K. A.,** ed.

Containing:

Die nordischen Tripyleen-Arten. **15**: 1–52, 56 figs.   (1901) **Borgert, A.**

Die nordischen Acantharien. Teil I: Acanthometriden. Teil II: Acanthophracten. **16**: 53–90, 29 figs.   (1905 and 1907) **Popofsky, A.**

Die nordischen Supmellarian. Teil I: Unterlegion Sphaerocollida, **16**: 91–120, 20 figs. Teil II: Unterlegion Sphaerellaria, **17**: 1–66, 37 figs.
    (1929 and 1909) **Schröder, O.**

Die nordischen Nassellarien. **17**: 67–146, 124 figs.   (1914) **Schröder, O.**

Ergebn. der Plankton-Expedition der Humboldt-Stiftung. Vol. 3. Kiel and Leipzig.
    (1904–1926) **Hensen, V. A. C.**, ed.

Containing:

Die Polycystinen der Plankton-Expedition. (Lde): 1–104, 4 figs, 3 pls.
    (1913) **Dreyer, F.**

Die Tripyleen-Familie der Aulacanthiden. (Lh1): 1–92, 8 pls.
    (1904) **Immerman, F.**

Die Tripyleen Radiolarien der Plankton-Expedition. Tuscaroridae. (Lh2): 93–113, 2 figs, 1 pl.   (1905) **Borgert, A.**

Atlanticellidae. (Lh3): 117–28, 1 pl.   (1905) **Borgert, A.**

Medusettidae. (Lh4): 133–92, 4 pls.   (1906) **Borgert, A.**

Concharidae. (Lh5): 195–232, 3 pls.   (1907) **Borgert, A.**

Castanellidae. (Lh6): 235–79, 4 pls.   (1908) **Schmidt, W. J.**

Phaeodinidae, Caementellidae, und Cannorrhaphidae. (Lh7): 283–316, 2 pls.   (1909) **Borgert, A.**

Circopordiae. (Lh8): 319–52, 3 pls.   (1909) **Borgert, A.**

Cannosphaeridae. (Lh9): 355–80, 2 pls.   (1909) **Borgert, A.**

Porospathidae und Cadiidae. (Lh10): 383–415, 2 pls.   (1910) **Borgert, A.**

Challengeridae. (Lh11): 419–536, 22 figs, 5 pls.   (1911) **Borgert, A.**

Atlanticellidae. II Teil. (Lh12): 539–610, 22 figs, 8 pls. (1913) **Borgert, A.**

Die Tripyleen Radiolarien der Plankton-Expedition. Coelodendridae (einschliesslich Coelographidae Haeckel). (Lh13): 1–101, 53 figs, 6 pls.
    (1926) **Popofsky, A.**

LABYRINTHOMORPHA

Labyrinthula. J. Protozool. **14**: 697–708.   (1967) **Pokorny, K. S.**

CILIOPHORA

Current status of the international collection of ciliate type-specimens and guide lines for future contributors. *Trans. Am. microsc. Soc.* **91**: 221–35.
    (1972) **Corliss, J. O.**

Proposition d'une classification du phylum Ciliophora Doflein, 1901 (Réunion
de Systématique, Clermont-Ferrand). *C. r. hebd. Séanc. Acad. Sci., Paris* **278**:
2799–802.   (1974) **Puytorac, P. de, Batisse, A., and Bohatier, J.**
The ciliated Protozoa: characterisation, classification, and guide to the literature, 2nd
edn. Pergamon, New York and Oxford. 455 pp, 37 pls
                                             (1979) **Corliss, J. O.**
Systematic descriptions of orders and families with classification of genera.

British and other freshwater ciliated Protoza, Part I Ciliophora: Kinetofragminophora. *Synopses Br. Fauna* (NS) **22**: 387 pp.   (1982) **Curds, C. R.**
Systematic descriptions and keys to genera.

British and other freshwater ciliated Protozoa. Part II Ciliophora: Oligo-
hymenophora and Polyhymenophora. *Synopses Br. Fauna* (NS) **23**: 474 pp.
        (1983) **Curds, C. R., Gates, M. A., and Roberts, D. McL.**
Keys and systematic descriptions of genera.

Étude monographique sue le groups des infusoires tentaculières. *Annls Soc.*
belge Microsc. **24**: 57–189; **25**: 7–205; **26**: 14–119.   (1899–1901) **Sand, R.**

Étude monographique sur les Acinétiens. I. Recherches expérimentales sur
l'étendue des variations et les facteurs tératogènes. *Arch. Zool. exp. gen.* **8**(5):
421–97, 29 figs, 2 pls.   (1911) **Collin, B.**
Études sur les infusoires d'eau douce. Georg, Geneva. 331 pp, 301 figs.
                                             (1921) **Penard, E.**

Tintinnidae. *Tierwelt N.-u. Ostsee* **2c**: 1–66, 33 figs.   (1927) **Jörgensen, E.**

Keys for the determination of the holotrichous ciliates. [Translated from the
Russian by Cecil A. Hoare.] *Proc. zool. Soc. Lond.* **1927**: 399–481.
                                             (1927) **Schewaikoff, V. T.**

Urtier oder Protozoa. I: Wimpertiere oder Ciliate (Infusoria). *Tierwelt Dtl.* **18,**
21, 25, 30: 886 pp, 3407 figs.   (1930–35) **Kahl, A.**
Keys to genera and species. Descriptions and figures of all known species of non-
parasitic ciliates except marine Tintinnidae.

Suctoria. *Tierwelt N.-u. Ostsee* **2c**: 184–226, 228 figs.   (1934) **Kahl, A.**

A study of spiraling in the ciliate *Frontonia* with a review of the genus and a
description of two new species. *Arch. Protistenk.* **92**: 10–66.
                                             (1939) **Bullington, W. E.**

The biology of Stentor. Pergamon, London. Keys to species in Ch. 18, 333–8 pp, 1
fig.   (1961) **Tartar, V.**

An illustrated key to the British freshwater ciliated Protozoa commonly found
in activated sludge. *Wat. Poll. Res. Tech. Pap.* No. 12: 89 pp, 89 figs. HMSO,
London.   (1969) **Curds, C. R.**

Marine ciliata from Plymouth: Peritricha, Vaginicolidae. *J. mar. biol. Ass. UK*
45: 229–39.   (1965) **Felinska, M.**

Order: Tintinnida. *Fich. Ident. Zooplancton* **117–27**: 70 pp, 263 figs.
                                             (1969) **Marshall, S. M.**

Orde Thigmotrichida (Ciliatea-Holotricha) 4. Familia Thigmophryidae. *Acta Protozool.* **9**: 121–70, 11 figs. (1971) **Raabe, Z.**

Szajkoszorous csillosok-peritricha. *Fauna Hung.* **105**: 1–245, 148 figs. [Hungarian]. (1971) **Stiller, J.**
Keys to species.

Ciliated Protozoa. An illustrated guide to the species used as biological indicators in freshwater biology. WHO, Geneva. 198 pp, 94 figs. (1972) **Bick, A. C.**

Blepharisma. The biology of a light-sensitive protozoan. Stanford University, Stanford. (1973) **Giese, A. C.**
Includes keys to species. Ch. 12, pp 304–32, classification, distribution, and evolution.

History, taxonomy, and evolution of species of *Tetrahymena*. In *Biology of Tetrahymena* (ed. A. M. Elliott), pp. 1–55, 46 figs. Dowden, Hutchinson, and Ross, Stroudsburg, Pennsylvania. (1973) **Corliss, J. O.**

Hypotrichida. *Fauna Hung.* **10**: 1–187. [Hungarian]. (1974) **Stiller, J.**
Keys to genera and species.

Morphology, taxonomy, and general biology of the genus *Paramecium*. In *Paramecium. A current survey* (ed. W. J. Van Wagtendonk), pp. 1–89, 44 figs. Elsevier, Amsterdam. (1974) **Vivier, E.**

A guide to the species of the genus *Euplotes* (Hypotrichida, Ciliatea). *Bull. Br. Mus. nat. Hist.* (Zool.) **28**: 1–61, 58 figs. (1975) **Curds, C. R.**

Opisthonectidae (Ciliata, Peritrichida) nov. fam. und revision der genera *Telotrochidium* (Kent) und *Opisthonecta* (Fauré–Fremiet). *Protistologica* **11**: 395–414. (1975) **Foissner, W.**

Revision der genera *Astylozoon* (Engelmann) und *Hastatella* (Erlanger) (Ciliata, Natantina). *Protistologica* **13**: 353–79. (1977) **Foissner, W.**

Marine flora and fauna of the Northeastern United States. Protozoa: Ciliophora. *NOAA Technical Report NMFS Circ-378.* 62 pp. (1978) **Borror, A. C.**
Keys to families.

Taxonomie und Phylogenie der Gattung *Colpidium* (Ciliophora, Tetrahymenidae) und Neubeschreibung von *Colpidium truncatum* Stokes, 1885. *Naturk. Jb. Stadt Linz* **24**: 21–40. (1978) **Foissner, W. and Schiffmann, H.**

A guide to the species of *Aspidisca. Bull. Br. Mus. nat. Hist.* (Zool.) **36**: 1–34. (1979) **Wu, I. C. H. and Curds, C. R.**

A taxonomic revision of the peritrich ciliate genera *Thuricola* and *Pseudothuricola. Beaufortia* **30**: 125–38. (1980) **Trueba, F. J.**

A taxonomic revision of the genus *Platycola* (Ciliophora: Peritrichida). *Bull. Br. Mus. nat. Hist.* (Zool.) **43**: 95–108. (1982) **Warren, A.**

Sand dwelling ciliates of south Wales. *Cah. Biol. mar.* **26**: 187–214. (1983) **Wright, J. M.**
Key to species of *Tracheloraphis*.

A revision of the genus *Epiclintes* (Ciliophora: Hypotrichida) including a
revision of *Epiclintes felis* comb. n. *Bull. Br. Mus. nat. Hist.* (Zool.) **45**: 41–54.
                                        (1983) **Carey, P. G. and Tatchell, E. C.**

A review of the Euplotidae (Hypotrichida, Ciliophora). *Bull. Br. Mus. nat. Hist.*
(Zool.) **44**: 191–247.   (1983) **Curds, C. R. and Wu, I. C. H.**

A revision of the genera *Trachelostyla* and *Gonostomum* (Ciliophora:
Hypotrichida), including redescriptions of *T. pediculiformis* (Cohn, 1866)
Kahl, 1932 and *T. caudata* Kahl, 1932. *Bull. Br. Mus. nat. Hist.* (Zool.) **47**: 1–
17.   (1984) **Carey, P. G. and Maeda, M.**

An illustrated guide to the species of the family Strombidiidae (Oligotrichida,
Ciliophora) free swimming Protozoa, common in the aquatic environment.
Bull. Ocean Res. Inst. Tokyo **19**: 1–68.   (1985) **Maeda, M. and Carey, P. G.**

A revision of the Suctoria (Ciliophora, Kinetofragminophora). 1. *Acineta* and
its morphological relatives. *Bull. Br. Mus. nat. Hist.* (Zool.) **48**: 75–129, figs.
                                        (1985*a*) **Curds, C. R.**
Key to species.

A revision of the Suctoria (Ciliophora, Kinetofragminophora). 2. An
addendum to *Acineta. Bull. Br. Mus. nat. Hist.* (Zool.) **49**: 163–65.
                                        (1985*b*) **Curds, C. R.**
Key to species.

A revision of the Suctoria (Ciliophora, Kinetofragminophora). 3. *Tokophrya*
and its morphological relatives. *Bull. Br. Mus. nat. Hist.* (Zool.) **49**: 167–93,
figs.   (1985*c*) **Curds, C. R.**
Key to species.

A revision of the Suctoria (Ciliophora, Kinetofragminophora). 4. *Podophrya*
and its morphological relatives. *Bull. Br. Mus. nat. Hist.* (Zool.) **50**: 59–91.
                                        (1986) **Curds, C. R.**
Key to species.

A revision of the genus *Vorticella* (Ciliophora, Peritrichida). *Bull. Br. Mus. nat.
Hist.* (Zool.) **50**: 1–57.   (1986) **Warren, A.**
Key to species.

Parasitic

GENERAL

Handbook of medical protozoology: for medical men, parasitologists, and zoologists.
Bailliére, Tindall, and Cox, London. 334 pp.   (1949) **Hoare, C. A.**

Protozoan parasites of domestic animals and man, 2nd edn. Burgess, Minneapolis,
Minnesota. 406 pp.   (1972) **Levine, N. D.**

The intestinal protozoa of man. Bale and Danielsson, London. 209 pp.
                                        (1921) **Dobell, C. and Connor, F. W.**

Précis de parasitologie, 6th edn. Masson, Paris. 1042 pp.   (1949) **Brumpt, E.**

Parasitic protozoa. Hutchinson, London. 176 pp.   (1969) **Baker, J. R.**

Medical and veterinary protozoology, an illustrated guide. Churchill Livingstone, Edinburgh and London. 200 pp.

(1971) **Adam, K. M. G., Paul, J., and Zaman, V.**

Parasites of laboratory animals. Iowa State University, Ames, Iowa. 884 pp.

(1973) **Flynn, R. J.**

Blood parasites found in mammals, birds and fishes in England. *Parasitology* 7: 17–61.   (1914) **Coles, A. C.**

Die parasitischen Protozoen der Pflanzen. *Handb. pathogen. Protozoen* 3(11): 1799–813.   (1925) **Nieschulz, O.**

Key-catalogue of parasites reported for Chiroptera with their possible public health importance. *Natn. Inst. Hlth. Bull.* no. **155**: 603–742.

(1931) **Stiles, C. W. and Nolan, M. O.**

Key-catalogue of parasites reported for Carnivora with their possible public health importance. *Natn. Inst. Hlth. Bull.* no. 163: 913–1223.

(1935) **Stiles, C. W. and Baker, C. E.**

Protozoan parasites of Orthoptera IV. Classified list of the protozoan parasites of Orthoptera of the world. *Ohio Sci.* **43**: 221–34, 271–6.

(1943) **Semans, F. M.**

An annotated list of protozoan parasites, hyperparasites and commensals of decapod Crustacea. *J. Protozool.* **18**: 526–37.

(1971) **Sprague, V. and Couch, J.**

Morphology and taxonomy of the intestinal protozoa of the guinea pig, *Cavia porcella. J. Morph.* **86**: 381–493.   (1950) **Nie, D.**

Parasites and epibionts of *Cladocera. Trans. Am. microsc. Soc.* **32**: 417–515.

(1974) **Green, J.**

Keys to families.

Studies on the parasitic protozoa of wild mice from Berkshire with a description of a new species of *Trichomonas. Proc. Zool. Soc. London* **132**: 381–401.

(1959) **Ring, M.**

New parasitic protozoa are recorded in *Annls. Parasit. hun. comp.* (1923 on). The *Zoological Record* has listed hosts since 1929.

Parasitic protozoa. Academic Press, New York. 4 vols (1977–8) **Kreier, J. P.**, ed.

Containing:

Taxonomy, kinetoplastids and flagellates of fish. Vol. 1 (1977) 441 pp.

Intestinal flagellates, histomonads, trichomonads, amoebae, opalinids and ciliates. Vol. 2 (1978) 730 pp.

Gregarines, Haemogregarines, Coccidia, Plasmodia, and Haemoproteids. Vol. 3 (1977) 563 pp.

Babesia, Theileria, Myxosporidia, Microsporidia, Bartonellaceae, Anaplasmataceae, Ehrlichia, and Pneumocystis. Vol. 4 (1977) 386 pp.

Atlas of rumen microbiology. Japan Scientific Societies, Tokyo. 231 pp, 378 figs.
(1981) **Ogimoto, K. and Imai, S.**
Descriptions and keys to species.

[*Piroplasmida: fauna, systematics, evolution.*] Nauka, Leningrad. 229 pp. [Russian].
(1981) **Krylov, M. V.**

Parasitic protozoa in British wild animals. Institute of Terrestrial Ecology, Cambridge. 24 pp.
(1982) **Baker, J. R.**

[*Coccidia of some domestic and economically important animals and of man.*] Academia, Prague. 144 pp, 132 figs. 120 pls. [Czech].
(1983) **Černá, Z.**

Laboratory guide to insect pathogens and parasites. Plenum, New York. 392 pp, 241 figs.
(1984) **Poinar, G. O. and Thomas, G. M.**

Atlas of diagnostic medical parasitology. Vol 2 Intestinal Protozoa. American Society of Clinical Pathologists, Chicago. 41 pp, 106 col. slides.
(1984) **Smith, J. W., McQuay, R. M., Ash, L. R., Melvin, D. M., Orihel, T. C., and Thompson, J. H. Jr.**

[Parasitic Protozoa.] In [*Key to the identification of parasites of freshwater fishes of the fauna of the USSR*] (ed. O. N. Bauer), Vol. 1, 426 pp, 609 figs. Nauka, Leningrad. [Russian].
(1984) **Shul'man, S. S.**
Keys to orders and genera.

SARCOMASTIGOPHORA

PHYTOMASTIGOPHORA

Les péridiniens parasites. Morphologies, reproduction, éthologie. *Archs. Zool. exp. gen.* **59**: 475 pp.
(1920) **Chatton, E.**
Parasitische Peridinea. *Tierwelt N.-u Ostsee* **2d**: 85–100, 10 figs.
(1930) **Reichenow, E.**
Algal flagellate symbiosis in the foraminifera. *Archaias. J. Protozool.* **16**: 71–81.
(1968) **Lee, J. S. and Zucker, W.**
Euglenoidina parasitic in Copepoda. (PWN–Polish Scientific, Warsaw. 224 pp.
(1972) **Michajlow, W.**
Cytology and taxonomy of parasitic flagellates. *Proc. int. Congr. Parasit.* **4**: 205–27.

(1978) **Honigberg, B. M., Vickerman, K., Kulda, J., and Brugerolle, G.**

ZOOMASTIGOPHOREA

Parasitische Flagellata (ausschliesslich Peridinea). *Tierwelt N.-u. Ostsee* **2e**: 1–18, 10 figs.
(1931) **Reichenow, E.**
Fish host trypanosomes.

Morphology of the intestinal trichomonad flagellates in man and of similar

forms in monkeys, cats, dogs and rats. *J. Morph.* **74**: 189–211.
(1944) **Wenrich, D. H.**

Evolutionary and systematic relationships in the flagellate order Trichomonadida Kirby. *J. Protozool.* **10**: 20–63. (1963) **Honigberg, B. M.**

A morphological study of trichomonads and related flagellates from the bovine digestive tract. *J. Protozool.* **11**: 386–94.
(1964) **Jensen, E. A. and Hammond, D. M.**

The trypanosomatid parasites of insects and arachnids. *Expl. Parasit.* **18**: 124–93.
(1966) **Wallace, F. G.**

The genus *Leishmania. Bull. Wld. Hlth. Org.* **44**: 477–89. (1971) **Garnham, P. C.**

The trypanosomes of mammals. Blackwell, Oxford and Edinburgh. 749 pp.
(1972) **Hoare, C. A.**

Taxonomy of trypanosomes (Protozoa: Trypanosomatidae) of *Spermophilus* spp. (Rodentia: Sciuridae). *Parasitology* **65**: 403–25.
(1972) **Hilton, D. F. J. and Mahrt, J. L.**

Leishmania. Ann. Rev. Microbiol. **28**: 189–217.
(1974) **Bray, R. S.**

The trypanosomes of Anura. *Adv. Parasitol.* **11**: 1–73.
(1973) **Bardsley, J. E. and Harmsen, R.**

The diversity of kinetoplastid flagellates. In *Biology of Kinetoplastida* (ed. W. H. R. Lumsden and O. A. Evans). Academic Press, New York and London. 1–34 pp.
(1976) **Vickerman, K.**

Trypanosomes et leishmaines africains. *Résult. scient. Explor. hydrobiol. Bassin Lac Bangeweulu & Luapula* **10**: 1–205.
(1978) **Jadin, J-M.**

Cytologie ultrastructurale systematic et evolution des Trichomonadida. *Annls Stn biol. Besse-en-Chandesse* **10**: 1–57.
(1977) **Brugerolle, G.**

The Trichomonads. *Publ. Hlth. Service Monogr. Ser.* No. 9. HMSO, London. 40 pp, 13 figs.
(1978) **Lowe, G. H.**

International reference strains of *Leishmania. Bull. Wld Hlth Org.* **63**: 43–5.
(1985) **Anon. (Scientific Working Group of the Leishmaniases)**

Flagellates of the genus *Trichonympha. Univ. Calif. Publs Zool.* **37**: 349–476.
(1932) **Kirby, H.**

The wood-feeding roach *Cryptocercus*, its protozoa and the symbiosis between protozoa and roach. *Mem. Am. Acad. Arts Sci.* **17**: 1–342.
(1934) **Cleveland, L. R., Hall, S. R., Saunders, E. P. and Collier, J.**

The flagellate subfamily Oxymonadinae. *Univ. Calif. Publs Zool.* **53**: 67–162.
(1946) **Cross, J. B.**

Flagellates of the caecum of ground squirrels. *Univ. Calif. Publs. Zool.* **53**: 315–66.
(1949) **Kirby, H. and Honigberg, B. M.**

OPALINATA

The opalinid ciliate infusorians. *Bull. US natn. Mus.* **120**: 1–471.
(1923) **Metcalf, M. M.**

Further studies on the opalinid ciliate infusorians. *Proc. US natn. Mus.* **87**: 465–634.
(1940) **Metcalf, M. M.**

The species problem in the opalinids (Protozoa, Opalinata) with special reference to *Protoopalina. Trans. Am. microsc. Soc.* **95**: 357–66.
(1976) **Sandon, H.**

SARCODINA (RHIZOPODA)

The amoebae living in man. Bale and Daniellson, London. 155 pp.
(1919) **Dobell, C.**

Entamoebae in farm animals. *J. Parasit.* **38**: 571–95.
(1952) **Noble, G. A. and Noble, E. R.**

Amoebic infections in animals. *Vet. Revs. Annot.* **5**: 91–102. (1959) **Hoare, C. A.**

Studies on pathogenic and non-pathogenic small free-living amoebae and the bearing of nuclear division on the classification of the order Amoebida. *Phil. Trans. R. Soc.* **B259**: 435–76, 12 pls.   (1970) **Singh, B. N. and Das, S. R.**

Small, free-living amoebas: cultivation, quantification, identification, classification, pathogenesis, and resistance. *Current Topics in Pathobiology* **1**: 205–54.
(1971) **Chang, S. L.**

The pathogenicity of soil amoebas. *A. Rev. Microbiol.* **25**: 231–54.
(1971) **Cuthbertson, C. G.**

A further study of the taxonomic criteria for *Limax* amoebae with descriptions of new species and a key to genera. *Arch. Protistenk.* **116**: 149–84.
(1974) **Page, F. C.**

Study of *Dientamoeba fragilis* Jepps & Dobell. I. Electronmicroscopic observations of the binucleate stages. II. Taxonomic position and revision of the genus. *J. Protozool.* **21**: 69–82.
(1974) **Camp, R. R., Mattern, C. F. T., and Honigberg, B. M.**

Pathogenic and non-pathogenic amoebae. Macmillan, London. ix + 235 pp, figs.
(1975) **Singh, B. N.**

CILIOPHORA

The ciliated Protozoa: characterisation, classification and guide to the literature, 2nd edn. Pergamon, New York and Oxford. 455 pp, 37 pls.   (1979) **Corliss, J. O.**
Systematic descriptions of orders and families with classification of genera.

Monographie der Familie Ophryoscolecidae. *Arch. Protistenk.* **59**: 1–288.
(1927) **Dogiel, V. A.**

Ciliates from *Bos indicus* Linn. I. The genus *Entodinium* Stein. *Univ. Calif. Publs Zool.* **33**: 471–544.   (1931) **Kofoid, C. A. and Maclennan, R. F.**

Ciliates from *Bos indicus* Linn. II. A revision of *Diplodinium* Scuberg. *Univ. Calif. Publs Zool.* **37**: 53–152.   (1933) **Kofoid, C. A. and Maclennan, R. F.**

Ciliates from *Bos indicus* Linn. III. *Epiplastron* gen. nov., *Epidinium* Crawley and *Ophryoscolex* Stein. *Univ. Calif. Publs Zool.* **39**: 1–34.
(1935) **Kofoid, C. A. and Maclennan, R. F.**

A monograph on the protozoa of the large intestine of the horse. *Iowa St. Coll. J. Sci.* **4**: 356–423.   (1930) **Hsiung, T.**

[Parasitic protozoa of the intestine of *Ungulata* belonging to the family Equidae.] *Uchen. Zap. pedegogich Inst. Gercena* **17**: 262 pp. [Russian].
(1939) **Streklow, A.**
Note. An English translation of this Russian monograph (11–133 pp) is available in the library of the Department of Zoology, University of Edinburgh.

Studies on ciliates from sea urchins. *Biol. Bull. mar. biol. Lab. Woods Hole.* **65**: 106–21.   (1933) **Powers, P. B. A.**

Studies on ciliates from sea urchins. *Pap. Tortugas Lab.* **29**: 293–326.
(1935) **Powers, P. B. A.**

The ciliates of *Strongylocentrotus dröbachiensis*: incidence, distribution in the host, and division. *Biol. Bull. mar. biol. Lab. Woods Hole* **94**: 99–112.
(1948) **Beers, C. D.**

Studies on the family Blepharocorythidae Hsiung. V. A review of genera and species. *Acta Protozool.* **9**: 23–40.   (1971) **Wolska, M.**

The entocommensal fauna of the pink sea urchin, *Allocentrotus fragilis*. *J. Parasit.* **47**: 417–18.   (1961) **Berger, J. and Profant, R. J.**

Morphology, systematics and demic variation of *Plagiopyliella pacifica* Poljansky, 1951 (Ciliatea: Philasterina), an entocommensal of strongylo-centroid echinoids. *Trans. Am. microsc. Soc.* **91**: 310–36.
(1972) **Lynn, D. H. and Berger, J.**

The Thyrophylacidae, a family of carvivorous philasterine ciliates entocom-mensal in strongylocentroitid echinoids. *Trans. Am. microsc. Soc.* **92**: 533–57.
(1973) **Lynn, D. H. and Berger, J.**

Ciliata ectocommensalia et parasitica. *Tierwelt N.-u Ostsee* **2c**: 184–226.
(1934) **Kahl, A.**

Untersuchungen an parasitischen Ciliaten aus Anneliden (Astomatida). *Arch. Protistenk.* **84**: 315–92.   (1935) **Heidenreich, E.**

Protozoa: Ciliophora. Fauna of British India. Taylor and Francis, London. 493 pp.
(1936) **Bhatia, B. L.**

Ciliates from Scandinavian molluscs. *Ophelia* **2**: 71–174.   (1968) **Fenchel, T.**

Symbiotic ciliates from solitary ascidians of the Pacific northwest with a description of *Parahypocoma rhamphisokarya* n. sp. *Trans. Am. microsc. Soc.* **92**: 517–22.   (1973) **Burreson, E. M.**

Sur un infusoire parasite des poils secréteurs des crustacés Edriophtalmes et la famille nouvelle des Pilisuctoridae. *C. r. hebd. Séanc. Acad. Sci., Paris* **199**: 696–99, 1 fig. (1934) **Chatton, E. and Lwoff, A.**

Les ciliés apostomes. Morphologie, cytologie, éthologie, évolution, systématique. *Archs. Zool. exp. gén.* **77**: 1–453, 217 figs, 21 pls. (1935) **Chatton, E. and Lwoff, A.**

Les Pilisuctoridae Ch. et Lw., ciliés parasites des poils secréteurs des Crustacés Edriophthalmes. Polarité, orientation et desmodexie chez les infusoires. *Bull. biol. Fr. Belg.* **70**: 86–144, 7 figs, 2 pls. (1936) **Chatton, E. and Lwoff, A.**

Beiträge zur Kenntnis der parasitischen Ciliaten aus Evertebraten IV. Neue Ciliaten aus der Familie Haptophryidae Cépède 1923, nebst einigen Bermerkungen zum heutigen Stand dieser Gruppe. *Arch Protistenk.* **104**: 133–54. (1959) **Lom, J.**

Resolution of persistent taxonomic and nomenclatural problems involving ciliate protozoa assignable to the astome family Haptophryidae Cépède, 1923. *J. Protozool.* **12**: 265–73. (1965) **Corliss, J. O., de Puytorac, P. and Lom, J.**

A contribution to the systematics and morphology of endoparasitic trichodinids from amphibians with a proposal of uniform specific characteristics (Urceolaridae). *J. Protozool.* **5**: 251–63. (1958) **Lom, J.**

Systematics of the family Urceolaridae Dujardin 1841. *Acta Protozool.* **1**: 121–38. (1963) **Raabe, Z.**

Recherches sur les ciliés thigmotriches, I. *Archs. Zool. exp. gén.* **86**: 169–253. (1949) **Chatton, E. and Lwoff, A.**

Recherches sur les ciliés thigmotriches, II. *Archs Zool. exp. gén.* **86**: 393–485. (1950) **Chatton, E. and Lwoff, A.**

Ordo Thigmotricha (Ciliata-Holotricha) I–V. *Acta Protozool.* **5**: 1–36; **7**: 117–80; **8**: 385–463; **9**: 121–70; **10**: 115–84. (1967–72) **Raabe, Z.**

APICOMPLEXA

GREGARINA

Die Gregarinen. *Handb. pathogen. Protoz.* **1**(4): 487–515. (1912) **Schellack, C.**

Die Gregarinen (Nachtrag). *Handb. pathogen. Protoz.* **3**(8): 1278–94. (1921) **Reichenow, E.**

Studies on gregarines. *Illinois biol. Monogr.* **2**: 211–468. (1916) **Watson, M. E.**

Studies on gregarines II. *Illinois biol. Monogr.* **7**: 1–104. (1922) **Kamm, M. W.**

Synopsis of genera and classification of haplocyte gregarines. *Parasitology* **22**: 156–67. (1930) **Bhatia, B. L.**

A new classification of Schizogregarina. *J. Protozool.* **2**: 6–12. (1955) **Weiser, J.**

Les Eugrégarines du genre *Gregarina* parasites de coléoptères ténébrionides. *Annls. Parasit. hum. comp.* **30**: 5–21.   (1955) **Théodoridès, J.**

Lankesteria barretti n. sp. (Eugregarinida, Diplocystidae) a parasite of the mosquito *Aedes triseriatus* and a review of the genus *Lankesteria* Mingazzini. *J. Protozool.* **16**: 546–70.   (1969) **Vavra, J.**

Taxonomy of the Archigregarinorida and Selenidiidae (Protozoa: Apicomplexa). *J. Protozool.* **18**: 704–17.   (1971) **Levine, N. D.**

Observations biologiques et ultra-structurales sur les Selenidiidae et leur conséquences sur la systématique des grégarinomorphes. *J. Protozool.* **18**: 448–70.   (1971) **Schrevel, J.**

Acephaline gregarines of earthworms—additions to the British records. *J. Protozool.* **18**: 313–17.   (1971) **Segun, A. O.**

Distribution of acephaline gregarines of British earthworms and their comparison with those of neighbouring European countries. *Parasitology* **65**: 47–53.   (1972) **Segun, A. O.**

New genera and higher taxa of septate gregarines (Protozoa, Apicomplexa). *J. Protozool.* **26**: 532–6.   (1979) **Levine, N. D.**

COCCIDIA

Die Haemogregarinen. *Handb. pathogen. Protoz.* **2**(5): 602–32.
                    (1912) **Reichenow, E.**

Die Coccidien. *Handb. pathogen. Protoz.* **3**(8): 1136–277 (1921) **Reichenow, E.**

A catalog and host-index of the species of the coccidian genus *Eimeria. Iowa St. Coll. J. Sci.* **8**: 402–5.   (1933) **Levine, N. D. and Becker, E. R.**

Coccidia and coccidiosis of domesticated game, laboratory animals and of man. *Iowa St. Coll. Monogr.* **1**: 147 pp.   (1934) **Becker, E. R.**

A check-list of the coccidia of the genus *Isospora. J. Parasit.* **20**: 195–6.
                    (1934) **Becker, E. R.**

Classification of Coccidia Eimeriidae in a periodic system of homologous genera. *Rev. brasil. Malar. Doenc. trop.* **8**: 197.   (1957) **Hoare, C. A.**

Catalogue of Eimeriidae in genera occurring in vertebrates and not requiring intermediate hosts. *Iowa St. Coll. J. Sci.* **31**: 85.   (1956) **Becker, E. R.**

Coccidiosis. Oliver and Boyd, Edinburgh. 264 pp.
                    (1963) **Davies, S. F. M., Joyner, L. P., and Kendall, S. B.**

Species of *Eimeria* of small wild rodents from the British Isles, with descriptions of two new species. *Syst. parasit.* **5**: 259–70.
                    (1983) **Lewis, D. C. and Ball, S. J.**

Coccidia and coccidiosis, 2nd edn. Akadémiai Kiadó, Budapest. 959 pp.
                    (1974) **Pellérdy, L. P.**

The coccidian parasites (Protozoa, Sporozoa) of rodents. *Illinois biol. Monogr.* **33**: 1–365.   (1965) **Levine, N. D. and Ivens, V.**

The coccidian parasites (Protozoa, Sporozoa) of ruminants. *Illinois biol. Monogr.* **44**: 278 pp.   (1970) **Levine, N. D. and Ivens, V.**

Coccidia of Leporidae. *J. Protozool.* **19**: 572–81.
(1972) **Levine, N. D. and Ivens, V.**

Diseases of wildlife. Iowa State University, Ames, Iowa.
(1970) **Todd, K. S. and Hammond, D. M.**
Coccidia of wild birds.

The Coccidia. University Park Press, Baltimore, Maryland and Butterworths, London. 482 pp.   (1973) **Hammond, D. M. and Long, P.**, eds.

Coccidia of Indian vertebrates. *Rec. zool. Surv. India* **70**: 39–120.
(1976) **Mandal, A. K.**
Keys to species.

Cryptosporidium wrairi sp. n. from the guinea pig *Cavia porcellus* with an emendation of the genus. *J. Protozool.* **18**: 242–7.
(1971) **Vetterling, J. M., Jervis, H. R., Merrill, T. G., and Sprinz, H.**

Advances in the biology of Sporozoa. *Z. Parasitenk.* **45**: 125–62.
(1974) **Frenkel, J. K.**
Toxoplasma, Sarcocystis, Frenkelia.

Sarcocystis infection in African antelopes. *Ann. trop. Med. Parasit.* **64**: 513–23.
(1970) **Mandour, A. M. and Keymer, I. F.**
Also infestation of deer in NW Europe.

Observations on the histomorphology of sarcosporidian cysts of some East African game animals (Artiodactyla). *Z. Parasitenk.* **46**: 13–23.
(1975) **Kaliner, G.**

Light and electron microscopic observations of the life cycle of *Sarcocystis orientalis* sp. n. in the rat (*Rattus norvegicus*) and the Malaysian python (*Python reticulatus*). *Z. Parasitenk.* **47**: 169–85. (1975) **Zaman, V. and Colley, F. C.**

The coccidial nature and life-cycle of *Sarcocystis. J. trop. Med. Hyg.* **77**: 248–59.
(1974) **Markus, M. B., Killick-Kendrick, R., and Garnham, P. C. C.**

A check-list and host-list of the genus *Haemoproteus. J. Parasit.* **22**: 88–105.
(1936) **Coatney, G. R.**

A catalog and host-index of the genus *Leucocytozoon. J. Parasit.* **23**: 202–12.
(1937) **Coatney, G. R.**

The identification of the avian malarias. *Am. J. trop. Med.* **18**: 565–75.
(1938) **Manwell, R. D.**

I plasmodi aviari. *Riv. Parasitol.* **3**: 1–46.   (1939) **Giovannola, A.**

Bird malaria. *Monographic series, Baltimore.* No. 15, 228 pp. (1940) **Hewitt, R.**

Report of the committee on terminology of strains of avian malaria. *J. Parasit.* **28**: 250–4.   (1942) **Huff, C. G., Boyd, G. H., and Manwell, R. D.**

Malaria parasites and other Haemosporidia. Blackwell, Oxford. 1114 pp.
(1966) **Garnham, P. C. C.**

Malaria of mammals excluding man. *Adv. Parasitol.* **5**: 139–204.
(1967) **Garnham, P. C. C.**

Recent research on malaria in mammals excluding man. *Adv. Parasitol.* **11**: 603–30.
(1973) **Garnham, P. C. C.**

Avian blood coccidians. *Adv. Parasitol.* **10**: 1–30.
(1972) **Baker, J. R., Bennett, G. F., Clark, G. W., and Laird, M.**

Studies on the exo-erythrocytic cycle of the genus *Plasmodium*. *Memoir Ser. Lond. Sch. Hyg. trop. Med.* **12**, 192 pp.
(1957) **Bray, R. S.**

Parasitic protozoa of the blood of rodents: a revision of *Plasmodium berghei*. *Parasitology* **69**: 229–37.
(1974) **Killick-Kendrick, R.**

A check-list of the genus *Haemoproteus* (Apicomplexa, Plasmodiidae). *J. Protozool.* **18**: 475–84.
(1971) **Levine, N. D. and Campbell, G. R.**

A check-list of the species of the genus *Leucocytozoon* (Apicomplexa, Plasmodiidae). *J. Protozool.* **20**: 195–203.
(1973) **Hsu, C. K., Campbell, G. R., and Levine, N. D.**

On species of *Leucocytozoon*. *Adv. Parasitol.* **12**: 1–67.
(1974) **Fallis, A. M. and Desser, S. S.**

Diversité des mécanismes assurant la pérennité de l'infection chez les sporozoaires coccidiomorphes. *Mém. Mus. natn. Hist. nat., Paris* (A) **77**: 1–62.
(1973) **Landau, I.**

On a new family of non-pigmented parasites in the blood of reptiles: Garniidae fam. nov. (Coccidia, Haemosporidiidae). Some species of the new genus *Garnia*. *Int. J. Parasitol.* **1**: 241–50.
(1971) **Lainson, R., Landau, I., and Shaw, J. J.**

The genera *Barrouxia*, *Defrentinella*, and *Goussia* of the coccidian family Barrouxiidae (Protozoa, Apicomplexa). *J. Protozool.* **30**: 542–7.
(1983) **Levine, N. D.**

PIROPLASMEA

Études sur les piroplasmoses bovines. Institute Pasteur d'Algérie, Algiers. 816 pp.
(1945) **Sergent, E., Donatien, A., Parrot, L., and Lestoquard, F.**

Classification, transmission, and biology of piroplasms of domestic animals. *Ann. NY Acad. Sci.* **64**: 56–111.
(1956) **Neitz, W. O.**

Taxonomy of the piroplasms. *Trans. Am. microsc. Soc.* **90**: 2–33.
(1971) **Levine, N. D.**

Babesioma gen. nov. and other babesioides in erythrocytes of cold blooded vertebrates. *Ann. NY Acad. Sci.* **64**: 112–27.
(1956) **Jakowska, S. and Nigrelli, R. F.**

Babesiosis. In *Infectious blood diseases of man and animals* (ed. D. Weinman and M. Ristic), Vol. 2, pp. 219–68. Academic Press, New York and London.
(1968) **Riek, R. F.**

Theileriosis, gonderioses, and cytauxzoonoses. *Onderstepoort J. vet. Res.* **27**: 275.
(1957) **Nietz, W. O.**

Theileriasis. In *Infectious diseases of man and animals* (ed. O. Weinman and M. Ristic), Vol. 2, pp. 269–328. Academic Press, New York and London.
(1968) **Barnett, S. F.**

MICROSPORA

A biologic and taxonomic study of the Microsporidia. *Illinois biol. Monogr.* **9**: 1–268.
(1924) **Kudo, R. R.**

Zur Kenntnis von in Oligochäten parasitierenden Microsporidien aus der Familie Mrazekidae. *Arch. Protistenk.* **87**: 314–44.   (1936) **Jirovec, O.**

Revision der Simulienlarven parasitierenden Microsporidien. *Zool. Anz.* **142**: 173–9.
(1943) **Jirovec, O.**

A list and brief description of Microsporidia infecting insects. *J. Insect. Path.* **2**: 346
(1960) **Thomson, H. M.**

Die Microsporidien als Parasiten des Insekten. *Monogrm. angew. Eng.* **17**.
(1961) **Weiser, J.**

Microsporidian parasites of *Simulium ornatum* Mg. in south of England. *Parasitology* **65**: 27–45.
(1972) **Gassouma, M. S. S.**

Two new species of *Plistophora* (Microsporidea) from North American fish with a synopsis of Microsporidea of freshwater euryhaline fishes. *J. Protozool.* **12**: 228–36.   (1965) **Putz, R. E., Hoffman, G. L., and Dunbar, C. E.**

[Studies on the microsporidian infection in salmonid fishes.] *Scient. Rep. Hokkaido Fish Hatch.* **29**: 1–95. [Japanese].   (1974) **Awakura, T.**

Revision of the Microsporida (Protozoa) close to Thelohania with descriptions of one new family, eight new genera and thirteen new species. *Tech. Bull. U.S. Dep. Agric.* No. 1530: 1–104. (1976) **Hazard, E. I. and Oldacre, S. W.** Keys to genera.

Contribution to the classification of microsporidia. *Věst. Čs. Spol. Zool.* **41**: 308–20.
(1977) **Weiser, J.**

Annotated list of species of microsporidia. In *Comparative pathobiology. 2 Systematics of the Microsporidia* (ed. L. A. Bulla and T. C. Cheng), pp. 31–335. Plenum, New York.
(1977) **Sprague, V.**

A new microsporidian *Berwaldia singularis* gen. et sp. nov. from *Daphnia pulex* and a survey of microsporidia described from Cladocera. *Parasitology* **83**: 325–42.
(1981) **Larsson, R.**

Characters of microsporidian genera. *Parasitology* **82**: 131–42.
(1981) **Vavra, J., Canning, E. U., Barker, R. J., and Desportes, I.**

Revision of the genus *Haplosporidium* and restoration of genus *Minchinia* (Haplosporidia, Haplosporidiidae). *J. Protozool.* **10**: 263–6.

(1963) **Sprague, V.**

ASCETOSPORA

Classification of the Halosporidia. *Mar. Fish Rev.* **41**: 40–4. (1979) **Sprague, V.**

MYXOZOA

MYXOSPOREA

Studies on Myxosporidia. A synopsis of genera and species of Myxosporidia. *Illinois biol. Monogr.* **5**: 1–265. (1920) **Kudo, R. R.**

A taxonomic consideration of Myxosporidia. *Trans. Am. microsc. Soc.* **52**: 195–216. (1933) **Kudo, R. R.**

Studies on some protozoan parasites of fishes of Illinois (Myxosporidia). *Illinois biol. Monogr.* **13**: 1–44. (1934) **Kudo, R. R.**

Some new Myxosporidia from Plymouth with a proposed new classification of the order. *Parasitology* **39**: 110–18. (1949) **Tripathi, Y. R.**

Five new species of Myxosporidia from marine fishes. *Parasitology* **52**: 177–86.

(1962) **Kabata, Z.**

Studies on *Myxosoma cartilaginis* n. sp. of centrarchid fish and a synopsis of the *Myxosoma* of North American freshwater fishes. *J. Protozool.* **12**: 319–32.

(1965) **Hoffman, G. L., Putz, R. E., and Dunbar, C. E.**

Myxidium macrocheilus n. sp. (Cnidospora, Myxidiidae) from the largescale sucker *Catostomus macrocheilus* and a synopsis of the *Myxidium* of North American freshwater vertebrates. *J. Protozool.* **14**: 415–24.

(1967) **Mitchell, L. G.**

ACTINOSPOREA

Actinomyxidia: morphology, ecology, history of investigations, systematics, development. *Acta parasit. pol.* **2**: 406–33, figs. (1955) **Janiszweska, J.**

Actinomyxidia II. New systematics, sexual cycle, descriptions of new genera and species. *Zoologica Pol.* **8**: 3–34. (1957) **Janiszweska, J.**

MESOZOA

A seldom reported group; the following references have been used to identify three species recorded from British seas.

Mesozoa. In *Traité de zoologie: anatomie, systématique, biologie* (ed. P-P. Grassé), Masson and Cie, Paris. Vol. 4(1), pp. 693–729.

(1961) **Grassé, P-P. and Caullery, N.**

DICYEMIDA

Protozoa through Ctenophora. *The invertebrates.* McGraw-Hill, New York. Vol. 1, 235 pp, 68 figs.   (1940) **Hyman, L. H.**
Contains a description of *Pseudicyema truncatum* infesting the kidney of *Sepia* specimens from Plymouth.

ORTHONECTIDA

Rhopalura granosa sp. nov., an orthonectid parasite of a Llamellibranch, *Heteranomia squamula* L., with a note on its swimming behaviour. *J. mar. biol. Ass. UK* **19**: 233–52.   (1933) **Atkins, D.**

Mesozoa. *Tierwelt N.-u Ostsee* **2h**, 10 pp.   (1933) **Neresheimer, E.**
Provides a description of *Rhopalura ophiocomae* Giard from the Irish Sea.

PORIFERA

General

Sponges. Hutchinson, London. 268 pp, 8 figs, 12 pls. (1978) **Bergquist, P. R.**
Biologie des spongiaires. *Colloques int. Cent. natn Rech. scient.* **291**: 533 pp.

(1979) **Lévi, C. and Boury-Esnault, N.**, eds.

The cell biology of sponges. Springer, New York. xix + 662 pp, 9 figs.

(1984) **Simpson, T. L.**

European contributions to the taxonomy of sponges. Sherkin Island Marine Station, County Cork. (In press).   (1987) Various authors

Freshwater

On the freshwater sponges of Denmark. *Vidensk. Meddr dansk naturh. Foren.* **130**: 173–8, 2 figs.   (1967) **Tendal, O. S.**
Freshwater Spongia. *Zoology Iceland* **2** (4a): 1–4.   (1976) **Tendal, O. S.**

Marine

Natural history of the marine sponges of southern New England. *Bull. Peabody Mus. nat. Hist.* **12**: 1–155, 12 pls.   (1958) **Hartman, W. D.**

Spongia. *Zoology Iceland* **2**(3–4): 1–71.   (1959) **Burton, M.**

Les Démosponges des côtes de France. I. Les Clathriidae. *Cah. Biol. mar.* **1**: 47–87, 25 figs.   (1960) **Lévi, C.**

Contribution à la connaissance de la faune des spongiaires de la Manche occidentale. Démosponges de la région de Roscoff. *Cah. Biol. mar.* **9**: 211–46, 12 figs, 2 pls.   (1968) **Cabioch, L.**

Structure et ultrastructure des papilles d'éponges du genre *Polymastia* Bowerbank. *Archs Zool. exp. gén.* **115**(1): 141–65, 7 figs, 5 pls.
(1974) **Boury-Esnault, N.**

Additions a l'inventaire de la faune marine de Roscoff. Spongiaires. *Trav. Stn biol. Roscoff* **20** (N.S.): 5–6.   (1976) **Cabioch, L.**

Marine and freshwater sponges (Porifera) of the Netherlands. *Zool. Meded. Leiden* **50**(16): 261–73, 5 figs, 3 pls.   (1977) **Soest, R. W. M. van**

Mycale micracanthoxea nov spec. (Porifera, Poecilosclerida) from the Netherlands. *Neth. J. Sea Res.* **11**(3–4): 297–304, 1 fig, 2 pls.
(1977) **Buizer, D. A. G. and Soest, R. W. M. van**

A note on the sponges and octocorals from Sherkin Island and Lough Ine, Co. Cork. *Ir. Nat. J.* **20**(1): 1–15, 12 figs.
(1980) **Soest, R. W. M. van and Weinberg, S.**

Sponges from Roaringwater Bay and Lough Ine. *J. Sherkin Is.* **1**(2) 1981: 35–49, 4 figs. (1983) **Soest, R. W. M. van, Guiterman, J. D., and Sayer, M.**

Calcareous sponges of the Netherlands (Porifera, Calcarea). *Bull. zool. Mus. Univ. Amsterdam* **8**(12): 89–98, 5 figs.   (1982) **Koolwijk, Th. van**

The genus *Thenea* (Porifera, Demospongia, Choristida) in the Norwegian Sea and adjacent waters; an annotated key. *Sarsia* **67**(4): 259–68, 4 figs.
(1982) **Steenstrup, E. and Tendal, O. S.**

Ecology and distribution of two sympatric, closely related sponge species, *Halichondria panicea* (Pallas, 1766) and *H. bowerbanki* Burton, 1930 (Porifera, Demospongiae), with remarks on their speciation. *Bijdr. Dierk.* **52**: 82–102.
(1982) **Vethaak, A. D., Cronie, R. J. A., and Soest, R. W. M. van**

Notes on some Mediterranean *Axinella* with descriptions of two new species. *Boll. Musei Ist. biol. Univ. Genova* **50–1**: 79–98, 10 figs.
(1982–1983) **Pansini, M.**

Présence de l'espèce *Esperiopsis fucorum* en Mediterranée. *Vie Milieu* **33**: 237–40.
(1983) **Uriz, M. J.**

The marine fauna of Lundy. Porifera (Sponges): a preliminary study. *Rep. Lundy Fld Soc.* **34** (1983): 16–35, 4 pls.
(1984) **Hiscock, K., Stone, S. M. K., and George, J. D.**

Sponges of the British Isles ('Sponge IV'). A colour guide and working document. Marine Conservation Society, Ross-on-Wye. 199 pp, 127 col. (*mini-prints*).
(1985) **Ackers, R. G., Moss, D., Picton, B. E., and Stone, S. M. K.**

A systematic revision of the North Eastern Atlantic shallow-water Haplo-sclerida (Porifera, Demospongiae). Part I: introduction, Oceanapiidae and Petrosiidae. *Beaufortia* **35**(5): 61–91, 6 figs.   (1985) **de Weerdt, W. H.**

CNIDARIA

About 290 British species.

General

INTRODUCTORY WORKS

Coelentérés. Traité de zoologie. Masson, Paris. 2 vols. (In press) **Bouillon, J.**, ed.
To become the standard systematic reference work.

Stamm Cnidaria, Nesseltiere. In *Lehrbuch der speziellen Zoologie* (co-ordinated by A. Kaestner), Band I: *Wirbellose Tiere* (ed. H-E. Gruner), 2 Teil: *Cnidaria, Ctenophora, Mesozoa, Plathelminthes, Nemertini, Entoprocta, Nemathelminthes, Priapulida.* G. Fischer, Stuttgart. 621 pp. (pp. 11–305). (1984) **Werner, B.**
A sound introduction to the group, partly replacing Hyman (1940); well illustrated; in German.

The invertebrates: Protozoa through Ctenophora. McGraw Hill, New York and London. 726 pp, 221 figs.   (1940) **Hyman, L.**

The open sea. Its natural history. Part one: The world of plankton. Collins, London. 335 pp, 93 figs, 24 pls.   (1956) **Hardy, A. C.**
Illustrations of many common planktonic forms occurring in British waters.

Synopsis of the medusae of the world. *J. mar. biol. Ass. UK* **40**: 1–469.
(1961) **Kramp, P. L.**
No illustrations: includes diagnoses of medusa stages of all species and higher taxa. Excludes the floating hydroid *Velella*, for which see Brinckmann-Voss (1970) on p. 37 and Kirkpatrick and Pugh (1984) on p. 38. Includes the sessile Stauromedusae.

Coelenterata, Cnidaria. In Erste Reihe: Deskriptive Morphogenese. *Eileitung zur Gesamwerk Morphogenetische Arbeitsmethoden und Begriffsysteme. Morphogenese der Tiere* (ed. F. Seidel). G. Fischer, Stuttgart. Lieferung **1**(A-1) 415 pp.
(1978) **Tardent, P.**

FAUNAL GUIDES

British coelenterates—a photographic identification guide. Marine Conservation Society, Ross-on-Wye. 8 pp, 24 col. slides.
(1986) **George, J. D. and Warren, L.**

Colour photographs of common British coelenterates in their natural environment with comprehensive systematic and ecological notes.

Flore et faune fixées sous-marines de Bretagne. Laboratoire Maritime, Concarneau and Laboratoire d'Océanographie Biologique, U.B.O., Brest. 100 pp, numerous figures.   (1982) **Castric, A. and Michel, C.**
An introductory guide to most groups.

Polypdyr (Coelenterata). II. Gopler. *Danm. Fauna* **43**: 1–233, 90 figs. [Danish].   (1937) **Kramp, P. L.**
Mostly planktonic forms: hydromedusae, scyphomedusae, stauromedusae, siphonophores, ctenophores.

Polypdyr (Coelenterata). III. Koraldyr. *Danm. Fauna* **51**: 1–167, 75 figs. [Danish].   (1945) **Carlgren, O.**
Anthozoans.

Coelentérés. *Faune Belge* pp. 283, 160 figs.   (1952) **Leloup, E.**

A survey of the marine benthonic macro-fauna along the Swedish west coast 1921–1938. *Acta R. Soc. scient. litt. gothoburg.* (Zool.) **6**: 1–146.
(1971) **Jägerskiöld, L. A.**
Species lists only. *See* p. 37: Rees and Rowe (1969).

HYDROZOA

Techniques

Nematocyst and hydranth morphology are becoming increasingly important in species identification. The following two references together with Bouillon (1985) on p. 33 summarize recent studies.

Taxonomic characters from the hydranths of thecate hydroids. In *Modern trends in the systematics, evolution and ecology of hydroids and hydromedusae* (ed. J. Bouillon, F. Boero, Cicogna, and P. F. S. Cornelius) Oxford University Press, Oxford. pp. 29–42.   (1987) **Cornelius, P. F. S.**
Techniques; identification of Campanulariidae.

New techniques and old problems in hydrozoan systematics. In *Modern trends in the systematics, evolution and ecology of hydroids and hydromedusae* (ed. J. Bouillon, F. Boero, Cicogna, and P. F. S. Cornelius), Oxford University Press, Oxford. pp. 67–82.   (1987) **Östman, C.**
Techniques and literature on nematocysts and electrophoresis in specific identification; identification of Campanulariidae.

General

Essai de classification de hydropolypes–hydroméduses (Hydrozoa–Cnidaria). *Indo-Malayan Zool.* **1**: 29–243.   (1985) **Bouillon, J.**
Extended definitions of all generic and suprageneric taxa; literature synopsis of all nematocyst types recorded in hydroid- and hydromedusa-like forms.

[Hydroids and hydromedusae of the USSR.] *Fauna SSSR* **70**, 660 pp, 463 figs,
30 pls. [Russian]. English translation (1969) by Israel program for scientific
translations, Jerusalem, No. 5108. [German translation (with amend-
ments) of identification keys: Thiel, M. E. (1962) Bestimmungstabellen der
Polypen und Medusen der russischen Gewässer. *Mitt. hamb. zool. Mus. Inst.*
60: 285–324.]   (1960) **Naumov, D. V.**
Some overlap with northern European fauna and only guide available to north
Norwegian waters. Many British species described.

'HYDRAS' (HYDRIDAE and ACTINULIDA)

A review of the Hydridae, and two new species from Natal. *Proc. Zool. Soc. Lond.*
118: 226–44.   (1948) **Ewer, R. F.**

Bestimmungsschlüssel der mitteleuropäischen *Hydra*–arten (Cnidaria,
Hydrozoa). *Archs Hydrobiol.* **107**: 529–43.   (1986) **Heitkamp, U.**

The freshwater hydras of Europe. 2. Description of *Hydra graysoni* sp. nov.
[from NE England]. *Archs Hydrobiol.* **69**: 547–56.   (1972) **Maxwell, T.**

Notes on the morphology and taxonomy of three hydras from a rheocrene
spring at Ashwell (Hertfordshire). *Naturalist, Hull* **1967**: 81–8.
                                    (1967) **Ball, I. R.**

The British freshwater hydras. *Countryside* **20**: 539–46.
                    (1968) **Grayson, R. F. and Hayes, A. D.**
Excellent guide to British freshwater hydras.

Some notes on hydras from south-Sweden and Denmark. *Kungl. Fysiog. Sälsk.*
Lund Förhandl. **18**: 170–7.   (1948) **Hansen-Melander, E.**

The Actinulida and their evolutionary significance. *Symp. Zool. Soc. Lond.* **16**:
119–33.   (1966) **Swedmark, B. and Teissier, G.**
For interstitial forms: useful bibliography.

Halammohydra vermiformis and *H. coronata* (Hydrozoa), new to the fauna of the
Netherlands. *Neth. J. Sea Res.* **8**: 407–9.
          (1974) **Wolff, W. J., Sandee, A. J. J., and Stegenga, H.**

Interstitial Cnidaria: Present status of their systematics and ecology. *Smithson.*
Contr. Zool. **76**: 1–8.   (1971) **Clausen, C.**
Review.

See also Delamere-Debouteville (1960) on p. 5 for interstitial forms and Clegg,
J. (1974) on p. 7 for coloured illustrations of *Hydra*.

HYDROIDS and HYDROMEDUSAE

No all-embracing guide available, though Ryland and Hayward (In press)
(see Cornelius and Ryland (1987), this section) will cover most British hydroid
stages. Meanwhile Barrett and Yonge (1958) on p. 6 is adequate for most of the
common and conspicuous intertidal Hydrozoa around British coasts. See also

Kramp (1935, *Danm. Fauna*), this section, Leloup (1952) on p. 33, and Russell (1970, appendix), this section, for literature between 1935 and 1970. Bouillon (1985), this section, gave the first acceptable classification of the group merging both hydroid and medusa stages. Velellids were treated by Kirkpatrick and Pugh (1984, listed here under Siphonophora, p. 38) and Brinckmann-Voss (1970), this section. Millard (1975), this section, provided keys to most families and genera of the hydroid stages. Nematocyst literature is indicated above.

Polypdyr (Coelenterata). I. Ferskvandspolypper og goplepolypper. *Danm. Fauna* **41**: 1–207, 81 figs. [Danish].   (1935) **Kramp, P. L.**

Monograph on the Hydroida of southern Africa. *Ann. S. Afr. Mus.* **68**: 1–513.
                                                    (1975) **Millard, N. A. H.**
Keys to and definitions of genera world-wide.

Essai de classification des hydropolypes–hydroméduses (Hydrozoa–Cnidaria). *Indo-Malayan Zool.* **2**: 29–243.   (1985) **Bouillon, J.**
Classification with definitions of all the genera; synopsis of occurrence of nematocyst types.

The hydromedusae of the Atlantic Ocean and adjacent waters. *Dana Rep.* **46**: 1–283, 335 figs, 2 pls.   (1959) **Kramp, P. L.**

The medusae of the British Isles. Anthomedusae, Leptomedusae, Limnomedusae, Trachymedusae, and Narcomedusae. Cambridge University Press, Cambridge. 530 pp, 319 figs, 35 pls.   (1953) **Russell, F. S.**
Supplement in Russell (1970) (see under Scyphomedusae on p. 39).

British thecate hydroids and their medusae. *Synopses Br. Fauna* (NS) (In prep.)
                                                    Cornelius, P. F. S.
Includes keys, descriptions, illustrations; NW France to mid-Norway.

Hydrozoa. In *An introduction to the marine fauna of the British Isles* (ed. J. S. Ryland and P. J. Hayward). Oxford University Press, Oxford.
                              (In press) **Cornelius, P. F. S. and Ryland, J. S.**
Includes all except minute species; keys, descriptions, illustrations; no medusa stages.

The hydroid species of *Obelia* (Coelenterata, Hydrozoa: Campanulariidae), with notes on the medusa stage. *Bull. Br. Mus. nat. Hist.* (Zool.) **28**: 249–93.
                                                    (1975) **Cornelius, P. F. S.**

A revision of the species of Lafoeidae and Haleciidae (Coelenterata: Hydroida) recorded from Britain and nearby seas. *Bull. Br. Mus. nat. Hist.* (Zool.) **28**: 373–426.   (1975) **Cornelius, P. F. S.**

A revision of the species of Sertulariidae (Coelenterata: Hydroida) recorded from Britain and nearby seas. *Bull. Br. Mus. nat. Hist.* (Zool.) **34**: 243–321.
                                                    (1979) **Cornelius, P. F. S.**

Hydroids and medusae of the family Campanulariidae recorded from the eastern North Atlantic, with a world synopsis of genera. *Bull. Br. Mus. nat. Hist.* (Zool.) **42**: 37–148.   (1982) **Cornelius, P. F. S.**

Redescription of *Laomedea exigua* M. Sars, a hydroid new to Scandinavia, with comments on its nematocysts, life cycle, and feeding movements. *Zool. Scripta* **16**: 1–8.   (1987) **Cornelius, P. F. S. and Östman, C.**
Details of a rarely reported species not treated fully elsewhere.

The British species of *Aglaophenia* (Coelenterata, Hydrozoa). *Bull. Br. Mus. nat. Hist.*   (In press) **Svoboda, A. and Cornelius, P. F. S.**

First records of the hydroid *Eudendrium glomeratum* in British and Irish waters. *Ir. Nat. J.* **22**: 244–6.   (1987) **Boero, F. and Cornelius, P. F. S.**

Hveldyr (Hydrozoa). *Natturufraedingurinn Reykjavik* **45**(1): 1–26. [Icelandic].
                                                 (1975) **Olafsson, E.**
Keys to 21 species from SW Iceland, recording *Sertularella tenella*. Hydroid stages only.

A history of the British hydroid zoophytes. J. Van Voorst, London. 2 vols. lxviii + 338 p., 67 pls + 45 woodcuts.   (1868) **Hincks, T.**
The standard work for many years. Plates still unsurpassed; taxonomy obsolescent.

A monograph of the gymnoblastic or tubularian hydroids. Ray Society, London. 2 vols (1871, 1872). 450 pp, 223 pls.   (1871–2) **Allman, G. J.**
Detailed and accurate; exceptionally well illustrated; taxonomy dated; omits some species, notably *Candelabrum (=Myriothela) phrygia*, for which see Hincks (1868) and Allman (1875), both this section, and the velellids, for which see Brinckmann-Voss (1970), this section, and Kirkpatrick and Pugh (1984) on p. 38.

On the structure and development of *Myriothela*. *Phil. Trans. R. Soc.* B**165**: 549–75.   (1875) **Allman, G. J.**

Hydrozoa I. *Tierwelt N.- und Ostsee* **3b**: 1–100, 105 figs.   (1928) **Broch, H.**
Key to species.

Hydrozoa (C I). A. Hydropolypen. *Fauna Ned.* **14**: 1–336, figs. [Dutch].
                                                 (1946) **Verwoort, W.**

Hydropoliepen (Hydroida). *Tabellenserie Strandwerkgemeens.* **27**: 1–22. [Dutch].
                                                 (1985) **Oosterbaan, A.**
Common intertidal species, Netherlands only.

Notes on the Hydrozoa of the Norfolk coast. *J. Linn. Soc.* (Zool.) **43**: 294–324, 26 figs, 1 pl.   (1957) **Hamond, R.**
Includes notes on and figs of many poorly known species.

Further notes on the Hydrozoa of the Norfolk coast. *Ann. Mag. nat. Hist.* (Ser. 13) **6**: 659–70.   (1963) **Hamond, R.**

The life-history of *Tubularia indivisa* (Hydrozoa: Tubulariidae) with observations on the status of *T. ceratogyne*. *J. mar. biol. Ass. UK* **63**: 467–79.
                                                 (1983) **Hughes, R. G.**

Corydendrium dispar, a new athecate hydroid from Scandinavian seas. *Göteborgs Vid. Samh. Handl.* (Ser. B) **5**, (4) **11**: 1–15.   (1935) **Kramp, P. L.**

Marine Hydrozoa. *Zoology Faroes* **1**(1:5): 1–59.   (1929) **Kramp, P. L.**
  Species list, distributions; no descriptions.

Marine Hydrozoa. a. Hydroida. *Zoology Iceland* **2**(5a): 1–82.
                                                          (1938) **Kramp, P. L.**
  Species list, distributions; no descriptions.

Observations on British and Norwegian hydroids and their medusae. *J. mar.
biol. Ass. UK* **23**: 1–42.   (1938) **Rees, W. J.**
  This paper and the next include many details of little known species not available
  elsewhere.

Notes on British and Norwegian hydroids and their medusae. *J. mar. biol. Ass.
UK* **25**: 129–41.   (1941) **Rees, W. J.**

On three northern species of *Hydractinia*. *Bull. Br. Mus. nat. Hist.* (Zool.) **3**: 351–
62.   (1956) **Rees, W. J.**

Hydroids of the Swedish west coast. *Acta R. Soc. scient. litt. gothoburg.* (Zool.) **3**:
1–24.   (1969) **Rees, W. J. and Rowe, M.**
  Species list only.

The hydroid fauna of the Oslo Fjord in Norway. *Norw. J. Zool.* **20**: 279–310.
                                                     (1972) **Christiansen, B. O.**
  Species list, distributions.

Bidrag til kundskaben om norges hydroider. *Forh. Vidensk. Selsk. Krist.* (**1873**):
91–150. [Old Norwegian].   (1874) **Sars, G. O.**
  A well illustrated and useful guide to most of the common and conspicuous southern-
  to mid-Norwegian species not treated by Hincks (1868), this section. Critique and
  list of contents in Hincks, T. (1874). Notes on Norwegian Hydroida from deep water.
  Ann. Mag. nat. Hist. (4) **13**: 125–37.

Anthomedusae/Athecatae (Hydrozoa, Cnidaria) of the Mediterranean. Part
I. Capitata. *Fauna Flora Golfo Napoli* **39**: 1–96, 106 text-figs, 11 pls.
                                                  (1970) **Brinckmann-Voss, A.**
  Includes several forms found further north.

Clavopsella quadranularia nov. spec. (Clavopsellidae nov. fam.), ein neuer
  Hydroidpolyp aus der Östsee und seine phylogenetische Bedeutung. *Z.
  Morph. Ökol. Tier* **51**: 88–90.   (1962) **Thiel, H.**
  Description and only published European record of the rare athecate hydroid
  Clavopsella navis. (For English-language description see Millard (1975), this section.)

Sur quelques hydroides de Roscoff. *Cah. biol. mar.* **13**: 323–64.
                                                          (1971) **Bouillon, J.**
  Detailed descriptions of *Dipurena simulans, Acauloides ammisatum, Stylactis claviformis,
  Thecocodium brieni, Lafoeina vilaevelebiti,* and *Helgicirrha schulzei.*

The hydroid and the medusa *Bougainvillia principis*, and a review of the British
species of *Bougainvillia. J. mar. biol. Ass. UK* **46**: 129–52. (1966) **Edwards, C.**

The hydroids and medusae *Podocoryne areolata, P. borealis and P. carnea. J. mar.
biol. Ass. UK* **52**: 97–144.   (1972) **Edwards, C.**

The hydroid *Trichydra pudica* and its medusa *Pochella polynema. J. mar. biol. Ass.
UK* **53**: 87–92.   (1973) **Edwards, C.**

The medusa *Modeeria rotunda* and its hydroid *Stegopoma fastigiatum*, with a
review of *Stegopoma* and *Stegolaria. J. mar. biol. Ass. UK* **53**: 573–600.
(1973) **Edwards, C.**

The medusa *Mitrocomella polydiademata* and its hydroid. *J. mar. biol. Ass. UK* **53**:
601–7.   (1973) **Edwards, C.**

The hydroids *Clava multicornis* and *Clava squamata. J. mar. biol. Ass. UK* **55**: 879–
86.   (1975) **Edwards, C. and Harvey, S. M.**

The hydroids and medusae *Sarsia occulta* sp.nov., *Sarsia tubulosa*, and *Sarsia
loveni. J. mar. biol. Ass. UK* **58**: 291–311.   (1978) **Edwards, C.**

Observations on the hydroids *Coryne pintneri* and *Thecocodium brieni* new to the
British list. *J. mar. biol. Ass. UK* **63**: 37–47.   (1983) **Edwards, C.**

The hydroids and medusae *Sarsia piriforma* sp.nov. and *Sarsia striata* sp.nov.
from the west coast of Scotland, with observations on other species. *J. mar.
biol. Ass. UK* **63**: 49–60.   (1983) **Edwards, C.**

On the occurrence of detached spherical colonies of the hydroids *Sarsia tubulosa*
and *Tubularia larynx* in Morecambe Bay. *J. mar. biol. Ass. UK* **51**: 495–503.
(1971) **Clare, J., Jones, D., and O'Sullivan, A. J.**
On aberrant colony forms of two common hydroids.

Branchiocerianthus norvegicus n. sp. from the Hardanger Fjord, western Norway.
Univ. Bergens Arb. **19**(6): 1–10.   (1957) **Brattström, H.**

A new record of *Branchiocerianthus norvegicus* (Hydrozoa) from western Norway.
Sarsia **48**: 99–104.   (1972) **Brattström, H.**

TRACHYMEDUSAE

A new, bottom-living trachymedusa from the Oslo Fjord. Description of the
species, and a general discussion of the life conditions and fauna of the fjord
deeps. *Nytt. Mag. Zool.* **6**: 121–43.   (1958) **Beyer, F.**

On the biology of a bottom-dwelling trachymedusa *Tesserogastria musculosa*
Beyer. *Norw. J. Zool.* **19**: 1–19.   (1971) **Hesthagen, I. H.**

SIPHONOPHORA

See also various sheets by A. K. Totton *Fiches d'identification du zooplankton* on p.
6. For complete list of these and other Totton siphonophore publications, see
Cornelius (1974), this section.

A synopsis of the Siphonophora. British Museum (Natural History), London. 230
pp, 153 figs, 40 pls.   (1965) **Totton, A. K.**
Descriptions and illustrations of all species, but excluding *Physalia* and also the
hydroidan velellids for both of which see next entry.

Bibliography of A. K. Totton. *Deep-Sea Res.* **21**: 787–9.
(1974) **Cornelius, P. F. S.**
Lists all A. K. Totton's siphonophore papers in letter to editor, Pugh, P. R.

Siphonophores and velellids. *Synopses Br. Fauna* (N.S.) **29**: i–vii + 1–154, figs
  1–60.   (1984) **Kirkpatrick, P. A. and Pugh, P. R.**

SCYPHOMEDUSAE

See also Hardy (1956) on p. 32 for shallow water British forms.

*The medusae of the British Isles. II. Pelagic Scyphozoa, with a supplement to the first
  volume on hydromedusae.* Cambridge University Press, Cambridge. 283 pp, 102
  figs, 15 pls.   (1970) **Russell, F. S.**
The most useful work but excluding Stauromedusae.

STAUROMEDUSAE

See also Barrett and Yonge (1958) on p. 6 for four of the British intertidal
species; Kramp (1961) on p. 32 for specific diagnoses (without illustrations);
and Ryland and Hayward (*in press*) on p. 6 for all seven British species.

A new species of the stauromedusan genus *Lucernariopsis* (Coelenterata:
  Scyphomedusae). *J. mar. biol. Ass. UK* **58**: 285–90.   (1978) **Corbin, P. G.**

Scyphomedusae of the seas of the U.S.S.R. *Fauna SSSR* **75**: 98 pp, text-figs 1–
  72, 3 pls. [Russian]. (Translation of Stauromedusa chapter in Marine
  Biological Association, Plymouth, and BM(NH), London).
                                              (1961) **Naumov, D. V.**
For *Lucernaria quadricornis, Lucernariopsis campanulata, Haliclystus auricula*, and *H. salpinx*
(as *Octomanus monstrosus*).

The biology of three New England stauromedusae, with a description of a new
  species. *Can. J. Zool.* **40**: 1249–62.   (1962) **Berrill, N. J.**
For *Lucernaria quadricornis*; and also *Haliclystus salpinx*, which occurs at Bergen,
possibly also in Orkney and Shetland, and perhaps on the north coast of mainland
Scotland.

Lucernariae and their allies. *Smithson. Contr. Knowl.* **2**(2): 1–30.
                                              (1878) **Clark, H. J.**
For *Haliclystus auricula*.

Lucernaria discoides, a new species from the Channel Islands. *J. mar. biol. Ass. UK*
  23: 167–70.   (1938) **Eales, N. B.**
For *Lucernariopsis campanulata* as *Lucernaria discoides*.

Report on the deep-sea medusae dredged by H.M.S. *Challenger* during the
  years 1873–1876. *Rep. scient. Results Voy. Challenger* (Zool.) **4**(12): cv + 154
  pp.   (1881) **Haeckel, E.**
For *Lucernaria bathyphila*.

A history of the British zoophytes. Lizars, Edinburgh. 2 vols. 341 pp. (2nd edn J.
  Van Voorst, London, 1847).   (1838) **Johnston, G.**
For *Craterolophus convolvulus*, as *Lucernaria campanulata*.

Fauna littoralis Norvegiae, oder Beschreibung und Abbildungen neuer oder wenig bekannten Seethiere, nebst Beobachtungen über die Organisation, Lebensweise u. Entwicklung derselben. Vol. 1. Christiania, Oslo.   (1846) **Sars, M.**
For *Depastruim cyathiforme* (as *Lucernaria cyathiformis*) and *Lucernaria quadricornis*.

INCERTAE SEDIS

Tetraplatia, a coronate scyphomedusan. *Proc. R. Soc.* **B152**: 263–81.
                                                          (1960) **Ralph, P. M.**

Account and figs of an aberrant cosmopolitan form recorded from British and Norwegian seas.

ANTHOZOA

General

British Anthozoa. *Synopses Br. Fauna* (N.S.) **18**: i–iii, 1–241, 2 pls, 83 figs.
                                                          (1981) **Manuel, R. L.**
Comprehensive and authoritative.

The Anthozoa of the British Isles, 2nd edn. Marine Conservation Society, Ross-on-Wye. 79 pp, 94 col. photographs.   (1983) **Manuel, R. L.**
Useful notes on identification, supplementing previous but less detailed; colour photographs of all species.

Anthozoans (Coelenterata: Anthozoa) new to Ireland and new records of some rarely recorded species. *Ir. Nat. J.* **21**: 484–8.   (1985) **Picton, B. E.**
Several species not in Manuel (1981), this section. Includes brief descriptive notes.

Anthozoa. *Tierwelt N.-u. Ostsee* **3e:** 1–317, 211 figs.   (1934) **Pax, F.**

Anthozoa. In *An introduction to the marine fauna of the British Isles* (ed. J. S. Ryland and P. J. Hayward). Oxford University Press, Oxford.
                                                          (In press) **Manuel, R. L.**
Keys, descriptions, illustrations; nearly all British species.

Alcyonarian and madreporarian corals in the Museum of Bergen, collected by the Fram-Expedition 1898–1902 and by the 'Michael Sars' 1900–1906. *Bergens Mus. Arb.* (Naturvid. række, Heft 2, 1915–1916) (6): 1–44.
                                                          (1916) **Jungersen, H. F. E.**
One of the few papers treating high-arctic species.

Actiniaria, Zoantharia, and Madreporaria. *Zool. Iceland* **2**(8): 1–20.
                                                          (1939) **Carlgren, O.**
Species lists, distributions, references; no descriptions or keys.

Actiniaria and Zoantharia. *Zool. Faroes* **7b**:K 1–5.   (1930) **Carlgren, O.**
Species lists, distributions, references; no descriptions or keys.

Die Gorgonaria, Pennatularia und Antipatharia des Tiefwassers der Biskaya (Cnidaria, Anthozoa). I. Allgemeiner Teil; II. Taxonomischer Teil. *Bull.*

Mus. natn. Hist. nat., Paris (Ser. 4) **3**(A) (1981): 731–66 + 941–78.
(1982) Grasshoff, M.
With keys and illustrations.

A survey of the Ptychodactiaria, Corallimorpharia and Actiniaria. *Kungl. Svensk. Vetenskakad. Handl.* (4) **1**(1): 1–121.   (1949) **Carlgren, O.**
Comprehensive guide to the anemone-like groups. Generic definitions and lists of species with distributions.

OCTOCORALLIA (=ALCYONARIA)

Key to the genera of Octocorallia exclusive of Pennatulacea (Coelenterata: Anthozoa), with diagnoses of new taxa. *Proc. biol. Soc. Wash.* **94**: 902–47.
(1981) Bayer, F. M.
World-wide; keys; figs of spicules.

Status of knowledge of octocorals of world seas. In *Seminarios de biologia marina* (ed. anon.), pp. 3–102. Academia Brasileira de Ciencias, Rio de Janeiro.
(1981) Bayer, F. M.
Nearly 1500 references to taxonomic and other works, listed geographically.

ZOANTHARIA (=ZOANTHIDEA)

Zoantharia. *Igolf Exped. Reps* **5**(4): 1–63.   (1913) **Carlgren, O.**
Northern forms.

CERIANTIPATHARIA

See also Manuel (1981, 1983) on p. 40 for adult forms.

Ceriantharia. *In golf Exped. Reps* **5**(3): 1–79.   (1912) **Carlgren, O.**
Northern species.

Arachnanthus sarsi Carlgren, 1912, a redescription of a cerianthid anemone new to the British Isles. *Zool. J. Linn. Soc.* **83**: 343–9.
(1985) Picton, B. E. and Manuel, R. L.
The Antipatharia of the Siboga expedition. *Siboga Rep.* **17**: 1–259.
(1914) Pesch, A. J. van
A major taxonomic work on the group.

GORGONACEA

Les gorgonaires de la Méditerranée. *Bull. Inst. océanogr. Monaco* **71**(1430): 1–140, 62 figs, 1 pl.   (1975) **Carpine, C. and Grasshoff, M.**
Keys to all genera and species. For 'Lusitanian' elements.

Die Gorgonaria des östlichen Nordatlantik und des Mittelmeeres I. Die Familie Ellisellidae (Cnidaria: Anthozoa). *Meteor Forsch.-Ergebnisse* (D) **10**: 73–87.   (1972) **Grasshoff, M.**

Die Gorgonaria des östlichen Nordatlantik und des Mittelmeeres II. Die
  Gattung *Acanthogorgia* (Cnidaria: Anthozoa). *Meteor Forsch.-Ergebnisse* (D)
  13: 1–10.   (1973) **Grasshoff, M.**

Die Gorgonaria des östlichen Nordatlantik und des Mittelmeeres III. Die
  Familie Paramuriceidae (Cnidaria, Anthozoa). *Meteor Forsch.-Ergebnisse* (D)
  27: 5–76.   (1977) **Grasshoff, M.**

ACTINARIA

See also Manuel (1981) on p. 40.

On the identity of the sea anemone *Scolanthus callimorphus* Gosse, 1853
  (Actiniaria: Edwardsiidae). *J. nat. Hist.* **15**: 265–76. (1981) **Manuel, R. L.**

British sea anemones. Ray Society, London. 2 vols, 574 pp, 33 pls.
                                            (1928–35) **Stephenson, T. A.**
  Includes coloured illustrations; taxonomy now amended but largely excellent;
  nomenclature partly obsolete.

Actiniaria I. *In golf Exped. Reps* **5**(9): 1–241.   (1921) **Carlgren, O.**
  Northern species.

SCLERACTINIA

Etude systématique et écologique des madréporaires de la région de Banyuls-
  sur-Mer (Pyrénées-Orientales). *Vie Milieu* (A) **20**: 293–325. (1969) **Best, M.**
  Includes many species recorded from British waters.

Les scléractiniares de la Méditerranée et de l'Atlantique nord-oriental. *Méms
  Inst. océanogr. Monaco* **11**: (in three volumes, unnumbered) 1–227 pp (text),
  1–107 pls, 229–84 (annexes + index).   (1980) **Zibrowius, H.**
  Detailed monograph on all species from Senegal to Iceland, Faroes, and north
  Norway.

PENNATULACEA

*Report on the Pennatulida collected in the Oban dredging excursion of the Birmingham
  Natural History and Microscopical Society, July 1881.* Herald Press, Birmingham.
  77 pp, 4 pls.   (1882) **Marshall, A. M. and Marshall, W. P.**

Pennatularia. *Das Tierreich* **43**: i–xv, 1–132. [German]. (1915) **Kükenthal, W.**
  Well illustrated; keys to and definitions of all taxa down to species.

CTENOPHORA

See also Newell and Newell (1977) on p. 6 for common British planktonic
forms.

Stamm Ctenophora, Rippenquallen, Kammquallen. In *Lehrbuch der speziellen Zoologie* (co-ordinated by A. Kaestner), Band I: *Wirbellose Tiere* (ed. H-E. Gruner), 2 Teil: *Cnidaria, Ctenophora, Mesozoa, Plathelminthes, Nemertini, Entoprocta, Nemathelminthes, Priapulida.* pp. 306–35. G. Fischer, Stuttgart.
(1984) **Werner, B.**
A sound introduction to the group, partly replacing Hyman (1940) on p. 32; well illustrated; in German.

Ctenophora. In *Synopsis and classification of living organisms* (ed. S. P. Parker), Vol. 1, pp. 707–15. McGraw Hill, New York.
(1982) **Harbison, G. R. and Madin, L. P.**
Orders and families defined, genera listed; most forms clearly illustrated. Best modern synopsis of classification.

On the classification and evolution of the Ctenophora. In *The origins and relationships of lower invertebrates* [*Syst. Ass. spec. Vol.* 28] (ed. S. C. Morris, J. D. George, R. Gibson, and H. M. Platt), pp. 78–100. Oxford University Press, Oxford.
(1985) **Harbison, G. R.**
Best source of literature for identification of extralimital species.

Ctenophora. *Fich. Ident. Zooplancton* **146**: 1–6, figs 1–5.
(1975) **Greve, W.**
English Channel to Spitzbergen. Figs, tabular key, lengthy reference list.

Ctenophora. *Tierwelt N.- u. Ostsee* **3f**: 1–50, 27 figs.
(1927) **Krumbach, T.**

The open sea. Its natural history. Part one: The world of plankton. Collins, London. 335 pp, 93 figs, 24 pls.
(1956) **Hardy, A. C.**
Illustrations of forms common in British waters.

PLATYHELMINTHES

Platyhelminthes and Rhynchocoela. The acoelomate Bilateria. *The invertebrates.* McGraw Hill, New York. viii + 550 pp, 208 figs.
(1951) **Hyman, L. H.**

Platyhelminthes. *Traité de zoologie, anatomie, systématique, biologie.* Masson and Cie, Paris. **4**(1): 21–692 and 8878–95.
(1961) **Grassé, P-P.**, ed.

Das Phylogenetische System der Plathelminthes. G. Fischer, Stuttgart. 317 pp.
(1985) **Ehlers, U.**

TURBELLARIA

Classe des turbellariés. Turbellaria (Ehrenberg, 1831). In *Traite de zoologie: anatomie, systématique, biologie* (ed. P-P. Grassé), Vol. 4, pp. 35–212. Masson, Paris.
(1961) **de Beauchamp, P.**

Turbellaria of the world. A guide to families and genera. Queensland Museum, Brisbane. 136 pp.
(1986) **Cannon, R. R. G.**

KALYPTORHYNCHIA

[Rostellar ciliated worms Kalyptorhynchia of the fauna of USSR and adjacent
countries. *Turbellaria.*] *Fauna SSSR* (N.S.) **115**. Izdatel'stvo 'Nauka', Leningrad. **1**(1) 399 pp. [Russian].   (1977) **Evodnin, L. A.**

RHABDOCOELIDA

GENERAL

Contributions to a knowledge of British marine Turbellaria. *Q. Jl microsc. Sci.*
34: 433–528, 3 pls.   (1893) **Gamble, F. W.**

Acoela. *Tierreich* **23**: 35 pp, 8 figs.   (1905) **Graff, L. von**

Rhabdocoela. *Tierreich* **35**: 484 pp, 394 figs.   (1913) **Graff, L. von**

Die Turbellarien Ostfennoskandiens. I. Acoela, Catenulida. *Fauna fenn.* **7**: 1–
155, 39 figs, 1 pl.   (1960) **Luther, A.**

Die Turbellarian Ostfennoskandiens. III. Neorhabdocoela 1. Dalyellioida,
Typhlopanoida: Brysophlebidae und Trgonostomidae. *Fauna fenn.* **12**: 1–
69, 29 figs.   (1962) **Luther, A.**

Die   Turbellarien   Ostfennoskandiens.   IV.   Neorhabdocoela   2.
Typhloplanoida: Typhloplanidae,   Solenopharyngidae   und   Carch-
arodopharyngidae. *Fauna fenn.* **16**: 1–163, 17 figs, 2 pls. (1963) **Luther, A.**

Die Turbellarien Ostfennoskandiens. V. Neorhabdocoela 3. Kalyptorhyn-
chia. *Fauna fenn.* **17**: 1–59, 126 figs.   (1962) **Karling, T. G.**

FRESHWATER

Süsswasserfauna Dtl. **19**: 94 pp, 120 figs.   (1909) **Graff, L. von**
Die Strudelwürmer. Mongraphien einheimischer Tiere. Klinkhardt, Leipzig. Vol. 5,
380 pp, 156 figs, 2 pls.   (1913) **Steinmann, P. and Bresslau, E.**

British and Irish freshwater Microturbellaria: historical records, new records
and a key for their identification. *Arch. Hydrobiol.* **67**(2): 210–41, 4 figs.
                                   (1970) **Young, J. O.**

MARINE

The interstitial Turbellaria Kalyptorhynchia from some North Wales
beaches. *Proc. zool. Soc., Lond.* **141**: 173–205, 41 figs. (1963) **Boaden, P. J. S.**

The interstitial fauna of some North Wales beaches. *J. mar. biol. Ass. UK* **43**:
79–96, 1 fig.   (1963) **Boaden, P. J. S.**

TRICLADIDA

Die Turbellarien Ostfennoskandiens. II. Tricladida. *Fauna fenn.* **2**: 1–42, 3 pls.
(1961) **Luther, A.**

British Planarians. *Synopses Br. Fauna* (N.S.) **19**: 141 pp.
(1981) **Ball, I. R. and Reynoldson, T. B.**

FRESHWATER

The freshwater triclads of Wicken Fen. *Guides to Wicken Fen.* National Trust Wicken Fen Local Committee, Cambridge. No. 6, 4 pp. (1968) **Ball, I. R.**

The taxonomy, habitat, and distribution of the freshwater triclad *Planaria torva* (Platyhelminthes: Turbellaria) in Britain. *J. Zool., Lond.* **157**(1): 99–123, 9 figs.   (1969) **Ball, I. R., Reynoldson, T. B., and Warwick, T.**

A key to the British species of freshwater triclads. *Scient. Publs. Freshwat. biol. Ass.* **23**, 28 pp, 7 figs.   (1967) **Reynoldson, T. B.**

MARINE

Tricladenstudien. I. Tricladida Maricola. *Z. wiss. Zool.* **81**: 344–504, 8 pls.
(1906) **Böhmig, L.**

All species described up to 1906.

Tricladen. *Fauna Fora Golf. Neapel* **32**, 405 pp, 16 pls. Friedländer, Berlin.
(1909) **Wilhelmi, J.**

All species described up to 1909.

Marine triclads of the Plymouth area. *J. mar. biol. Ass. UK* **48**: 209–23, 2 figs.
(1968) **Hartog, C. den**

TERRESTRIAL

On the name of the ground fluke *Fasciola terrestris* O. F. Müller, on *Othelosoma symondsi* Gray, and on the genus *Amblyplana* von Graff. *J. Lin. Soc. Lond.* (Zool.) **42**(285): 207–18, 1 fig, 3 pls.   (1953) **Pantin, C. F. A.**

A taxonomic review of the terrestrial planarians of Europe. *Boll. Zool.* **44**: 399–419.   (1977) **Minelli, A.**

POLYCLADIDA

Die Polycladen (Seeplanarien) des Golfes von Neapel. . . . *Fauna Flora Golf. Neapel* **2**: 688 pp, 39 pls.   (1884) **Lang, A.**

Studien über Polycladen. *Zool. Bidr. Upps.* **2**: 31–343, 8 pls.   (1913) **Bock, S.**

British polyclad turbellarians. *Synopses Br. Fauna* (N.S.) **26**: 77 pp.
(1982) **Prudhoe, S.**

A monograph on polyclad Turbellaria. British Museum (Natural History), London and Oxford University Press, Oxford. 259 pp.   (1985) **Prudhoe, S.**

MONOGENEA

GENERAL

Monogenetic trematodes. Their systematics and phylogeny (ed. W. J. Hargis; translated by P. C. Oustinoff). American Institute of Biological Sciences, Washington, DC. xx + 627 pp, 315 figs.   (1961) **Bychowsky, B. E.**

Bibliography of the monogenetic trematode literature of the world, 1758 to 1969. *Spec. Sci. Rep. Virginia Inst. mar. Sci.* **55**, v + 195 pp. Supplement I with errata. iv + 13 pp (1970).   (1969) **Hargis, W. J.** *et al.*

A synopsis of the monogenetic trematodes. *Trans. zool. Soc., Lond.* **25**(4): 185–600.   (1947) **Sproston, N. G.**
Diagnoses, keys to suborders, families, genera. Lists species with references to literature. World host-list, distribution. List British species, with hosts.

Monogenea and Aspidocotylea. *Systema helminthum IV.* Interscience Publishers, New York and London. viii + 699 pp, 134 pls.
          (1963) **Yamaguti, S.**

FRESHWATER

[*Key to parasites of freshwater fish.*] Israel program for scientific translations, Jerusalem. 919 pp, 1628 figs.
          (1964) **Bychowskaya-Pavlovskaya, I. E.** *et al.*

[Parasitic metazoa (first part)] In [*Keys to the parasitic Monogenea of freshwater fishes in the fauna of the USSR*] (ed. O. N. Bauer) (*Keys to the fauna of the USSR*, 143).] **2**: 425 pp. [Russian].   (1985) **Gusev, A. V.**

Species of *Gyrodactylus* von Nordmann, 1832 (Monogenea: Gyrodactylidae) from freshwater fishes in southern England, with a description of *Gyrodactylus rogatensis* sp. nov. from the bullhead *Cottus gobio* L. *J. nat. Hist.* **19**: 791–809.   (1985) **Harris, P. D.**

A checklist of British and Irish freshwater fish parasites with notes on their distribution. *J. Fish. Biol.* **6**: 613–44.   (1974) **Kennedy, C. R.**

Les monogenes Monopisthocotylea parasites des poissons d'eau douce de la France méditerranée. *Bull. Mus. natn. Hist. nat. Paris* (Ser. 3) no. 429, (Zool.) **299**: 177–214.   (1977) **Lambert, A.**

OTHER VERTEBRATES

The Trematoda of British fishes. Ray Society, London. viii + 364 pp.
          (1947) **Dawes, B.**
Provides keys to families and higher groups, diagnoses. Short descriptions of many species. Host list.

A checklist of monogenean (platyhelminth) parasites of Plymouth hosts. *J. mar. biol. Ass. UK* **64**: 881–7.
(1984) **Llewellyn, J., Green, J. E., and Kearn, G. C.**

Platyhelminth parasites of the Amphibia. British Museum (Natural History), London and Oxford University Press, Oxford. 217 pp + 371 pp on microfiche.
(1982) **Prudhoe, S. and Bray, R. A.**

TREMATODA (DIGENEA and ASPIDOCOTYLEA)

GENERAL

The Trematoda, with special reference to British and other European forms. Cambridge University Press, Cambridge. xvi + 644 pp, 81 figs.
(1946) **Dawes, B.**

[*Trematodes of animals and man. Fundamentals of trematodology.*] Izdatlst'vo 'Nauka', Moscow. 1–26 [Russian]. English translations: Amerind, Delhi, Vols 1, 2, 23, 24; Israel program for scientific translations, Jerusalem, Vols. 17, 18.
(1947–78) **Skrjabin, K. I.**, ed.

Keys to the trematodes of animals and man (ed. H. P. Arai; translated by R. W. Dooley). University of Illinois Press, Urbana. xvi + 351 pp, 919 figs.
(1964) **Skrjabin, K. I.** *et al.*

Synopsis of digenetic trematodes of vertebrates. Keigaku Publishing, Tokyo. Vol. 1, 1074 pp Vol. 2, 1796 pp.
(1971) **Yamaguti, S.**

A synoptical review of life histories of digenetic trematodes of vertebrates. Keigaku Publishing, Tokyo. 590 pp, 219 pls.
(1975) **Yamaguti, S.**

FRESHWATER HOSTS

[*Key to parasites of freshwater fish.*] Israel program for scientific translations, Jerusalem. 919 pp, 1628 figs.
(1964) **Bychowskaya-Pavlovskaya, I. E.** *et al.*

The Trematoda of British fishes. Ray Society, London. viii + 364 pp.
(1947) **Dawes, B.**

A checklist of British and Irish freshwater fish parasites with notes on their distribution. *J. Fish. Biol.* **6**: 613–44.
(1974) **Kennedy, C. R.**

British freshwater cercariae. Universidad de Oriente, Cumana, Venezuela. 345 pp.
(1984) **Nasir, P.**

MARINE HOSTS

The Trematoda of British fishes. Ray Society, London. viii + 364 pp.
(1947) **Dawes, B.**

The Accacoeliidae (Digenea) of fishes from the north-east Atlantic. *Bull. Br. Mus. nat. Hist.* (Zool.) **31**: 51–99.   (1977) **Bray, R. A. and Gibson, D. I.**

The Azygiidae, Hirudinellidae, Ptychogonimidae, Sclerodistomidae and Syncoeliidae of fishes from the north-east Atlantic. *Bull. Br. Mus. nat. Hist.* (Zool.) **32**: 167–245.   (1977) **Gibson, D. I. and Bray, R. A.**

The Hemiuroidea: terminology, systematics and evolution. *Bull. Br. Mus. nat. Hist.* (Zool.) **36**: 35–146.   (1979) **Gibson, D. I. and Bray, R. A.**

The Fellodistomidae (Digenea) of fishes of the north-east Atlantic. *Bull. Br. Mus. nat. Hist.* (Zool.) **37**: 199–293. (1980) **Bray, R. A. and Gibson, D. I.**

The Hemiuridae (Digenea) of fishes of the north-east Atlantic. *Bull. Br. Mus. nat. Hist.* (Zool.) **51**: 1–125.   (1986) **Gibson, D. I. and Bray, R. A.**

The Zoogonidae (Digenea) of fishes of the north-east Atlantic. *Bull. Br. Mus. nat. Hist.* (Zool.) **51**: 127–206.   (1986) **Bray, R. A. and Gibson, D. I.**

AMPHIBIANS AND REPTILES

Platyhelminth parasites of the Amphibia. British Museum (Natural History), London and Oxford University Press, Oxford. 217 pp + 371 pp on 4 microfiche.   (1982) **Prudhoe, S. and Bray, R. A.**

[*Helminths of amphibians in the fauna of the USSR.*] Izdat. 'Nauka', Moscow. 278 pp, 115 figs. [Russian].
    (1980) **Ryzhikov, K. M., Sharpilo, V. P. and Shevchenko, N. N.**

[*Parasitic worms of reptiles in the fauna of the USSR.*] Izdat. 'Naukova Dumka', Kiev. 287 p, 116 figs. [Russian].   (1976) **Sharpilo, V. P.**

BIRDS AND MAMMALS

Synopsis des Strigeidae et des Diplostomatidae (Trematoda). *Mém. Soc. neuchât. Sci. nat.* **10**(1): 1–258, 270 figs; **10**(2): 259–727, 480 figs.
    (1968–70) **Dubois, G.**

The helminth fauna of mustelids and the ways of its formation. Amerind, New Delhi. xvi + 607 pp.   (1985) **Kontramavichus, V. L.**

[*Helminths of diving and marsh birds of the fauna of the Ukraine.*] Izdat 'Naukova Dumka', Kiev. 415 pp, 74 figs. [Russian]. (1976) **Smogorzhevskaya, L. A.**

[*Identification key to the helminths of rodents of the fauna of the USSR. Cestodes and trematodes.*] Izdat. 'Nauka', Moscow. 231 pp, 45 figs. [Russian].
    (1978) **Ryzhikov, K. M.** *et al.*

[*Keys to the identification of helminths in domestic and wild pigs.*] Izdat. 'Nauka', Moscow. 168 pp, 75 figs. [Russian].   (1983) **Ryzhikov, K. M.** *et al.*

[*Helminths of deer.*] Izdat. 'Nauka', Alma-Ata. 223 pp, 50 figs. [Russian].
    (1976) **Pryadko, E. I.**

[*Key to the helminths of carnivores in the USSR.*] Izdat. 'Nauka', Moscow. 275 pp, 123 figs. [Russian]. (1970) **Gvozdev, E. V.** *et al.*

[*Helminths of waterfowl. Key to their identification, morphology, biology and laboratory diagnosis.*] Academia, Prague. 239 pp, 100 figs. [Czech].
(1982) **Ryšavÿ, B.** *et al.*

Key to trematodes reported in waterfowl. United States Departments of the Interior, Fish and Wildlife Service, Washington DC. Resource Publication no. 142, 156 pp. (1981) **McDonald, M. E.**

MAN AND DOMESTIC ANIMALS

A checklist of the helminth parasites of domestic animals of the United Kingdom. Hoechst UK Ltd, Milton Keynes. 147 pp. (1983) **Scofield, A. M.**

Trematode zoonoses. *Handbook series in zoonoses* (Section C). CRC Press, Boca Raton, Florida. Vol. 3, 337 pp. (1982) **Hillyer, G. V. and Hopla, C. E.**, eds

Helminths, arthropods, and protozoa of domestic animals, 7th edn. Baillière Tindall, London. 809 pp. (1982) **Soulsby, E. J. L.**

Worms and diseases. A manual of medical helminthology. William Heinemann Medical Books, London. 161 pp. (1975) **Muller, R.**

CESTODA

GENERAL

Biology of the Eucestoda. Academic Press, London. Vol. 1, 296 + lxxi pp; Vol. 2, 297–628 + i–xxi pp. (1980) **Arme, C. and Pappas, P. W.**

Études critiques sur les Tetrarhynches du Museum de Paris. *Arch. Mus. natn. Hist. nat. Paris* **19**: 466 pp. (1942) **Dollfus, R. P.**

[*Essentials of cestodology.*] Izdat. 'Nauka', Moscow. Vols. 1–11. [Russian] English translations by the Israel program for scientific translations, Vols. 1, 4, 5. (1951–69) (Vols. 1–7) **Skrjabin, K. I.**, ed.
(1977–78) (Vols. 8, 9) **Ryzhikov, K. M.**, ed.
(1981) (Vol. 10) **Freze, V. I.**, ed.
(1985) (Vol. 11) **Sudarikov, V. E.**, ed.

Handbook of tapeworm identification. CRC Press, Boca Raton, Florida. 675 pp.
(1986) **Schmidt, G. D.**

FRESHWATER HOSTS

[*Key to parasites of freshwater fish.*] Israel program for scientific translations, Jerusalem. 919 pp, 1628 figs.
(1964) **Bychowskaya-Pavlovskaya, I. E.** *et al.*

A checklist of British and Irish freshwater fish parasites with notes on their distribution. *J. Fish. Biol.* **6**: 613–44. (1974) **Kennedy, C. R.**

MARINE HOSTS

[*Helminthofauna of marine mammals (ecology and phylogeny).*] Israel program for scientific translations, Jerusalem. ix + 522 pp, 240 figs.
(1968) **Delyamure, S. L.**

AMPHIBIANS AND REPTILES

Platyhelminth parasites of the Amphibia. British Museum (Natural History), London and Oxford University Press, Oxford. 217 pp + 371 pp on 4 microfiche.
(1982) **Prudhoe, S. and Bray, R. A.**

[*Helminths of amphibians in the fauna of the USSR.*] Izdat. 'Nauka', Moscow. 278 pp, 115 figs. [Russian].
(1980) **Ryzhikov, K. M., Sharpilo, V. P., and Shevchenko, N. N.**

[*Parasitic worms of reptiles in the fauna of the USSR.*] Izdat. 'Naukova Dumka', Kiev. 287 pp, 116 figs. [Russian].
(1976) **Sharpilo, V. P.**

BIRDS AND MAMMALS

The helminth fauna of mustelids and the ways of its formation. Amerind Publishing, New Delhi. xvi + 607 pp.
(1985) **Kontramavichus, V. L.**

[*Helminths of diving and marsh birds of the fauna of the Ukraine.*] Izdat. 'Naukova Dumka', Kiev. 415 p, 74 figs. [Russian]. (1976) **Smogorzhevskaya, L. A.**

[*Identification key to the helminths of rodents of the fauna of the USSR. Cestodes and trematodes.*] Izdat. 'Nauka', Moscow. 231 pp, 45 figs. [Russian].
(1978) **Ryzhikov, K. M.** *et al.*

[*Keys to the identification of helminths in domestic and wild pigs.*] Izdat. 'Nauka', Moscow. 168 pp, 75 figs. [Russian].
(1983) **Ryzhikov, K. M.** *et al.*

[*Helminths of deer.*] Izdat. 'Nauka', Alma-Ata. 223 pp, 50 figs. [Russian].
(1976) **Pryadko, E. I.**

[*Key to the helminths of carnivores in the USSR.*] Izdat. 'Nauka', Moscow. 232 pp, 72 figs. [Russian].
(1979) **Gvozdev, E. V.** *et al.*

[*Helminths of waterfowl. Key to their identification, morphology, biology, and laboratory diagnosis.*] Academia, Prague. 239 pp, 100 figs. [Czech].
(1982) **Ryšavý, B.** *et al.*

MAN AND DOMESTIC ANIMALS

A checklist of the helminth parasites of domestic animals of the United Kingdom. Hoechst UK, Milton Keynes, 147 pp.
(1983) **Scofield, A. M.**

Cestode zoonoses. *Handbook series in zoonoses*, Section C. CRC Press, Boca Raton, Florida. Vol. 1, 361 pp. (1982) **Jacobs, L. and Arambulo, P.**, eds

Helminths, arthropods, and protozoa of domestic animals, 7th edn. Baillière Tindall,
London. 809 pp.   (1982) **Soulsby, E. J. L.**

Worms and diseases. A manual of medical helminthology. William Heinemann
Medical Books, London. 161 pp.   (1975) **Muller, R.**

NEMERTINEA

Embranchement des nemertiens. In *Traité de zoologie, anatomie, systématique,
biologie* (ed. P-P. Grassé), Masson, Paris. Vol. 4(1), pp. 783–886.
                                                  (1961) **Gontcharoff, M.**

Nemerteans. Hutchinson, London. 224 pp, 33 figs.   (1972) **Gibson, R.**

British Nemerteans. *Synopsis Br. Fauna* **24**: 212 pp, 50 figs. (1982) **Gibson, R.**

ROTIFERA

About 600 British species

General and freshwater

Synopsis of the Rotatoria. *Bull. US natn. Mus.* **81**: 226 pp.
                                                  (1913) **Harring, H. K.**

Rotatorien. *Bronn's Kl. Ordn. Tierreichs* **4**(2)1, 1–4: 1–576.
                                                  (1929–33) **Remane, A.**

Introduction to the study of rotifers. I–XVII. *Microscope* (London) **5**(10): 253–6;
   (11): 292–5; (12): 300–6; **6**(1): 3–9; (2): 31–4; (3): 78–82; (4): 98–100; (6):
   160–4; (8): 205–11; (9): 237–44; (12): 309–11; **7**(1): 2–6; (3): 65–8; (7): 169–
   74; (8): 197–9; (11): 287–90; (12): 322–4.   (1945–50) **Hollowday, E. D.**
   A simple account.

The Czechoslovak Rotatoria of the Order Bdelloidea. *Vestn. csl. zool. Spol.* **15**:
   241–500, 66 figs.   (1951) **Bartos, E.**
   Keys to genera and species, keys in English relevant to most of Europe.

Matériaux relatifs à la nomenclature et à la bibliographie de rotifères. *Polskie
   Archwm Hydrobiol.* **2**(15): 7–249.   (1954) **Wisniewski, J.**

Beiträge zur ökologie und Systematik der Bodenrotatorien. *Zool. Jb. (Syst.)* **82**:
   551–617.   (1954) **Schulte, H.**

Rädertiere (Rotatorien). Frankh'sche Verlangshandlung, Stuttgart. 56 pp. [Ger-
   man] English translation: Wright, H. G. S. (1966). *Rotifers.* Warne,
   London. 80 pp.   (1956) **Donner, J.**

How to begin the study of rotifers. Parts 1–7. *Countryside* (N.S.) **19**(4): 150–6;
(5): 188–94; (6): 246–50; (7): 291–4; (8): 334–9; (9): 382–8; (10): 424–30.
                                                         (1961–63) **Galliford, A. L.**
A simple account. Part 7 gives a review of the literature.

Ordnung Bdelloidea (Rotatorien, Rädertiere). *Bestimm. Büch. Bodenfauna Europ.* **6**: 297 pp, 203 figs.   (1965) **Donner, J.**

Kolovratki fauny SSSR (Rotatoria), podklass Eurotatoria, otryady Ploimida, Monimotrochida, Paedotrochida. *Opred. Faune SSSR* **104**: 1–744. [Russian].
                                                         (1970) **Kutikova, L. A.**
Useful keys, relevant to most of Europe.

Das Zooplankton der Binnengewässer. 3. Rotatoria. *Binnengewässer* Suppl. **26**(1): 146 pp.   (1972) **Ruttner-Kolisko, A.**

Rotatoria—Die Rädertiere Mitteleuropas Ein Bestimmungswerk, begründet von Max Voigt. überordnung Monogononta, 2nd edn. Borntraeger, Berlin and Stuttgart. Vol. 1, viii + 673 pp; Vol. 2, 234 pls.   (1978) **Koste, W.**
Contains keys to species level, with brief descriptions for the Monogononta.

Rotatoria. In *Limnofauna europaea*, 2nd edn (ed. J. Illies), pp. 54–91. Fischer, Stuttgart and Swets and Zeitlinger, Amsterdam.   (1978) **Berzins, B.**

A key to the freshwater planktonic and semi-planktonic Rotifera of the British Isles. *Scient. Publs Freshwat. Biol. Ass.* **38**: 178 pp.   (1978) **Pontin, R. M.**

Remarques taxonomiques sur quelques rotifères des tourbières avec la description d'une espèce et d'une sous-espèce nouvelles. *Hydrobiologia* **109**(2): 125–30.   (1984) **Francez, A. J. and Pourriot, R.**
Rotifers in peat bogs.

Marine

Rotatoria. *Tierwelt N.-u. Ostsee* **73e**: 1–156.   (1929) **Remane, A.**

A preliminary report on the Plymouth marine and brackish-water Rotifera. *J. mar. biol. Ass. UK* **28**: 239–53.   (1949) **Hollowday, E. D.**

Contributions to the knowledge of the marine Rotatoria of Norway. *Univ. i Bergen Arbok, Nat. rekke* No. 6, 11 pp.   (1952) **Berzins, B.**

Zooplankton sheets. *Fich. Ident. Zooplancton.* **84**, Synchaetidae, 7 pp; **85**, Trichocercidae, 3 pp; **86, 87** Brachionidae, 9 pp; **88**, Asplanchnidae and Synchaetidae, 4 pp; **89**, Testudinellidae, Conochilidae, and Collothecidae, 5 pp.   (1960) **Berzins, B.**
Short descriptions, figures, and references for marine species.

A simple key to the genera of marine and brackish-water rotifers. *Ophelia* **5**: 299–311.   (1968) **Thane-Fenchel, A.**

Inventaire de la faune marine de Roscoff. Gastrotriches, kinorhynches, rotiferes, tardigrades. *Cah. Biol. mar. Paris* Suppl. (No. 10, Editions de la Station Biologique de Roscoff) 28 pp.   (1970) **d'Hondt, J-L.**

The rotifer fauna of rock-pools in the Tvärminne Archipelago, southern Finland. *Acta zool. fenn.* **135**: 3–30.   (1972) **Bjorklund, B. G.**

Marine Rotatoria aus dem Insterstitial der Nordseeinsel Sylt. *Mikrofauna Meeresbodens* **71**: 1–64. [English summary].   (1979) **Tzschaschel, G.**

Geographic

A list of the Rotatoria known to occur in Ireland, with notes on their habitats and distribution. *Irish Fish. Invest.* (ser. A) **21**: 3–24. (1981) **Horkan, J. P. K.**

The Rotifera; or wheel-animalcules. Longmans, Green and Co, London. 2 vols + suppl. (text and plates). 144 + 64 pp, 34 pls.
(1886–89) **Hudson, C. T. and Gosse, P. H.**

A checklist and bibliography of records of Rotifera (Rotatoria) in Britain. British Museum (Natural History), London. 8 pp + 10 microfiche [of 910 pp + 2 maps].   (1981) **Hussey, C. G.**

Present knowledge of the non-marine invertebrate fauna of the outer Hebrides. *Proc. R. Soc. Edinb.*(b) **79**(4): 215–321. (1981) **Waterston, A. R.**

Investigations on rotifers in the brackish waters of eastern Lorraine (France). *Hydrobiologia* **32**: 340–83.   (1968) **Ridder, M. de**

[Morphological and ecological observations on some rotifers in Belgium.] *Natuurwelt. Tijdschr.* **61**(4–5): 108–131. [Flemish: English and French summaries.]   (1979) **De Maeseneer, J.**

[Contribution to the knowledge of the rotifer fauna of the brackish water of the Belgian coastal area.] *Natuurwelt. Tijdschr.* **62**(4–6): 129–38. [Flemish: English and French summaries.] (1980) **Ridder, M. de and Verhaye, H.**

Bijdrage tot de kennis van de Bdelloidea (Rotatoria van Belgie, met de Bechrijving van een nieuwe soort. *Verh. K. vlaam. Acad. Wet.* (Klasse Wet.) **50**: 63 pp.   (1954) **Schepens, D.**

Cartographie des invertébrés européens. Atlas provisoire des rotifères de Belgique. Faculté des Sciences Agronomiques, Gembloux. No pagination.
(1973) **Ridder, M. de**

Raderdieren—Rotifera. *Wet. Meded. K. ned. natuurh. Veren.* **59**: 1–36.
(1965) **Horsthuis, F.**

Über die Rädertiere (Rotatoria, Phylum Aschelminthes) des Dümmen, NW Deutschland. *Osnabrücker naturwiss. Mitt.* **11**: 91–125.
(1984) **Koste, W. and Poltz, J.**

[The rotifers of the submerged mosses and other biotopes in the dammed regions of the Danube at the German–Austrian frontier.] *Arch. Hydrobiol.* (Suppl.) **44**(1): 49–114. [German: English summary.]   (1972) **Donner, J.**

Taxonomical and ecological studies on planktonic Rotatoria from central Sweden. *K. svenska Vetensk Akad. Handl.* (Ser. 4) **6**(7): 1–52. (1957) **Pejler, B.**

On the taxonomy and ecology of benthic and periphytic Rotatoria investigations in northern Swedish Lapland. *Zool. Bidr. upps.* **33**: 327–422.
(1962) Pejler, B.

Rotifers from two tarns in southern Finland, with a description of a new species, and a list of rotifers previously found in Finland. *Acta zool. Fenn.* **125**: 1–36.
(1969) Eriksen, B. G.

Rotatoria. *Zool. Iceland* **2**(13): 106 pp.
(1972) Ridder, M. de

GASTROTRICHA

General

An introduction to the taxonomy of the Gastrotricha, with a study of eighteen species from Michigan. *Trans. Am. microsc. Soc.* **69**: 325–52.
(1950) Brunson, R. B.

Freshwater

Gastrotricha. *Süsswasserfauna Dtl.* **14**: 240–65, 33 figs.   (1912) **Collin, A.**
Gastrotricha Gastrotrichen. *Tierwelt Mitteleur.* **1**(4a): 1–74, 12 pls.
(1958) Voigt, M.

Marine

Gastrotricha. *Tierwelt N.-u. Ostsee* **7d**: 56 pp, 62 figs.   (1927) **Remane, A.**
Marine Gastrotricha from the interstitial fauna of some North Wales beaches. *Proc. zool. Soc., Lond.* **140**: 485–502, 8 figs.   (1963) **Boaden, P. J. S.**
Key to marine genera.

Beiträge zur Gastrotrichenfauna der Nord- u. Oustsee. *Kieler Meeresforsch.* **17**: 206–18, 4 pls.   (1961) **Forneris, L.**
Descriptions of species, with checklist and bibliography.

Clés tabulaires de déterination des genres marins de Gastrotriches. *Bull. Soc. zool. Fr.* **99**: 645–65, 7 figs.   (1974) **Hondt, J. L. d'**
Gastrotricha. *Oceanogr. mar. Biol. Ann. Rev.* **9**: 141–92, 18 figs.
(1971) Hondt, J. L. d'

KINORHYNCHA

About 25–30 British species known
The classification of the phylum Kinorhyncha. *Proc. Int. Congr. Zool.* **15**: 941–3.
(1958) Chitwood, B. G.

Indian Ocean Kinorhyncha 2. Neocentrophyidae; a new Homalorhagid family. *Proc. biol. Soc. Wash.* **82**, 113–28, 5 figs.   (1969) **Higgins, R. P.**
Includes a key to genera of adults of the Kinorhyncha.

Monograph der Echinodera. Leipzig. 396 pp, 73 figs, 27 pls. (1928) **Zelinska, K.**

The class Kinorhyncha (Echinoderida) in British waters. *J. mar. biol. Ass. UK.* **42**: 503–9, 2 figs.   (1962) **McIntyre, A. D.**

Occurrence of *Echinoderella* (Echinoderida) off Plymouth. *Ann. Mag. nat. Hist.* **5**(13): 61–3, fig.   (1962) **Krishnaswamy, S.**

NEMATODA

General

[*Essentials of nematodology*]. Izdatel'stvo 'Nauka', Moscow. 29 vols. [Russian]. English translations: Israel program for scientific translations, Jerusalem. Vols. 2–5, 7, 8, 13, 16, 17, 21, 22. (1949–73) (Vols. 1–23) **Skrjabin, K. I.**, ed. (1975–79) (Vols. 24–9) **Ryzhikov, K. M.**, ed.

Nemathelminthes (Nematodes–Gordiaces), Rotiferes–Gastrotriches, Kinorhynques. *Traite de zoologie: anatomie, systematique, biologie.* Masson, Paris. Vol. 4 (2–3), 1497 pp.   (1965) **Grassé, P-P.**

Evolution as a basis for the systemization of nematodes. Pitman Publishing, London. 287 pp, 40 figs.   (1976) **Andrassy, I.**

The natural history of nematodes. Prentice-Hall, Englewood Cliffs, New Jersey. 323 pp.   (1983) **Poinar, G. O. Jr**

Free-living

FRESHWATER AND SOIL

Freilebende Nematoden. *Tierwelt. Mitteleur.* **1**(5a): 164 pp, 54 pls.
    (1960) **Meyl, A. H.**
Keys to genera of soil and freshwater nematodes; brief descriptions of many species from Central Europe.

Soil and freshwater nematodes, 2nd edn. Methuen, London. 544 pp, 298 figs.
    (1963) **Goodey, T.** revised by **Goodey, J. B.**
One of the few texts available covering most of the genera but somewhat dated taxonomically especially for plant parasitic forms.

C.I.H. descriptions of plant-parasitic nematodes. Commonwealth Agricultural Bureaux, Farnham Royal. 8 sets.

(1972–85) **Muller, R. I. J., Gooch, P. S., Siddiqi, M. R., and Hunt, D. J.**, eds
Each set contains illustrated descriptions of 15 nematode species with details of their
taxonomy, biology, and control.

Plant nematology, 3rd edn. HMSO, London. MAFF Ref. Book 407. 440 pp, 101
figs, 20 pls.   (1978) **Southey, J. F.**, ed.
Includes a general classification of plant and soil nematodes with a key to superfamilies; descriptions of common free-living forms are included together with
detailed information on the identity, biology, and control of plant parasitic forms.

Tylenchida parasites of plants and insects. Commonwealth Agricultural Bureaux,
Farnham Royal. 645 pp, 150 figs.   (1986) **Siddiqi, M. R.**
Keys plus detailed diagnoses are given for each category from order down to genus/
subgenus level and a full, well-illustrated description is given for the type or a
representative species of each genus.

MARINE

Free-living marine nematodes. (Reports of the Lund University Chile Expedition 1948–49) I–IV. *Acta Univ. Lund* (N.F.) **49**(6): 1–155; **50**(16): 1–148;
52(13): 1–115; **55**(5): 1–111, 253 figs.   (1953–59) **Wieser, W.**

A taxonomic hierarchy and checklist of the genera and higher taxa of marine
nematodes. *Smithson. Contr. Zool.* **137**: 101 pp.
                                        (1972) **Hope, W. D. and Murphy, D. G.**

The Bremerhaven checklist of aquatic nematodes. *Veröff. Inst. Meeresforsch.
Bremerh.* (Suppl.) **4**(1, 2): 736 pp. (1973–74) **Gerlach, S. and Riemann, F.**

An illustrated guide to the marine nematodes. Florida Institute of Food and
Agricultural Sciences. 135 pp, 226 figs.   (1980) **Tarjan, C.**

Entwurf eines phylogenetischen Systems der freilebenden Nematoden. *Veröff.
Inst. Meeresforsch. Bremerh.* (19 Suppl.) **7**: 472 pp, 36 figs. (1981) **Lorenzen, S.**

The systematics and ecology of free-living marine nematodes. *Helminth. Abstr.*
(Ser. B) **51**: 1–31. (1982) **Help, C., Vincx, M., Smol, N., and Vranken, G.**

Free-living marine nematodes. Part 1. British enoplids. *Synopses Br. Fauna* **28**:
307 pp, 137 figs.   (1983) **Platt, H. M. and Warwick, R. M.**
Pictorial keys to world genera and British species. Part 2, British chromadorids, and
Part 3, British monhysterids, will complete descriptions of all known British species.

Parasitic

OF INSECTS AND OTHER INVERTEBRATES

CIH key to the groups and genera of nematode parasites of invertebrates. Commonwealth
Agricultural Bureaux, Farnham Royal. 43 pp.   (1977) **Poinar, G. O. Jr**

Entomophagus nematodes. A manual and list of insect-nematode associations. Brill,
Leiden. 317 pp.   (1975) **Poinar, G. O. Jr**

[*Aquatic Mermithidae of the fauna of the USSR.*] (1977) Vol. 1, Amerind Publishing, New Delhi. [Translated from Russian]; (1974) Vol. 2, 222 pp. Izdat 'Nauka', Leningrad. [Russian]. (1974–77) **Rubsov, I. A.**

Tylenchida. Parasites of plants and insects. Commonwealth Agricultural Bureaux, Farnham Royal. 645 pp. (1986) **Siddiqi, M. R.**

OF VERTEBRATES

CIH keys to the nematode parasites of vertebrates. Commonwealth Agricultural Bureaux, Farnham Royal. 10 vols.
(1974–83) **Anderson, R. C., Chabaud, A. G., and Willmott, S.**, eds.

The nematodes of vertebrates. *Systema Helminthum.* Interscience, New York. Vol. 3 (2 parts), 1261 pp, 909 figs. (1961) **Yamaguti, S.**

OF FISH

Nematodes of fishes in Czechoslovakia. *Acta Scientarum Naturalium Academiae Scientarum Bohemoslovacae. Brno* **5**: 1–49. (1971) **Moravec, F.**

Key to parasites of freshwater fish of the USSR. Israel program for scientific translations, Jerusalem. (1964) 919 pp, 1628 figs. [Translation of: *Akad. Nauk SSSR. Zool. Inst., Leningrad* 1962].
(1964) **Bychowskaya-Pavlovsksaya, I. E.** *et al.*

Nematodes from some Norwegian fishes. *Sarsia* **2**: 1–50. (1961) **Berland, B.**

OF AMPHIBIANS AND REPTILES

[*Helminths of amphibians in the fauna of the USSR.*] Izdat. 'Nauka', Moscow. 278 pp, 115 figs. [Russian].
(1980) **Ryzhikov, K. M., Sharpilo, V. P., and Sihevchenko, N. N.**
[*Parasitic worms of reptiles in the fauna of the USSR.*] Izdat. 'Naukova Dumka', Kiev, 287 pp, 116 figs. [Russian.] (1976) **Sharpilo, V. P.**

OF BIRDS AND MAMMALS

Helminths of fish-eating birds of the palaearctic region. Nematoda. Academia, Prague. Vol. 1, 318 pp. (1978) **Barus, V.** *et al.*

Helminthofauna of marine mammals. (Ecology and Phylogeny). Israel program for scientific translations, Jerusalem. 522 pp, 240 figs. (1968) **Delyamure, S. L.**

Key to nematodes reported in waterfowl. *Resource Publication* no. 22. United States Departments of the Interior, Bureau of Sport Fisheries and Wildlife, Washington DC. 44 pp. (1974) **McDonald, M. E.**

The helminth fauna of mustelids and the ways of its formation. Amerind Publishing, New Delhi. xvi + 607 pp.   (1985) **Kontramavichus, V. L.**

[*Helminths of diving and marsh birds of the fauna of the Ukraine.*] Izdat. 'Naukova Dumka', Kiev. 415 pp, 74 figs. [Russian]. (1976) **Smogorzhevskaya, L. A.**

[*Keys to the identification of helminths in domestic and wild pigs.*] Izdat. 'Nauka', Moscow. 168 pp, 75 figs. [Russian].   (1983) **Ryzhikov, K. M.** *et al.*

[*Helminths of deer.*] Izdat. 'Nauka', Moscow. 223 pp, 50 figs. [Russian].
                                                          (1976) **Pryadko, E. I.**

[*Keys to the helminths of carnivores in the USSR.*] Izdat. 'Nauka', Moscow. 275 pp, 123 figs. [Russian].   (1977) **Kozlov, D. P.**

[*Key to the helminths of lagomorphs of the USSR.*] Izdat. 'Nauka', Moscow. 232 pp, 72 figs. [Russian].   (1970) **Gvozdev, E. V.** *et al.*

[*Helminths of waterfowl. Key to their identification, morphology, biology and laboratory diagnosis.*] Academia, Prague. 239 pp, 100 figs. [Czech].
                                                          (1982) **Ryšavÿ, B.** *et al.*

OF MAN AND DOMESTIC ANIMALS

A checklist of the helminth parasites of domestic animals of the United Kingdom. Hoechst UK, Milton Keynes. 147 pp.   (1983) **Scofield, A. M.**

Parasitic zoonoses. Handbook series in zoonoses, Section C. CRC Press. Boca Raton, Florida. Vol. 2, 360 pp.   (1982) **Schultz, M. G.**, ed.

Helminths, arthropods, and protozoa of domestic animals, 7th edn. Baillière Tindall, London. 809 pp.   (1972) **Soulsby, E. J. L.**

Worms and diseases. A manual of medical helminthology. William Heinemann Medical Books, London. 161 pp.   (1975) **Muller, R.**

OF PLANTS

(*See* 'Free-living—Freshwater and soil' above.)

NEMATOMORPHA

General

Monografia dei Gordii. *Memorie R. Accad. Sci. Torino.* **47** (Ser. 2): 339–419, 3 pls.   (1897) **Camerano, L.**

Revisione dei Gordii. *Memorie R. Accad. Sci. Torino.* **66** (Ser. 2): 1–66.
                                                          (1915) **Camerano, L.**

Freshwater and soil

Some observations on the Nematomorpha (Gordiacea) of the British Isles. *Parasitology* **84**(4): lxix–lxx.   (1982) **Harris, E. A.**
Includes a key.

Marine

Nectonema munidae Brinkmann (Nematomorpha) parasitizing *Munida tenuima*
   G.O. Sars (Crust. Dec.) with notes on the host-parasite relations and new
   host species. *Sarsia* **38**: 91–110.   (1969) **Nielsen, S. O.**

ACANTHOCEPHALA

Acanthocephala. *Tierwelt Mitteleur.* **1**(6): 1–40, 46 figs.   (1938) **Meyer, A.**
Systematique des Acanthocephales (Acanthocephala Rudolphi 1801).
   Première partie. L'ordre des Palaeacanthocephala Meyer 1931, premier
   fascicule la super-famille des Echinorhynchoidea (Cobbold 1876) Golvan et
   Houin 1963. *Mém. Mus. natn. Hist. nat. Paris* (Zool.) **57**: 5–373, 260 figs.
                                                            (1969) **Golvan, Y. J.**
The biology of the Acanthocephala. Cambridge University Press, Cambridge. 519
   pp.   (1985) **Crompton, D. W. T. and Nickol, B. B.**
Acanthocephala. *Systema Helminthum.* Interscience, New York. Vol. 5, 423 pp,
   855 figs.   (1963) **Yamaguti, S.**
   Contains keys to orders, families, genera, etc. of freshwater hosts.

Of freshwater fish

Keys to parasites of freshwater fish of the USSR. [English translation] Israel
   program for scientific translations, Jerusalem. 919 pp, 1628 figs.
                                    (1964) **Bychowskaya-Pavlovskaya, E.** *et al.*
A checklist of British and Irish freshwater fish parasites with notes on their
   distribution. *J. Fish. Biol.* **6**: 613–44.   (1974) **Kennedy, C. R.**

Of amphibians and reptiles

[*Helminths of amphibians in the fauna of the USSR.*] Izdat. 'Nauka', Moscow. 278
   pp, 115 figs. [Russian].
          (1980) **Ryzhikov, K. M., Sharpilo, V. P, and Shevchenko, N. N.**
[*Parasitic worms of reptiles in the fauna of the USSR.*] Izdat. 'Naukova Dumka',
   Kiev. 287 pp, 116 figs. [Russian].   (1976) **Sharpilo, V. P.**

Of birds and mammals

The helminth fauna of mustelids and the ways of its formation. Amerind Publishing,
   New Delhi. xvi + 607 pp.   (1985) **Kontramavichus, V. L.**

[*Acanthocephala of domestic and wild animals.*] English translation by Israel
   program for scientific translations, Jerusalem. Vol. 1, iii + 465 pp, 182 figs;
   Vol. 2, iv + 478 pp, 178 figs.   (1971) **Petrochenko, V. I.**

[*Helminths of diving and marsh birds of the fauna of the Ukraine.*] Izdat. 'Naukova
   Dumka', Kiev. 415 pp, 74 figs. [Russian]. (1976) **Smogorzhevskaya, L. A.**

[*Key to the helminths of carnivores in the USSR.*] Izdat. 'Nauka', Moscow. 275 pp,
   123 figs. [Russian].   (1977) **Kozlov, D. P.**

[*Key to the helminths of lagomorphs of the USSR.*] Izdat. 'Nauka', Moscow. 232 pp,
   72 figs. [Russian].   (1970) **Gvozdev, E. V.** *et al.*

[*Helminths of waterfowl. Key to their identification, morphology, biology and laboratory
   diagnosis.*] Academia, Prague. 239 pp, 100 figs. [Czech].
                                                     (1982) **Rysavy, B.** *et al.*

[*Helminthofauna of marine mammals (Ecology and phylogeny).*] English translation
   by Israel program for scientific translations, Jerusalem. ix + 522 pp, 240
   figs.   (1968) **Delyamure, S. L.**

Of man and domestic animals

A checklist of the helminth parasites of domestic animals of the United Kingdom. Hoechst
   UK, Milton Keynes. 147 pp.   (1983) **Scofield, A. M.**

Helminths, arthropods, and protozoa of domestic animals, 7th edn. Baillière Tindall,
   London. 809 pp.   (1982) **Soulsby, E. J. L.**

Worms and diseases. A manual of medical helminthology. William Heinemann
   Medical Books, London. 161 pp.   (1975) **Muller, R.**

ENDOPROCTA (KAMPTOZOA)

Approximately 75 species, at least 30 reported from Britain and Europe.

Bryozaires (première partie), Entoproctes, Phylactolèmes, *Faune Fr.* **60**: 1–
   398, 151 figs. [French].   (1956) **Prenant, M. and Bobin, G.**
   Includes some Bryozoa; see p. 198.

Three species of Kamptozoa new to Britain. *Proc. zool. Soc. Lond.* **133**(3): 423–
   33, 3 figs, 1 pl.   (1960) **Ryland, J. S. and Austin, A. P.**

Three new species of Entoprocta from West Norway. *Sarsia* **1**: 39–45, 3 figs.
                                                     (1961) **Nielsen, C. and Ryland, J. S.**
   Key to some species of the genus *Loxosomella*.

Entoprocta from the Bergen area. *Sarsia* **17**: 1–6, 2 figs.   (1964) **Nielsen, C.**

Studies on Danish Entoprocta. *Ophelia* **1**: 1–76, 48 figs.   (1964) **Nielsen, C.**
Key to species.

The Loxosomatidae of the Isle of Man. *Proc. zool. Soc. Lond.* **145**(4): 529–47, 11
figs.   (1965) **Eggleston, D.**

A nomenclatural index to *A History of the British Marine Polyzoa* by T. Hincks
(1880). *Bull. Br. Mus. (Nat. Hist.)* Zool. **17**(6): 205–60. (1969) **Ryland, J. S.**
Gives annotated check-list of British and European species including Bryozoa, see
p. 198.

Loxomespilon perezei—ein Entoproctenfund im Mittelatlantik Überlegungen
zur Benthosbesiedlung der Grossen Meteorbank. *Mar. Biol. Berlin* **9**: 51–62,
6 figs.   (1971) **Emschermann, P.**

Entoproct life-cycles and the Entoproct/Ectoproct relationship. *Ophelia* **9**:
209–341, 82 figs.   (1971) **Nielsen, C.**

Cuticular pores and spines in the Pedicellinidae and Barentsiidae
(Entoprocta), their relationship, ultrastructure and suggested function and
their phylogenetic evidence. *Sarsia* **51**: 7–16, 11 pls.
                                                  (1972) **Emschermann, P.**

Les Kamptozoaires. État actuel de nos connaissances sur leur anatomie, leur
dévelopment, leur biologie et leur position phylogénétique. *Bull. soc. zool. Fr.*
107(2): 317–44, 6 figs, 3 pls.   (1982) **Emschermann, P.**
Summarizes and reviews the present state of knowledge and is a valuable source of
references.

Urnatella gracilis, a freshwater Kamptozoan, occurring in Japan. *Annot. zool.*
Japon. **55**(3): 151–66, 11 figs.   (1982) **Oda, S.**
Describes the only freshwater genus of Entoprocta, which is also reported from
Europe.

PRIAPULIDA

Systematics, zoogeography, and ecology of the Priapulida. *Zool. Verh. Leiden*
112, 118 pp, 5 pls, 89 figs.   (1970) **Land, J. van der**
Sipunculiens, Echiuriens, Priapuliens. *Faune Fr.* **4**: 1–30, 14 figs.
                                                       (1922) **Cuénot, L.**

British Echiurids (Echiuroidea), Sipunculids (Sipunculoidea), and Priapulids
(Priapuloidea) with keys and notes for the identification of the species.
Synopses Br. Fauna (Ser. 1) **12**, 27 pp, 18 figs.   (1960) **Stephen, A. C.**

SIPUNCULA

About 15 species are known from the shores and shelf seas of north-western
Europe.

The phyla Sipuncula and Echiura. British Museum (Natural History), London. vii + 528 pp, 60 figs.   (1972) **Stephen, A. C. and Edmonds, S. J.**

A classification of the phylum Sipuncula. *Bull. Brit. Mus. nat. Hist.* (Zool.) **52**: 43–58.   (1987) **Gibbs, P. E. and Cutler, E. B.**

Gephyrea of the coasts of Ireland. *Scient. Invest. Fish Brch Ire.* **1912** (3) 46 pp, 7 pls.   (1913) **Southern, R.**

British Sipunculans. Keys and notes for the identification of the species. *Synopsis Br. Fauna* (N.S.) **12**, 35 pp, 13 figs.   (1977) **Gibbs, P. E.**

Sipunculiens, Échuriens, Priapuliens. *Faune Fr.* **4**, 29 pp, 14 figs.
                    (1922) **Cuénot, L.**

Northern and arctic invertebrates in the collection of the Swedish State Museum (Riksmuseum). 1. Sipunculids. *K. svenska VetenskAkad. Handl.* **39**(1), 130 pp, 15 pls.   (1905) **Théel, H.**

On the genus *Phascolion* (Sipuncula) with particular reference to the north-east Atlantic species. *J. mar. biol. Ass. UK* **65**: 311–23.   (1985) **Gibbs, P. E.**

The taxonomy of some little-known Sipuncula from the north-east Atlantic region including new records. *J. mar. biol. Ass. UK* **66** 335–41.
                    (1986) **Gibbs, P. E.**

ECHIURA

Five species are known from the shores and shelf seas of north-western Europe.

The phyla Sipuncula and Echiura. British Museum (Natural History), London. vii + 528 pp, 60 figs.   (1972) **Stephen, A. C. and Edmonds, S. J.**

Gephyrea of the coasts of Ireland. *Scient. Invest. Fish Brch Ire.* 1912 (3), 46 pp, 7 pls.   (1913) **Southern, R.**

British echiurids (Echiuroidea), sipunculids (Sipunculoidea), priapulids (Priapuloidea) with keys and notes for the identification of the species. *Synopses Br. Fauna* (Ser. 1) **12**, 27 pp, 18 figs.   (1960) **Stephen, A. C.**

Sipunculiens, Échuriens, Priapuliens. *Faune Fr.* **4**, 29 pp, 14 figs.
                    (1922) **Cuénot, L.**

Northern and arctic invertebrates in the collection of the Swedish State Museum (Riksmuseum) II. Priapulids, echiurids, etc. *K. svenska VetenskAkad. Handl.* **40**(4), 28 pp, 2 pls.   (1906) **Théel, H.**

On the structure and affinities of 'Thalassema' lankesteri Herdman and the classification of the Group Echiuroidea. *Göteborgs K. Vetensk.-o. VitterhSamh. Handl.* (B2) **6**(6), 94 pp, 11 figs, 6 pls.   (1942) **Bock, S.**

POGONOPHORA

General

Pogonophora. Academic Press, London. 479 pp, 325 figs. (1963) **Ivanov, A. V.**
  Includes a key to all species known at that date.

Recent researches on the Pogonophora. *Oceanogr. mar. Biol.* **9**: 193–220.
                                               (1971) **Southward, E. C.**
  Lists all new species described 1963–71.

The distribution of Pogonophora in the Atlantic Ocean. *Symp. zool. Soc. Lond.*
  19: 145–58.   (1967) **Southward, E. C. and Southward, A. J.**

Systematic

Description of a new species of *Oligobrachia* (Pogonophora) from the North
  Atlantic, with a survey of the Oligobrachiidae. *J. mar. biol. Ass. UK* **58**: 357–
  65.   (1978) **Southward, E. C.**

A new species of *Lamellisabella* (Pogonophora) from the North Atlantic. *J. mar.
  biol. Ass. UK* **58**: 713–18.   (1978) **Southward, E. C.**

On some Pogonophora from the North-East Atlantic, including two new
  species. *J. mar. biol. Ass. UK* **37**: 627–32, 3 figs.
                              (1958) **Southward, E. C. and Southward, A. J.**

Two new species of Pogonophora from the North-East Atlantic. *J. mar. biol.
  Ass. UK* **38**: 439–44, 2 figs.   (1959) **Southward, E. C.**

On a new species of *Siboglinum* (Pogonophora), found on both sides of the
  North Atlantic. *J. mar. biol. Ass. UK* **43**: 513–17, 1 fig.
                                               (1963) **Southward, E. C.**

Siboglinum fiordicum sp. nov. (Pogonophora) from the Raunefjord, Western
  Norway. *Sarsia* **13**: 33–44.   (1963) **Webb, M.**

A redescription of *Siboglinum ekmani* Jägersten (Pogonophora). *Sarsia* **15**: 37–
  47.   (1964) **Webb, M.**

A new bitentaculate pogonophoran from Hardangerfjorden, Norway. *Sarsia*
  15: 49–55.   (1964) **Webb, M.**

Additional notes on *Sclerolinum brattstromi* (Pogonophora) and the establish-
  ment of a new family, Sclerolinidae. *Sarsia* **16**: 47–58.   (1964) **Webb, M.**

A new species of multitentaculate Pogonophora from northern Norway. *Sarsia*
  22: 55–63.   (1966) **Brattegard, T.**

Geographic

Siboglinum holmei Southward (Frenulata: Pogonophora) from Galway Bay,
  west coast of Ireland. *Ir. Nat. J.* **20**(8): 350.   (1981) **O'Connor, B.**

Horizontal and vertical distribution of Pogonophora in the Atlantic Ocean. *Sarsia* **64**: 51–55.　　　　　　　　　　　　(1979) **Southward, E. C.**

Pogonophora. In *Peuplements profonds du Golfe de Gascogne* (ed. L. Laubier and C. Monniot), pp. 369–73. IFREMER, Brest.　　　　(1985) **Southward, E. C.**

Pogonophora in the Skagerrak. *Sarsia* **67**: 211–12.

(1982) **Flügel, H. J. and Langhof, I.**

A new hermaphroditic pogonophore from the Skagerrak. *Sarsia* **68**: 131–8.

(1983) **Flügel, H. J. and Langhof, I.**

Pogonophora and associated fauna in the deep basin of Sognefjorden. *Sarsia* **29**: 299–306.　　　　　　　　　　　　(1967) **Brattegard, T.**

Notes on the distribution of Pogonophora in Norwegian fjords. *Sarsia* **18**: 11–15.　　　　　　　　　　　　　　　　　(1965) **Webb, M.**

POLYCHAETA

General

Polychètes errantes. *Faune Fr.* **5**: 1–488, 181 figs.　　　(1923) **Fauvel, P.**
Polychètes sédentaires. Addenda aux errantes, archiannélides, myzostomaires. *Faune Fr.* **16**: 1–494, 152 figs.　　　　　　　(1927) **Fauvel, P.**
Still the two most useful volumes for identifying British polychaetes in general, although the nomenclature is dated.

Catalogue of the polychaetous annelids of the world. Parts 1 and 2. *Occ. Pap. Allan Hancock Fdn.* **23**: 1–628.　　　　　　　　(1959) **Hartman, O.**

Catalogue of the polychaetous annelids of the world. Supplement 1960–1965 and index. *Occ. Pap. Allan Hancock Fdn.* **23** (suppl): 1–197.

(1965) **Hartman, O.**
Lists the world's species and gives senior synonyms. Very useful but now somewhat out of date.

A monograph on the Polychaeta of Southern Africa. **1** Errantia; **2** Sedentaria. British Museum (Natural History), London. 878 pp.　　　　(1967) **Day, J. H.**
Useful for identification at the family level. Some species found in European waters.

The polychaete worms. Definitions and keys to the orders, families, and genera. *Nat. Hist. Mus. Los Ang. Cty. Sci. Ser.* **28**: 190 pp. (1977) **Fauchald, K.**
Orders are defined (not keyed as the title would suggest) but there are keys with a world-wide application to families and genera.

Systematic

Sphaerodoridae (Polychaeta: Errantia) from world-wide areas. *J. nat. Hist.* **8**: 257–89.　　　　　　　　　　　　　　　　(1974) **Fauchald, K.**

Interstitial Dorvilleidae (Annelida, Polychaeta) from Europe, Australia and
New Zealand. *Zoologica Scr.* **14**(3): 183–99.
                         (1985) **Westheide, W. and Nordheim, H. von**

Ophelia celtica, (Annélide Polychète), nouvelle espèce avec quelques remarques
sur les diverses espèces du genre. *Bull. Soc. zool. France* **106**(2): 189–94.
                         (1981) **Amoureux, L. and Dauvin, J-C.**

Scionella lornensis sp.n., a new terebellid (Polychaeta: Annelida) from the west
coast of Scotland, with notes on the genus *Scionella* Moore, and a key to the
genera of the Terebellidae recorded from European waters. *J. nat. Hist.* **3**:
509–16, 3 figs.   (1969) **Pearson, T. H.**

Polychaeta Terebellomorpha. Mar. Invert. Scand. **7**: 1–194.   (1986) **Holthe, T.**
    Includes Pectinariidae, Ampharetidae and Trichobranchidae in addition to the
    Terebellidae.

Sur un nouveau cas de salissures biologiques favorisées par le chlore. *Tethys*
1(2): 375–82.   .   (1969) **Zibrowius, H. and Bellan, G.**
    Pomatoceros triqueter and *P. lamarckii* differentiated.

Geographic

Polychaetes: British Amphinomida, Spintherida, and Eunicida. *Synopses Br.*
Fauna (N.S.) **32**: 1–221.
                         (1985) **George, J. D. and Hartmann-Schröder, G.**
    Contains a key to all orders and families of Polychaeta and with keys and figures to
    species on the Continental Shelf around the British Isles; useful notes on life histories
    and ecology. (Other volumes by various authors *in prep.*)

Polychaeta. Families: Aphroditidae, Phyllodocidae, and Alciopidae. *Fich.*
Ident. Zooplancton **52**: 1–6.   (1953) **Muus, B. J.**
    Provides details of pelagic stages in the Atlantic north of 40 °N.

Polychaeta (contd.). Families: Tomopteridae and Typhloscolecidae. *Fich.*
Ident. Zooplancton **53**: 1–5.   (1953) **Muus, B. J.**
    Provides details of pelagic stages in the Atlantic north of 40 °N waters.

Polychaeta: larvae. Families: Spionidae, Disomidae, Poecilochaetidae. *Fich.*
Ident. Zooplancton **91**: 1–12.   (1961) **Hannerz, L.**
    Provides details of pelagic late larval stages in European coastal waters.

Polychaeta. Family: Syllidae. Subfamily: Autolytinae. *Fich. Ident. Zooplancton*
113: 1–4.   (1967) **Hamond, R.**
    Provides details of pelagic stages from North European waters.

The family Paraonidae (Polychaeta) in British waters: a new species and new
records with a key to species. *J. mar. biol. Ass. UK* **61**(1): 133–49.
                         (1981) **Hartley, J. P.**

Taxonomy and ecology of British Spirorbidae (Polychaeta). *J. mar. biol. Ass.*
UK **57**: 453–99.   (1977) **Knight-Jones, P. and Knight-Jones, E. W.**

Polychaetes from Scottish waters. Part 2. Families Aphroditidae, Sigalionidae and Polyodontidae. *R. Scot. Mus. Stud.* **1985**: 1–38.   (1985) **Chambers, S.**

The polychaetes of Lewis and Harris with notes on other marine invertebrates. *Proc. R. Soc. Edinb.* **77B**: 189–216.   (1979) **George, J. D.**
Contains useful identification and natural history notes.

Nephtys pente sp. nov. (Polychaeta: Nephtyidae) and a key to *Nephtys* from Northern Europe. *J. mar. biol. Ass. UK* **64**: 899–907.   (1984) **Rainer, S. F.**

De Familie Nephtyidae (Polychaeta). *Tabellenser. Strandwerkgemeens.* **22**, 1–6.
(1968) **Wolff, W. J.**
Key to species from Netherlands waters.

Annelida, Borstenwürmer, Polychaeta. *Tierwelt Dtl.* **58**: 1–594, 191 figs.
(1971) **Hartmann-Schröder, G.**
Keys and figures to species from the North Sea, Skagerrak, and Kattegat, with useful notes on life histories and ecology.

Neue Nerillidae (Archiannelida) aus dem Sublitoral der Nordsee und des Mittelatlantik (Nordwest-Afrika). *Zoologica Scr.* **7**(4): 257–62.
(1978) **Faubel, A.**

Polychaeten von Expeditionen der 'Anton Dohrn' in Nordsee und Skagerrak. *Veröff. Inst. Meeresforsch. Bremerh.* **14**: 169–274.
(1974) **Hartmann-Schröder, G.**

Zur Kenntnis der Maldaniden der Nord- und Ostsee. *Wiss. Meeresunters. Kiel* **15**(1): 1–94.   (1912) **Nolte, W.**

Octobranchus floriceps sp. nov. (Polychaeta: Trichobranchidae) from the northern North Sea with a re-examination of *O. antarcticus* Monro. *Sarsia* **65**: 249–54.   (1980) **Kingston, P. F. and Mackie, A. S. Y.**

Polychaeta. *Zoology Iceland* **2**(19): 1–182.   (1951) **Wesenberg-Lund, E.**

The use of methyl green as an aid in species discrimination in Onuphidae (Annelida, Polychaeta). *Zoologica Scr.* **14**(1): 19–23.   (1985) **Winsnes, I.**
Deals with species from Norway to Iceland.

The Glyceridae (Polychaeta) of the North Atlantic and Mediterranean, with descriptions of two new species. *J. nat. Hist.* **21**: 167–89.
(1987) **O'Connor, B. D. S.**

Les ophéliens Norvégiens. *Meddr. zool. Mus., Oslo* **52**: 21–61.
(1945) **Stöp-Bowitz, C.**

Les scalibregmiens de Norvège. *Meddr. zool. Mus., Oslo* **55**: 63–87.
(1945) **Stöp-Bowitz, C.**

Les flabelligériens Norvégiens. *Bergens Mus. Arb.* **1946/1947**(2): 1–59.
(1948) **Stöp-Bowitz, C.**

Nephtyidae (Polychaeta) from Norwegian waters. *Sarsia* **13**: 1–32, 9 figs.
(1963) **Fauchald, K.**

A revision of Autolytinae (Syllidae, Polychaeta) with special reference to Scandinavian species, and with notes on external and internal morphology,

reproduction, and ecology. *Ark. Zool.* Ser. 2 **19**: 157–213, 31 figs.
(1966) Gidholm, L.

Studien über die skandinavischen und arktischen Maldaniden nebst Zusammenstellung der übrigen bisher bekannten Arten dieser Familie. *Zool. Jb.* Suppl. **9**(1): 1–308.
(1907) Arwidsson, I.

Arctic and Scandinavian Oweniidae (Polychaeta) with a description of *Myriochele fragilis* sp. n., and comments on the phylogeny of the family. *Sarsia* **70**(1): 17–32.
(1985) Nilsen, R. and Holthe, T.

OLIGOCHAETA

About 400 European species of which c. 150 are British.

General

Oligochaeta. *Tierwelt Mitteleur.* **1**(7a), 162 pp, 22 figs. [German].
(1958) Wilcke, D. E.

Aquatic

Aquatic Oligochaeta of the World. Oliver and Boyd, Edinburgh. 860 pp, many figures, 4 pl.   (1971) **Brinkhurst, R. O. and Jamieson, B. G. M.**
Enchytraeidae omitted.

A guide for the identification of British aquatic Oligochaeta. *Scient. Publs Freshwat. biol. Ass.* **22**: 1–55, 15 figs, 1 pl.   (1971) **Brinkhurst, R. O.**

British and other marine and estuarine Oligochaetes. *Synopses Br. Fauna* (NS) **21**, 127 pp, 39 figs.   (1982) **Brinkhurst, R. O.**

A guide for the determination of European Naididae. *Zool. Bidr. Upps* **29**: 45–78, 28 figs, 3 pls.   (1952) **Sperber, C.**

Taxonomy and phylogeny of the gutless Phallodrilinae (Oligochaeta, Tubificidae) with descriptions of one new genus and twenty-two new species. *Zoologica Scr.* **13**: 239–272, 31 figs.   (1984) **Erseus, C.**
A world-wide group that has not yet been reported from the NE Atlantic. Provides key to species.

[*Aquatic Oligochaeta of the USSR (Keys Fauna USSR).*] Amerind Publishing, New Delhi. Vol. 78, 513 pp, 255 figs. [English translation of Vodnye Maloshchetinkovye Chervy Fauny SSSR.]   (1981) **Chekanovskaya, O. V.**
Provides identification to Branchiobdellidae.

A new species of *Lumbricillus* with a revised checklist of British Enchytraeidae
(Oligochaeta). *J. mar. biol. Ass. UK* **46**: 89–95, 1 fig.   (1966) **Tynen, M. J.**

A key to the European littoral Enchytraeidae (Oligochaeta). *Ann. Zool. Fenn.* **6**:
150–5, 4 figs.   (1969) **Tynen, M. J. and Nurminen, M.**

For other Enchytraeidae, *see* Terrestrial section.

Terrestrial

Earthworms. *Synopses Br. Fauna* (NS) **31**, 171 pp, 52 figs.
                                    (1985) **Sims, R. W. and Gerard, B. M.**

Lombriciens de France, ecologie et systematique. *Annls Zool. Ecol. anim.* **72–2**:
671 pp, 100 figs.   (1972) **Bouché, M. B.**
Classification and nomenclature often differ from those of other authors.

Die Regenwürmer Deutschlands. *Schrift Forsch. land Braunschweig-Völk.* **7**, 81
pp, 25 pl.   (1953) **Graff, O.**
Includes photographs of all species.

A contribution to our knowledge of the systematics and zoogeography of
Norwegian earthworms (Annelida Oligochaeta: Lumbricidae). *Nytt Mag.
Zool.* **17**: 169–280, 65 figs.   (1969) **Stöp-Bowitz, C.**

The Enchytraeidae. Critical revision and taxonomy of the European species.
Natura jutl. **8–9**, 160 pp, 177 figs.
                                    (1959) **Neilsen, C. O. and Christensen, B.**

HIRUDINEA

About 70 European species (c. 50 marine and c. 20 freshwater or terrestrial
habitats) of which some 40 occur in the British area (26 marine and 14
freshwater).

General

Leeches (Hirudinea). Their structure, physiology, ecology and embryology. Pergamon
Press, London. 201 pp, 112 figs, 1 pl.   (1962) **Mann, K. H.**
Includes keys to European terrestrial and freshwater species and marine genera.

Leech biology and behaviour. Oxford University Press, Oxford. **2**, 430 pp, 170 figs.
                                    (1986) **Sawyer, R. T.**
Includes keys to the genera and species of the world.

Marine

Identification key to the leech (Hirudinoidea) genera of the world, with a
catalogue of species. I. Family: Piscicolodae. *Acta zool. hung.* **11**: 417–63.
                                    (1965) **Soos, A.**

De Nederlandse Bloedzuigers (Hirudinea). *Wet. Mededm K. ned. natuurh. Veren.*
39, 60 pp, 47 figs. [Dutch].
(1960) **Dresscher, G. N., Engel, H., and Middelhoek, A.**
Three new leeches (Piscicolidae) from marine shore fishes (Cottidae) in
British waters. *J. Zool., Lond.* **150**: 297–318, 18 figs. (1966) **Srivastva, L. P.**

Freshwater and terrestrial

A key to the British freshwater Leeches. *Scient. Publs Freshwat. biol. Ass.* **40**, 72
pp, 52 figs, 1 pl. (1979) **Elliott, J. M. and Mann, K. H.**
De Nederlandse Bloedzuigers (Hirudinea). *Wet. Mededm K. ned. natuurh. Veren.*
39, 60 pp, 47 figs. [Dutch].
(1960) **Dresscher, G. N., Engel, H., and Middelhoek, A.**
Hirudinea. *Tierwelt Mitteleur.* **1**(7b), 30pp, 56 figs. [German].
(1958) **Autrumn, H.**
Contributions to the ecology and biology of the Danish freshwater leeches
(Hirudinea). *Folia limnol. scand.* **2**, 109 pp, 29 figs, 16 tables.
(1953) **Bennike, S. A.**
Identification key to the leech (Hirudinoidea) genera of the world, with a
catalogue of species. III. Family: Erpobdellidae. *Acta zool. hung.* **12**: 371–
407. (1966) **Soos, A.**
Identification key to the leech (Hirudinoidea) genera of the world, with a
catalogue of species. VI. Family: Glossihoniidae. *Acta zool. hung.* **15**: 397–
454. (1969) **Soos, A.**
Identification key to the species of the genus *Dina* R. Blanchard, 1892
(emended. Mann, 1952) (Hirudinea: Erpobdellidae). *Acta biol. Szeged.* **9**:
253–61. (1963) **Soos, A.**
On the genus *Glossiphonia* Johnston, 1816, with a key and catalogue to the
species (Hirudinoidea: Glossiphoniidae). *Annls hist. -nat. Mus. natn. hung.* **58**:
271–9. (1966) **Soos, A.**
On the genus *Batracobdella* Viguier, 1879, with a key and catalogue to the
species (Hirudinoidea: Glossiphoniidae). *Annls hist. -nat. Mus. natn. hung.* **59**:
243–57, 43 figs. (1967) **Soos, A.**
Identification key to the species of the genus *Erpobdella* de Blainville, 1818
(Hirudinoidea: Erpobdellidae). *Annls hist. -nat. Mus. natn. hung.* **60**: 141–5.
(1968) **Soos, A.**

CHELICERATA (ARACHNIDA)
SCORPIONES

On the occurrence of the scorpion *Euscorpius flavicaudis* (De Geer) at Sheerness
Port, Isle of Sheppey, Kent. *Bull. Br. arachnol. Soc.* **4**: 74–6, 2 pls.
(1977) **Wanless, F. R.**

La répartition, en France métropolitaine, des scorpions appartenant au genre
  Euscorprius Thorell 1876 (Famille des Chactidae). *Bull. sci. Bourg.* **36**(1): 25–
  41.   (1983) **Vachon, M.**

ARANEAE

Katalog der Araneae von 1758 bis 1940. Bremen and Brussels. Vol. 1 (1942) viii +
  1040 pp; Vol. 2 (1954) 1751 pp.   (1942–54) **Roewer, C. F.**

Bibliographia Araneorum. Douladoure, Toulouse. Vol. 1 (1945) 832 pp, 28 pls;
  Vol. 2 (1955–59) 5508 pp; Vol. 3 (1961) 591 pp.   (1945–61) **Bonnet, P.**

The comity of spiders. Ray Society, London. Vol. 1 (1939) x + 228 pp, 15 figs, 19
  pls; Vol. 2 (1941) xiv + 229–560 pp, 81 figs, 3 pls.

                                              (1939–41) **Bristowe, W. S.**

The world of spiders, revised edn. Collins, London. xvi + 304 pp, 116 figs, 32 pls.

                                                    (1971) **Bristowe, W. S.**

Les Araignées. Société Nouvelle des Editions Boubée, Paris. 277 pp.

                                                        (1979) **Hubert, M.**

Die schönsten Spinnen Europas. Fauna-Verlag, Karlsfeld. 94 pp.

                                (1982) **Sauer, F. and Wunderlich, J.**

The Country Life guide to spiders of Britain and Northern Europe. Country Life Books,
  Feltham. 320 pp, 350 col. photographs.   (1983) **Jones, D.**
  Harvestmen included.

The spiders of Great Britain and Ireland. Vol. 1 (text): Atypidae – Theridio-
  somatidae. Vol. 3 (plates): Atypidae – Linyphiidae. Harley Books, Colchester. Vol. 1 (1985), 229 pp, 100 figs, 3 pls; Vol. 2 (1987); Vol. 3 (1985), 256
  pp, 237 pls.   (1985, 1987) **Roberts, M. J.**

A history of the spiders of Great Britain and Ireland. Ray Society, London. 2 vols, 384
  pp, 29 pls.   (1861–64) **Blackwall, J.**

British spiders. Ray Society, London. Vol. 1 (1951) ix + 310 pp., 142 figs, 1 pl;
  Vol. 2 (1953), vii + 449 pp, 254 figs.

                            (1951–53) **Locket, G. H. and Millidge, A. F.**

British spiders. Ray Society, London. Vol. 3, ix + 315 pp, 75 figs, 612 maps.

                  (1974) **Locket, G. H., Millidge, A. F., and Merrett, P.**

A check list of British spiders. *Bull. Br. arachnol. Soc.* **6**(9): 381–403.

              (1985) **Merrett, P., Locket, G. H., and Millidge, A. F.**
  Useful bibliography to recent literature.

Zodarion italicum (Araneae: Zodariidae), a species newly recorded in Britain.
  Newsl. Br. arachnol. Soc. **44**: 4.   (1985) **Harvey, P. and Murphy, J.**

Les arachnides de France. Roret, Libraire Encyclopédique, Paris. Vol. 1 (1874)
  350 pp, 3 pls; Vol. 2 (1875) 358 pp, 4 pls; Vol. 3 (1876), 360 pp, 4 pls; Vol. 4
  (1878) 334 pp, 5 pls; Vol. 5 (1881) 885 pp, 1 pl; Vol. 6 (1914–37) 1298 pp,
  2028 figs.   (1874–1937) **Simon, E.**

Spinnentiere oder Arachnoidea II: Lycosidae s.lat. (Wolfspinnen im weiteren Sinne). *Tierwelt Dtl.* **5**: 1–80, 192 figs. (1927) **Dahl, F. and Dahl, M.**

Spinnentiere oder Arachnoidea: I, Springspinnen (Salticidae); VI, 2d Familie, Agelenidae; VIII, 19 Familie, Hahniidae, 20 Familie, Argyronetidae. *Tierwelt Dtl.* **3**: 1–55, 159 figs; **23**: 1–136, 218 figs; **33**: 100–18, 38 figs. (1926, 1931, and 1937) **Dahl, M.**

Spinnentiere oder Arachnoidea. VIII, 16 Familie, Gnaphosidae oder Plattbauchspinnen; 17 Familie, Anyphaenidae oder Zartspinnen; 18 Familie, Clubionidae oder Röhrenspinnen. *Tierwelt Dtl.* **33**: 1–99, 104 figs.

(1937) **Reimoser, E.**

Spinnentiere oder Arachnoidea. VI, 27 Familie, Araneidae, VIII, 26 Familie, Theridiidae oder Haubennetzspinnen (Kugelspinnen); IX, Orthognatha–Cribellatae–Haplogynae–Entelegyinae (Pholcidae, Zodariidae, Oxyopidae, Mimetidae, Nesticidae); 28 Familie Linyphiidae Baldachinspinnen; XI, Micryphantidae–Zwergspinnen; XII, Tetragnathidae–Streckspinnen und Dickkiefer. *Tierwelt Dtl.* **23**: 47–136, 218 figs; **33**: 119–222, 286 figs; **42**: 1–150, 305 figs; **44**: viii + 1–337, 551 figs; **47**: xi + 1–620, 1147 figs; **49**: viii + 1–76, 124 figs. (1931–63) **Wiehle, H.**

Egentliga spindlar. Araneae. Fam. 1–4. Salticidae, Thomisidae, Philodromidae och Eusparrassidae. In *Svensk Spindelfauna* Vol. 3, pp. 1–138, 48 figs, 18 pls. Entomologiska, Föreningen, Stockholm.

(1944) **Tullgren, A.**

Egentliga spindlar. Araneae. Fam. 5–7. Clubionidae, Zoridae och Gnaphosidae. In *Svensk Spindelfauna* Vol. 3, pp. 1–141, 39 figs, 10 pls. Entomologiska Föreningen, Stockholm. (1946) **Tullgren, A.**

Egentliga spindlar. Araneae. Fam. 8–10. Oxyopidae, Lycosidae och Pisauridae. In *Svensk Spindelfauna* Vol. 3, pp. 1–48, 20 figs, 10 pls. Entomologiska Föreningen, Stockholm. (1947) **Holm, A.**

PSEUDOSCORPIONES (=CHELONETHIDA)

Ordnung Pseudoscorpionidea (Afterskorpione). *Bestimm. Büch. Bodenfauna Europ.* **1**: 313 pp, 300 figs. (1963) **Beier, M.**

The arachnid order Chelonethida. *Stanf. Univ. Publs Biol. Sciences* **7**(1): 284 pp, 71 figs. (1931) **Chamberlin, J. C.**
Detailed account of morphology. Systematic review with keys to families, genera, and some species. Extensive bibliography.

A new species of pseudoscorpion from Britain. *J. Zool. Lond.* **150**: 165–81, 5 figs. (1966) **Gabbutt, P. D.**
Keys European species of *Chthonius (Ephippiochthonius)*.

Pseudoscorpiones. *Synopses Br. Fauna* (In press). (1987) **Legg, G.**

A key to all stages of the British species of the family Neobisiidae (Pseudoscorpiones: Diplosphyronida). *J. nat. Hist.* **3**: 183–95, 13 figs.
(1969) **Gabbutt, P. D.**

A synopsis of the false-scorpions of Britain and Ireland. *Proc. R. Ir. Acad.* **29B**: 38–64, 3 pls; **33B**: 71–85, 3 figs.   (1911, 1916) **Kew, H. W.**
The classic work on British pseudoscorpions.

The genitalia and associated glands of five British species belonging to the family Chthoniidae (Pseudoscorpiones: Arachnida). *J. Zool., Lond.* **177**: 99–121, 15 figs.   (1975) **Legg, G.**

The genitalia and associated glands of five British species belonging to the family Neobisiidae (Pseudoscorpiones: Arachnida). *J. Zool., Lond.* **177**: 123–51, 21 figs.   (1975) **Legg, G.**
Both above papers by Legg include keys to species based on genitalia.

Remarques sur les Chernetidae (Pseudoscorpiones) de la faune britannique. *Ann. Mag. nat. Hist.* (12) **10**: 389–94, 9 figs.   (1957) **Vachon, M.**
Keys genera: illustrates spermathecal types in Chernetidae.

OPILIONES (=PHALANGIDA)

Spinnentiere, Arachnida—Weberknechte, Opiliones. *Tierwelt Dtl.* **64**: xx pp, 815 figs, 38 maps.   (1978) **Martens, J.**

Spinachtigen-Arachnida III De Hooiwagens (Opilionida) van Nederland. *Wet. Meded. K. ned. natuurh. Veren.* **4**: 28 pp, 13 figs.   (1964) **Spoek, G. L.**

Key to the determination of the British harvestmen (Arachnida, Opiliones). *Entomologist's mon. Mag.* **84**: 109–13, 11 figs.   (1948) **Todd, V.**

Opiliones. *Synopses Br. Fauna* (NS). (In press).
(1987) **Hillyard, P. D. and Sankey, J. H. P.**

ACARI

GENERAL

Acari. In *Synopsis and classification of living organisms* (ed. S. Parker), Vol. 2, pp. 111–69. McGraw-Hill, New York.
(1982) **Johnston, D. E., Kethley, J., and OConnor, B. M.**

How to know the mites and ticks. Wm C. Brown Co., Dubuque, Iowa. 335 pp, 682 figs.   (1979) **McDaniel, B.**

An introduction to acarology. Macmillan, New York. 465 pp, 377 figs, 1 pl.
(1952) **Baker, E. W. and Wharton, G. W.**

Introduction and biology. *The terrestrial Acari of the British Isles. An introduction to their morphology, biology, and classification.* British Museum (Natural History),

London. Vol. 1, 219 pp, 216 figs.
                    (1961) **Evans, G. O., Sheals, J. G., and Macfarlane, D.**
Clare Island Survey, part 39 ii. Acarinida ii. Terrestrial and marine Acarina.
   Proc. R. Ir. Acad. **31**(39): 45–136, 5 pls.   (1915) **Halbert, J. N.**
A manual of acarology, 2nd edn. Oregon State University, Corvallis, Oregon. 509
   pp, 163 pls.   (1978) **Krantz, G. W.**
The Acari. A practical manual. University of Nottingham School of Agriculture,
   Sutton Bonington. Vol. 1, 500 pp, illustrated.   (1985) **Evans, G. O.** *et al.*

MORPHOLOGY AND CLASSIFICATION

Acarina. *Bronn's Kl. Ordn. Tierreichs* **5**(4): 5, xi + 1011 pp, 522 figs.
                                             (1940–43) **Vitzthum, H. G.**
The Acarina of the seashore. *Proc. R. Ir. Acad.* **35B**: 106–52, 3 pls.
                                                    (1920) **Halbert, J. N.**
Mites. In *Parasites of endothermal laboratory animals* (ed. R. J. Flynn), pp. 425–92
   (illustrated). Iowa State University Press, Ames, Iowa. (1973) **Yunker, C.**
Medical and veterinary parasites. *The Acari. A practical manual.* University of
   Nottingham School of Agriculture, Sutton Bonington. Vol. 2, 150pp.
                                        (1985) **Fain, A. and Till, W. M.**
Mites of the rodent fauna of the USSR. *Opred. Faune SSSR* **59**: 459 pp, 984 figs.
   [Russian].   (1955) **Bregetova, N. G.** *et al.*
Mites of moths and butterflies. Comstock Publishing Associates, Ithaca, New
   York.   (1975) **Treat, A. E.**
Mites injurious to economic plants. University of California Press, Berkeley. xxiv +
   614 pp, 138 figs, 74 pls.
                    (1975) **Jeppson, L. R., Keifer, H. H., and Baker, E. W.**
The mites of stored food and houses. *Tech. Bull. Ministr. Agric. Fish Fd.* no. 9:
   400 pp, 437 figs.   (1976) **Hughes, A. M.**
Biological control of pests and mites. University of California Press, Berkeley. 185
   pp, illustrated.
                    (1983) **Hoy, M. A., Cunningham, G. L., and Knutson, L.**, eds.
A synonymic catalogue of British Acari. *Ann. Mag. nat. Hist.* (12) **6**: 1–26, 88–
   99.   (1953) **Turk, F. A.**

METASTIGMATA or IXODOIDEA

*Bibliography of ticks and tickborne diseases from Homer (about 800 BC) to 31 December
   1969.* US Naval Medical Research Unit 3, Cairo. Vol. 1 (1970) 499 pp; Vol.
   2 (1970) 495 pp; Vol. 3 (1971) 435 pp; Vol. 4 (1972) 355 pp; Vol. 5 (1):
   (1974) 492 pp; Vol. 3 (1971) 435 pp; Vol. 4 (1972) 355 pp; Vol. 5 (1): (1974)
   492 pp; Vol. 5 (2): (1978) 455 pp; Vol. 6 (1981) 407 pp; Vol. 7 (1982) 219 pp.
   After Vol. 4 the title was modified progressively to end with 31 December
   1981 at Vol. 7.   (1970 on) **Hoogstraal, H.**

Ticks, a monograph of the Ixodoidea. Vol. 1 (1) The Argasidae, (2) The Ixodidae, (3) The genus *Haemaphysalis.* Cambridge University Press, Cambridge. 550 pp, 449 figs, 13 pls.

(1908–15) **Nuttall, G. H. F., Warburton, C., Cooper, W. F., and Robinson, L. E.**

Also Vol. 1. *Bibliography of the Ixodoidea* (in the same series) 68 pp.

(1911) **Nuttall, G. H. F., Robinson, L. E., and Cooper, W. F.**

And Vol. 2. *Bibliography of the Ixodoidea II.* 32 pp.

(1915) **Nuttall, G. H. F. and Robinson, L. E.**

And Vol. 3 (4) *The genus Amblyomma.* 302 pp, 130 figs, 7 pls.

(1926) **Robinson, L. E.**

Ticks, a monograph of the Ixodoidea. Part V. On the genera of *Dermacentor, Anocentor, Cosmiomma, Boophilus & Margaropus.* Cambridge University Press, Cambridge. xviii + 251 pp, 510 figs, 4 pls. (1960) **Arthur, D. R.**

British ticks. Butterworth, London. ix + 213 pp, 362 figs. (1963) **Arthur, D. R.**

Ticks and tickborne diseases. *Index catalogue of Medical and veterinary zoology.* Special Publication. US Department of Agriculture, Washington DC. Vol. 3, I. Genera and species of ticks: Part 1, 429 pp; Part 2, 593 pp; Part 3, 329 pp. II. Hosts: Parts 1–3, 1268 pp. (1974) **Doss, M. A.** *et al.*

Identification of larval ticks found on small mammals in Britain. Mammal Society, Reading. 14 pp, illustrated. (1978) **Snow, K. R.**

MESOSTIGMATA

Manual of mesostigmatid mites parasitic on vertebrates. *Contr. Inst. Acarol Univ. Md.* **4**: xi + 330 pp, 69 pls.

(1958) **Strandtmann, R. W. and Wharton, G. W.**
Keys families and genera and lists species. Host list and excellent bibliography.

Pytoseiidae et Aceosejidae (Acarina, Gamasina) d'Algerie IV Genre *Typhlodromus* Scheuten 1857. *Bull. Soc. Hist. nat. Afr. N.* **51**: 62–107, 41 figs.

(1960) **Athias-Henriot, C.**
Key species of *Typhlodromus.*

Contribution à l'étude des *Amblyseius* paléarctiques (Acariens actinotriches, Phytoseiidae). *Bull. scient. Bourgogne* **24**: 181–230, 133 figs, 1 pl.

(1966) **Athias-Henriot, C.**
All palaearctic species reviewed.

Observations sur les *Pergamasus* 1–5. *Acarologia* **9**: 669–761, 334 figs, 3 pls; **9**: 762–800, 152 figs, 3 pls; **10**: 181–90, 1 fig. *Bull. scient. Bourgogne* **25**: 175–228, 162 figs. (1967–68) **Athias-Henriot, C.**
Keys to species.

Die Familie Ascaidae (Oudemans, 1905) Bernhard nov. comb. In *Beitrage zur Systematik und Okologie mitteleuropaischer Acarina* (ed. H. J. Stammer), Vol. 2

(Mesostigmata 1), part III, pp. 3–177, 90 figs. Geest and Portig, Leipzig.
(1963) **Bernhard, F.**

A revision of the British mites of the genus *Pergamasus* Berlese *s.lat.* (Acari: Mesostigmata). *Bull. Br. Mus. nat. Hist.* (Zool.) **11**: 1–112, 313 figs, 8 pls.
(1963) **Bhattacharyya, S. K.**

Zerconidae (Acari, Mesostigmata) Polski. *Monografie Fauny Polski*, Vol. 3, 315 pp, 173 figs. Polish Academy of Science, Warsaw. [Polish; figure captions and keys in English.]   (1974) **Blaszak, C.**
Keys to genera and species.

[Gamasid mites (Gamasoidea).] *Opred. Faune SSSR* **61**: 1–246, 562 figs. [Russian].   (1956) **Bregetova, N. G.**

Immature and adult stages of some British Phytoseiidae Berl., 1916 (Acarina). *J. Linn. Soc., Zool.* **43**: 599–643, 26 figs.   (1958) **Chant, D. A.**

Phytoseiid mites (Acarina: Phytoseiidae). Part I. Bionomics of seven species in southeastern England. Part II. A taxonomic review of the family Phytoseiidae with descriptions of 38 new species. *Can. Ent. Suppl.* **2**: 1–166, 306 figs.   (1959) **Chant, D. A.**

A revision of the family Epicriidae (Acarina–Mesostigmata). *Bull. Br. Mus. nat. Hist.* (Zool.) **3**: 169–200, 41 figs, 2 pls.   (1955) **Evans, G. O.**

British mites of the genus *Veigaea* Oudemans (Mesostigmata – Veigaiaidae). *Proc. zool. Soc. Lond.* **125**: 569–86, 21 figs.   (1955) **Evans, G. O.**

An introduction to the British Mesostigmata (Acarina) with keys to families and genera. *J. Linn. Soc. Zool.* **43**: 203–59, 92 figs.   (1957) **Evans, G. O.**

A revision of the British Aceosejinae (Acarina: Mesostigmata). *Proc. zool. Soc. Lond.* **131**: 177–229, 73 figs.   (1958) **Evans, G. O.**

British mites of the subfamily Macrochelinae Trägordh (Gamasina–Macrochelidae). *Bull. Br. Mus. nat. Hist.* (Zool.) **4**: 1–55, 85 figs, 4 pls.
(1956) **Evans, G. O. and Browning, E.**

British mites of the genus *Pachylaelaps* Berlese (Gamasina–Pachylaelaptidae). *Entomologist's mon. Mag.* **92**: 118–29, 35 figs.
(1956) **Evans, G. O. and Hyatt, K. H.**

A revision of the Platyseiinae (Mesostigmata: Aceosejidae) based on material in the collection of the British Museum (Natural History). *Bull. Br. Mus. nat. Hist.* (Zool.) **6**: 25–101, 204 figs.   (1960) **Evans, G. O. and Hyatt, K. H.**

Studies on the British Dermanyssidae. Part I, external morphology. *Bull. Br. Mus. nat. Hist.* (Zool.) **13**: 247–94, 21 figs. Part II, classification. *Bull. Br. Mus. nat. Hist.* (Zool.) **14**: 107–370, 101 figs.
(1965–66) **Evans, G. O. and Till, W. M.**

Les acariens parasites nasicoles des oiseaux de Belgique. I–V. *Bull. Annls Soc. r. ent. belg.* **98**: 252–70, 16 figs; **99**: 168–81, 20 figs; **99**: 471–85, 24 figs; **100**: 55–

61, 14 figs; **102**: 117–22, 8 figs.   (1962–66) **Fain, A.**
Contains many descriptions of Rhinonyssidae.

[*Key to soil-inhabiting mites.*] Mesostigmata. Nauka, Leningrad. 718 pp, 538 figs.
[Russian].   (1977) **Gilyarov, M. S. and Bregetova, N. G.**

Gangsystematik der Parasitiformes. In *Acarologie, Schriftenreihe für vergleichende
Milbenkunde.* Hirschmann, Fürth-i-Bayern. Teile 1–232, Folge 1–12, 14–22;
Teile 234–489, Folge 23–32.   (1957–85) **Hirschmann, W.**
A most important series of well-illustrated generic revisions.

Die Familie Parasitidae Oudemans 1901. In *Acarologie, Schriftenreihe für
vergleichende Milbenkunde.* Hirschmann, Fürth-i-Bayern. Folge 13, 55 pp, 23
pls.   (1969) **Holzmann, C.**

British mites of the genera *Halolaelaps* Berlese and Trouessart and *Saprolaelaps*
Leitner (Gamasina: Rhodacaridae). *Entomologist's Gaz.* **7**: 7–26, 55 figs.
                                                    (1956) **Hyatt, K. H.**

British mites of the genus *Pachyseius* Berlese, 1910 (Gamasina:
Neoparasitidae). *Ann. Mag. nat. Hist.* (12) **9**: 1–6, 10 figs.
                                                    (1956) **Hyatt, K. H.**

Acari (Acarina), Milben, Unterordnung Anactinochaeta (Parasitiformes).
Die freilebenden Gamasina (Gamasides), Raubmilben. *Tierwelt Dtl.* **59**: 1–
475, 516 figs.   (1971) **Karg, W.**

Die europäischen Arten der Gattungen *Macrocheles* Latreille 1829 u. *Geholaspis*
Berlese 1918. In *Acarologie, Schriftenreihe für vergleichende Milbenkunde.* Hirschmann, Fürth-i-Bayern. Folge **14** 2–43 pp., 16 pls.   (1970) **Krauss, W.**

Taxonomic concepts in the Ascidae with a modified setal nomenclature for the
idiosoma of the Gamasina. *Mem. ent. Soc. Can.* **47**: 1–64, 70 figs.
                          (1965) **Lindquist, E. E. and Evans, G. O.**
Key to genera.

Die Familie Parasitidae Oudemans 1901. Pánstwowe Wydawnictwo Naukowe,
Cracow. 690 pp, 425 figs.   (1969) **Micherdzinski, W.**

A revision of the mites of the family Spinturnicidae (Acarina). *Univ. Calif.
Publs Ent.* **17**: 157–284, 31 pls.   (1960) **Rudnick, A.**

Podřád Čmelikovci—Mesostigmata. *Klič Zvířeny ČSSR* **4**: 313–52, 175 figs.
[Czech].   (1971) **Samšiňák, K. and Dusbábek, F.**
Keys to families and genera.

Die Familie Zerconidae Berlese. *Acta zool. hung.* **3**: 313–68, 58 figs.
                                                    (1958) **Sellnick, M.**
Keys to genera and species.

Die Familie Podocinidae Berlese 1916. In *Beiträge zur Systematik und Ökologie
mitteleuropäischer Acarina* (ed. H. J. Stammer), Vol. 2 (Mesostigmata 1), Part
IV, pp. 179–450, pp. 778–804, 177 figs. Geest and Portig, Leipzig.
                                                    (1963) **Westerboer, I.**
Keys to genera and species. Immature stages and both sexes described and figured.

Die Familie Phytoseiidae Berlese 1916. In *Beiträge zur Systematik und Ökologie*

mitteleuropäischer Acarina (ed. H. J. Stammer), Vol. 2, Part V, pp. 451–777, 231 figs (pp. 778–804, references and index for Vol. 2). Geest and Portig, Leipzig.   (1963) **Westerboer, I. and Bernhard, F.**
Keys to genera and species. Immature stages and both sexes described and figured.

Eine Taxonomische Analyse der Familie Macronyssidae Oudemans, 1936. I. Subfamilie Ornithonyssinae Lange, 1958 (Acarina, Mesostigmata). Pánstwowe Wydawnictwo Naukowe, Warsaw and Cracow. 264 pp, 132 figs.
                                        (1980) **Micherdzinski, W.**

Die Raubmilbengattung *Proctolaelaps* Berlese 1923. *Zool. Jb. Syst.* **112**: 185–206, 9 figs.   (1985) **Karg, W.**

Mites of the subfamily Parasitinae (Mesostigmata: Parasitidae) in the British Isles. *Bull. Br. Mus. nat. Hist.* (Zool.) **38**: 237–378.   (1980) **Hyatt, K. H.**

Mites of the genus *Holoparasitus* Oudemans, 1936 (Mesostigmata: Parasitidae) in the British Isles. *Bull. Br. Mus. nat. Hist.* (Zool.) **52**: 139–64.
                                                (1987) **Hyatt, K. H.**

ASTIGMATA

House dust biology for allergists, acarologists, and mycologists. Johanna E. M. H. van Bronswijk, Zoelmond. 316 pp, illustrated.
                                        (1981) **Bronswijk, J. E. M. H. van**

The feather mite genus *Proctophyllodes (Sarcoptiformes: Proctophyllodidae). Bull. Univ. Nebraska St. Mus.* **5**: 1–354, 313 figs.
                                        (1966) **Atyeo, W. T. and Braasch, N. L.**
Keys to species.

Nadkohorta Acaridiae. *Klíč Zvířkeny ČSSR* **4**: 496–529, 106 figs. [Czech].
                                        (1971) **Černy, V. and Samsinák, K.**
Keys to families and genera.

[Feather mites (Analgesoidea). Part II, Epidermoptidae and Freyanidae; part III, Pterolichidae.] *Faune SSSR*, Arachnida **6**(6): 1–411, 167 figs; **6**(7): 1–813, 398 figs. [Russian].   (1953) **Dubinin, V. B.**

A review of the family Epidermoptidae Trouessart parasitic on the skin of birds (Acarina: Sarcoptiformes). *Verh. K. vlaam. Acad. Wet.* (Klasse der Wetenschappen) **27**(84)7: (pt. 1) 1–176; (pt. 2) 1–144, 185 figs.
                                                (1965) **Fain, A.**

Les hypopes parasites des tissus cellulaires des oiseaux (Hypodectidae: Sarcoptiformes). *Bull. Inst. r. Sci. nat. Belg.* **43**(4): 1–139, 150 figs.
                                                (1967) **Fain, A.**

Le genre *Dermatophagoides* Bogdanov 1864, son importance dans les allergies respiratoires et cutanées chez l'homme (Psoroptidae: Sarcoptiformes). *Acarologia* **9**: 179–225, 48 figs.   (1967) **Fain, A.**
Key to species of *Dermatophagoides.*

Étude de la variabilité de *Sarcoptes scabiei* avec une revision des Sarcoptidae. *Acta zool. path. antverp.* **47**: 1–196, 209 figs.   (1968) **Fain, A.**

Les deutonymphes hypopiales vivant en association phoretique sur les mammifères (Acarina: Sarcoptiformes). *Bull. Inst. r. Sci. nat. Belg.* **45**(33): 1–262, 367 figs.   (1969) **Fain, A.**

Morphologie et cycle evolutif des Glycyphagidae commensaux de la taupe *Talpa europaea* (Sarcoptiformes). *Acarologia* **11**: 750–795, 44 figs.
                                                                    (1969) **Fain, A.**

Acariens récoltés par le Dr J. Travé sur îles subantarctiques. I. Familles Saproglyphidae et Hyadesidae (Astigmates). *Acarologia* **16**: 684–708, 51 figs.   (1975) **Fain, A.**
Keys to species of *Hyadesia*.

Les acariens de la famille Knemidokoptidae producteurs de gale chez les oiseaux. *Acta zool. path. antverp.* **45**: 1–142, 74 figs.
                                                      (1967) **Fain, A. and Elsen, P.**

Les Myocoptidae parasites des rongeurs en Hollande et en Belgique (Acarina: Sarcoptiformes). *Acta zool. path. antverp.* **50**: 67–172, 80 figs.
                                    (1970) **Fain, A., Munting, A. J., and Lukoschus, F.**

A further systematic study of the genus *Acarus* L., 1758 (Acaridae, Acarina), with a key to species. *Bull. Br. Mus. nat. Hist.* (Zool.) **19**: 83–118, 40 figs, 4 pls.   (1970) **Griffiths, D. A.**

[*Key to soil-inhabiting mites. Sarcoptiformes.*] Nauka, Leningrad. 491 pp, 1196 figs. [Russian].   (1975) **Gilyarov, M. S. and Krivolutskii, D. A.**

Tyrophagus neiswanderi, a new acarid mite of agricultural importance. *Res. Bull. Ohio agric. exp. Stn* **977**: 17 pp, 4 figs.
                                              (1965) **Johnston, D. E. and Bruce, W. A.**
Key to species of *Tyrophagus*.

British Tyroglyphidae. Ray Society, London. Vol. 1, xv + 291 pp, 19 pls; Vol. 2, xi + 183 pp, 20 pls.   (1901–03) **Michael, A. D.**

Systematik und Ökologie der deutschen Anoetinen. In *Beiträge zur Systematik und Ökologie mitteleuropäischer Acarina* (ed. H. J. Stammer), Vol. 1, pp. 233–384, 79 figs. Geest and Portig, Leipzig.   (1957) **Scheucher, R.**

Systematik und Ökologie der Tyroglyphiden Mitteleuropas. In *Beiträge zur Systematik und Ökologie mitteleuropäischer Acarina* (ed. H. J. Stammer), Vol. 1, pp. 1–231, 182 figs. Geest and Portig, Leipzig.
                                                          (1957) **Turk, E. and Turk, F.**

[Tyroglyphoidea (Acari).] *Faune SSSR* Arachnoidea **6**(1): 475 pp, 705 figs. [Russian] [English translation: *Dokl. (Proc.) Acad. Sci. USSR.*]
                                                          (1941) **Zachvatkin, A. A.**

CRYPTOSTIGMATA (ORIBATEI)

The Oribatid genera of the world. Akadémiai Kiàdó, Budapest. 188 pp, 16 figs, 71 pls.   (1972) **Balogh, J.**
Keys to genera. Setal diagnoses.

Primitive oribatids of the Palaearctic region. In *The soil mites of the World*. Elsevier, Amsterdam, Oxford, New York. Vol. 1, 372 pp, 133 pls.
(1983) **Balogh, J. and Mahunka, S.**

[*Key to soil-inhabiting mites. Sarcoptiformes.*] Nauka, Leningrad. 491 pp, 1196 figs. [Russian].   (1975) **Gilyarov, M. S. and Krivolutskii, D. A.**

Morphologische und entwicklungsgeschichtliche Untersuchungen zum phylogenetischen System der Acari: Acariformes Zach. 1, Oribatei, Malaconothridae. *Mitt. zool. Mus. Berl.* **33**: 97–213, 41 figs.   (1957) **Knülle, W.**
Keys to genera and species. Comprehensive bibliography.

British Oribatidae. Ray Society, London. 2 vols. 657 pp, 54 pls.
(1884–88) **Michael, A. D.**
Notes of life histories, habits, etc. Still useful but nomenclature out of date.

British mites of the genus *Brachychthonius* Berl. 1910. *Ann. Mag. nat. Hist.* (12)**5**: 227–39, 8 figs.   (1952) **Evans, G. O.**

Mites of the genus *Malaconothrus* (Acari: Cryptostigmata) from the British Isles. *J. nat. Hist.* (In press)   (1987) **Luxton, M.**

Sur les *Hydrozetes* (Acariens) de l'Europe occidentale. *Bull. Mus. Hist. nat. Paris* (2)**20**: 328–35, 3 figs.   (1948) **Grandjean, F.**

The Oribatei (Acari) of the Netherlands. *Zool. Verh. Leiden* **17**: 1–139, 12 figs.
(1952) **Hammen, L. van der**

Berlese's primitive oribatid mites. *Zool. Verh. Leiden* **40**: 1–93.
(1959) **Hammen, L. van der**
Key to families of 'lower' Cryptostigmata.

Studies on the Oribatei (Acari) of Norway. *Suom. hyönt Aikak.* **37**: 30–53, 14 maps.   (1971) **Karppinen, E.**

Die Gattung *Carabodes* C. L. Koch in der schwedischen Bodenfauna (Acar. Oribat.). *Ark. Zool.* (2)**4**: 367–90, 13 figs.
(1953) **Sellnick, M. and Forsslund, K.-H.**

Die Camisiidae Schwedens (Acar. Oribat.). *Ark. Zool.* (2)**8**: 473–530, 47 figs.
(1955) **Sellnick, M. and Forsslund, K.-H.**

Formenkreis: Hornmilben, Oribatei. *Tierwelt Mitteleur* **3**(4), 9, 42 pp, 91 figs; pp. 45–134, 2 pls. (Ergänzung).   (1929) **Sellnick, M.**
Keys to genera and species.

Revision of the family Oppiidae Grandjean 1953 (Acarina, Oribatei). *Acarologia* **17**: 331–45, 4 figs.   (1975) **Seniczak, M.**

Bestimmungstabelle der holsteinischen *Suctobelba*-Arten. *Arch. Hydrobiol.* **45**: 340–8.   (1951) **Strenzke, K.**

Die norddeutschen Arten der Oribatiden-Gattung *Suctobelba*. *Zool. Anz.* **147**: 147–66, 20 figs.   (1951) **Strenzke, K.**

Die Jugendstadien der nordischen Camisiiden (Acar. Orib.) und etwas über die Systematik der Erwachsenen I. *Ent. Meddr.* **26**: 392–403, 6 pls.

(1952) **Tuxen, S. L.**

Keys to genera and species of immature Camisiidae.

Moosmilben oder Oribatiden (Cryptostigmata). *Tierwelt Dtl.* **22**: 79–200, 364 figs.

(1931) **Willmann, C.**

Keys to genera and species. Descriptions. Nomenclature now substantially out of date.

PROSTIGMATA (including ENDEOSTIGMATA, HETEROSTIGMATA and HYDRACHNELLAE)

A proposed classification of the trombidiform mites (Acarina). *Proc. ent. Soc. Wash.* **57**: 209–18.

(1955) **Cunliffe, F.**

Keys to families.

Podřád Sametkovci—Trombidiformes. *Klíč Zvířeňy ČSSR* **4**: 357–422, 194 figs. [Czech].

(1971) **Daniel, M.**

A modern comprehensive key to families and genera.

Acarina. Bdellidae, Nicoletiellidae, Cryptognathidae. *Tierreich* **56**: 1–84, 93 figs.

(1931) **Thor, S.**

Acarina Prostigmata 6–11. Eupodidae, Penthalodidae, Penthaleidae, Rhagidiidae, Pachygnathidae, Cunaxidae. *Tierreich* **71a**: 1–186, 252 figs.

(1941) **Thor, S. and Willmann, C.**

Anystoidea

Neues über Anystidae (Acari). *Arch. Naturgesch.* (N.F.) **5**: 364–446, 28 pls.

(1935) **Oudemans, A. C.**

Bdelloidea (Cunaxoidea)

The Icelandic Bdellidae (Acarina). *J. Kans. ent. Soc.* **35**: 281–98, 29 figs.

(1962) **Atyeo, W. T. and Tuxen, S. L.**

Acaros de la familia Cunaxidae. *An. Esc. nac. Cienc. Biol. Mex.* **5**: 229–73, 100 figs. [Spanish].

(1948) **Baker, E. W. and Hoffmann, H. G.**

Keys to genera and species.

Cheyletoidea

A review of the mites of the family Cheyletidae in the United States National Museum. *Proc. US natn. Mus.* **99**: 267–320, 17 pls.   (1949) **Baker, E. W.**

Observations sur les Myobiidae parasites des rongeurs. Evolution parallele hôtes-parasites (Acariens: Trombidiformes). *Acarologia* **16**: 441–75, 17 figs.

(1975) **Fain, A.**

Keys to genera and species. Host lists.

A revision of the family Syringophilidae (Prostigmata: Acarina). *Contr. Am. ent. Inst.* **5**(6): 1–76, 38 figs.

(1970) **Kethley, J. B.**

New classification of the mites of the superfamilies Cheyletoidea W. Dub. and
Demodicoidea W. Dub. *Mag. Parasit.* **17**: 71–136, 51 figs. [Russian].
(1957) Dubinin, V. B.

Beschreibung neuer *Psorergates*-Arten (Psorergatidae: Trombidiformes).
Tidschr. Ent. **110**: 133–81, 84 figs.
(1967) Lukoschus, F., Fain, A., and Beaujean, M. M. J.
Key to species.

Review of the mite family Cheyletidae. *Univ. Calif. Publs Ent.* **6**: 1–153, 59 figs.
(1970) Summers, F. M. and Price, D. W.
Keys to genera and species.

[Mites of the family Cheyletidae of the world fauna.] *Opred. Faune SSSR* **101**: 1–
431, 536 figs. [Russian].   (1969) **Volgin, V. I.**
Keys to genera and species.

Cheyletiella (Acari: Cheyletiellidae) of dog, cat and domesticated rabbit, a
review. *J. med. nt.* **13**: 315–27, 12 figs.
(1976) Bronswijk, J. E. M. H. and Kreek, E. J. de

Demodicoidea

Studies on Acari I. The genus *Demodex* Owen. British Museum (Natural
History), London. 44 pp, 4 figs, 13 pls.   (1919) **Hirst, S.**

Endeostigmata (Pachygnathoidea)

Quelques genres d'acariens appartenant au groupe des Endeostigmata. *Ann.
Sci. nat.* (11) **2**: 1–122, 25 figs; **4**: 81–135; **5**: 137–95, 18 figs.
(1939–43) Grandjean, F.

Eriophyoidea

Catalogue of eriophyid mites (Acari: Eriophyoidea). Warsaw Agricultural University Press, Warsaw. 254 pp.   (1982) **Davis, R.** *et al.*

The pocket encyclopaedia of plant galls in colour, 2nd edn. Blandford Press, Poole.
191 pp, 293 figs.   (1975) **Darlington, A.**

Bestimmungstabellen der Gallen an Pflanzen Mittel- und Nordeuropas. G. Fischer,
Jena. 2 vols, xvi + 1572 pp, 25 pls.   (1964–65) **Burh, H.**
Keys species on each plant genus according to gall type.

Familie Eriophyidae. Gallmilben. *Tierwelt Mitteleur.* **3**(3) (Neubearb.): 1–155,
115 figs.   (1965) **Farkas, H.**

Eriophyiden-Gallenmilben. II. Teil. Systematik der Gallmilben. Beschreibung der Gallmilben Deutschlands. *Zoologica, Stuttg.* **24**: 205–93, 6 pls.
(1910) Nalepa, A.
Descriptions and illustrations.

Eriophyidocecidien, die durch Gallmilben verursachten Pflanzengallen.
Zoologica, Stuttg. **24**: 295–498, 34 figs, 24 pls.
(1916) Schlechtendal, D. H. R. von
Excellent illustrations of galls, many in colour.

Erythraeoidea

Die bisjetzt bekannten Larven von Trombidiidae und Erythraeidae. *Zool. Jahrb.*, Suppl. **14**: 1–230, 271 figs.   (1912) **Oudemans, A. C.**

Die Landmilben des schweizerischen Nationalparkes. II Trombidiformes Reuter 1909. *Ergebn. widd. Unters. schweiz. Nat.-Parks* (N.F.) **3**: 119–64, 71 figs.   (1951) **Schweizer, J.**
Keys to species. Excellent illustrations.

Studies on the systematics and biology of the Erythraeoidea (Acarina), with a critical revision of the genera and subfamilies. *Aust. J. Zool.* **9**: 367–610, 25 figs, 1 pl.   (1961) **Southcott, R. V.**

Eupodoidea

Soil mites of the family Rhagidiidae (Actinedida: Eupodoidea). Morphology, systematics, ecology. *Acta Univ. Carol.* (Biol.) **1978**: 489–785, 134 figs.
(1980) **Zacharda, M.**

Prostigmatic mites from the experimental farm in Etzdorf/Saalkreis, G.D. R. *Abhandlungen Ber. Naturk. Mus.-ForschStelle, Görlitz* **50**(2): 1–33, illustrated.
(1976) **Strandtmann, R. W. and Prasse, J.**
Keys to species of *Eupodes* given.

The eupodoid mites of Alaska (Acarina, Prostigmata). *Pacif. Insects* **13**: 75–118, 15 figs.   (1971) **Standmann, R. W.**
Keys to genera and known species.

Hydracarina (Hydrachnellae, Halacaridae, Limnochalacaridae)

A systematic and ecological study of the Halacaridae of eastern North America. *Bull. Bingham oceanogr. Coll.* **10**(3): 1–232, 331 figs.
(1947) **Newell, I.**
Keys to families, genera, and species many of which occur in N.W. Europe.

The taxonomy of water mite larvae. *Mem. Am. ent. Inst.* **18**: ii + 326 pp, 126 figs.
(1972) **Prasad, V. and Cook, D. R.**
Illustrated keys and diagnoses for families and genera.

The British Hydracarina. Ray Society, London. Vol. 1, x + 216 pp, 20 pls. Vol. 2, viii + 215 pp, 20 pls. Vol. 3, viii + 184 pp, 20 pls.
(1925) **Soar, C. D. and Williamson, W.**

Water mite genera and subgenera. *Mem. Am. ent. Inst.* **21**: vii + 860 pp, 1942 figs.   (1974) **Cook, D. R.**
Illustrated keys and diagnoses for families, subfamilies, and genera.

A synonymic and bibliographic check-list of the freshwater mites (Hydrachnellae and Limnohalacaridae, Acari) recorded from Great Britain and Ireland. *Occ. Publs freshwat. biol. Ass.* **1**: 59 pp.
(1976) **Gledhill, T. and Viets, K. O.**

A key to the water mites (Hydracarina) of the Flatford area. *Field Studies* **1**: 1–20, illustrated.   (1961) **Hopkins, C. C.**

A check-list of British marine Halacaridae (Acari) with notes on two species of the subfamily Rhombognathinae. *J. mar. biol. Ass. UK* **39**: 63–9, 4 figs.
(1960) **Green, J.**

Clare Island Survey, part 39. Acarinida: I, Hydracarina; II, Terrestrial and marine Acarina. *Proc. R. Ir. Acad.* **31**(39): 1–44, 45–136.
(1911–15) **Halbert, J. N.**

A taxonomic key to the British marine mites of the Halacaridae. *Porcupine Newsl.* **2**: 92–8, figs. (1982) **Bamber, R. N.**

Halacariens marins. *Faune Fr.* **46**: 152 pp, 83 figs. (1946) **André, M.**

Les Porohalacaridae de la faune française. *Annls Limnologie* **1**: 213–20, 8 figs.
(1965) **Angelier, E.**

De nederlandse Watermijten (Hydrachnellae Latreille 1802). *Monographieen ned. ent. Vereen* **1**: 199 pp, 333 figs. [Dutch]. (1964) **Besseling, A. J.**
Keys to genera and species.

Abteilung: Wassermilben, Hydracarina. *Tierwelt Mitteleur.* **3**(4): 8, 57 pp, 135 figs (with Viets, K. Q.) pp. 1–44, 6 pls. (Ergänzung). (1928, 1960) **Viets, K.**
Keys to families, genera, and species.

Spinnentiere oder Arachnoidea VII: Wassermilben oder Hydracarina (Hydrachnellae und Halacaridae). *Tierwelt Dtl.* **31–32**: x + 574 pp, 652 figs.
(1936) **Viets, K.**

Die Milben des Süsswassers und des Meeres. Hydrachnellae und Halacaridae (Acari). Gustav Fischer, Jena. Vol. 1, Bibliographie, 476 pp; Vol. 2, Katalog und Nomenclator, 870 pp, 140 figs. (1955–56) **Viets, K.**

Hydracarinen Schwedens I–III. *Zool. Bidr. Uppsala* **11**: 185–540, 5 pls; *Ark. Zool.* **14**: 1–635, 534 figs; **21**: 1–633, 15 figs, 7 pls.
(1927, 1962, 1968) **Lundblad, O.**

Raphignathoidea

The genus *Raphignathus* Duges (Acarina, Raphignathidae) in the United States with notes on the Old World species. *Acarologia* **3**: 14–20, 7 figs.
(1961) **Atyeo, W. T.** *et al.*

A taxonomic study of the genera *Mediolata*, *Zetzellia*, and *Agestemus* (Acarina: Stigmaeidae). *Univ. Calif. Publs Ent.* **41**: 1–64, 116 figs.
(1965) **Gonzalez-Rodriguez, R. H.**
Key to genera of Stigmaeidae.

Eupalopsis and eupalopsellid mites (Acarina: Stigmaeidae, Eupalopsellidae). *Florida Ent.* **43**: 119–38, 22 figs. (1960) **Summers, F. M.**

Mites of the family Caligonellidae (Acarina). *Hilgardia* **23**: 539–61, 9 pls.
(1955) **Summers, F. M. and Schlinger, F. I.**

New and redescribed species of *Ledermuelleria* from North America (Acarina: Stigmaeidae). *Hilgardia* **31**: 369–81, 6 pls.
(1961) **Summers, F. M. and Price, D. W.**

The genus *Stigmaeus* (Acarina: Stigmaeidae). *Hilgardia* **33**: 491–537, 55 figs.
(1962) **Summers, F. M.**

Tarsonemina (=Heterostigmata)

A revision of the Tarsonemidae of the western hemisphere (Order Acarina).
Kans. Univ. Sci. Bull. **36**: 1091–387, 25 pls.   (1954) **Beer, R. E.**

Revisione dei generi della famiglia Tarsonemidae (Acarina). *Boll. Zool. agric.
Bachic. (2)* **7**: 19–43, 77 figs. [Italian]. (1965) **Beer, R. E. and Nucifora, A.**

The generic relationships of the family Pyemotidae (Acarina: Trombidiformes). *Kans. Univ. Sci. Bull.* **45**: 29–275, 100 figs.   (1965) **Cross, E. A.**
Keys to genera and species lists.

A new dimorphic species of *Pyemotes* and a key to previously described forms
(Acarina: Tarsonemoidea). *Ann. ent. Soc. Amer.* **68**: 723–32, 19 figs.
(1975) **Cross, E. A. and Moser, J. C.**

Systematik und Ökologie der Scutacariden. In *Beiträge zur Systematik und
ökologie mitteleuropäischer Acarina* (ed. H. J. Stammer), Vol. 1, Part 2, pp. 627–
712, 42 figs. Geest and Portig, Leipzig.   (1959) **Karafiat, H.**
Keys to genera and species.

Systematik und Ökologie der Pyemotiden. In *Beiträge zur Systematik und
Ökologie mitteleuropäischer Acarina* (ed. H. J. Stammer), Vol. 1, Part 2, pp.
385–625, 85 figs. Geest and Portig, Leipzig.   (1959) **Krczal, H.**
Keys to genera and species.

Identification keys for the species of the family Scutacaridae (Acari: Tar-
sonemini). *Acta zool. Bpest.* **11**: 353–401, 29 figs.   (1965) **Mahunka, S.**

Beiträge zur Tarsonemini-Fauna Ungarns, VI. (Acari, Trombidiformes).
Opusc. zool. Bpest. **9**: 363–72, 13 figs.   (1969) **Mahunka, S.**
Key to species of *Siteroptes.*

Considerations on the systematics of the Tarsonemina and the description of
new European taxa (Acari: Trombidiformes). *Acta zool. hung.* **16**: 137–74, 23
figs.   (1970) **Mahunka, S.**
Keys to families and to genera of Pyemotoidea and Pygmephoroidea.

Bibliographica gefundene Arten der Gattung *Pygmephorus* (Acarina, Pyg-
mephoridae). *Mitt. Hamburg zool. Mus. Inst.* **72**: 157–76, 26 figs.
(1975) **Rack, G.**
Key to species of *Pygmephorus s.str.*

Systematik und ökologie der Tarsonemiden. In *Beiträge zur Systematik und
Ökologie mitteleuropäischen Acarina* (ed. H. J. Stammer), Vol. 1, Part 2, pp.
713–823, 55 figs. Geest and Portig, Leipzig.   (1959) **Schaarschmidt, L.**

Tetranychoidea

A revision of the spider mite family Tetranychidae. *Mem. Pacif. Coast ent. Soc.* **2**:
472 pp, 391 figs, 1 pl.   (1955) **Pritchard, A. E. and Baker, E. W.**

The false spider mites (Acarina: Tenuipalpidae). *Univ. Calif. Publs Ent.* **14**: 175–275, 51 figs.

Trombidioidea

Trombidiidae. Prospetto dei generi e delle specie finora noti. *Redia* **8**: 1–291, 137 figs, 1 pl. [Italian]. (1912) **Berlese, A.**

Acarina Trombidoidea. *Fauna Rep. pop. Romina* **5**(1): 1–190, 110 figs. [Rumanian]. (1955) **Feider, F.**

Die bisjetzt bekannten Larven von Trombidiidae und Erythraeidae. *Zool. Jahrb.* (Suppl.) **14**:J 1–230, 271 figs. (1912) **Oudemans, A. C.**

Die Landmilben des schweizerischen Nationalparkes. II Trombidiformes Reuter 1909. *Ergebn. wiss. Unters. schweiz. Nat.-Parks* (N.F.) **3**: 119–64, 71 figs. (1951) **Schweizer, J.**
Keys to species. Excellent illustrations.

Acarina. Trombidiidae. *Tierreich* **71b**: 187–541, 347 figs.
(1947) **Thor, S. and Willmann, C.**

Tydeoidea

A review of the genera of the family Tydeidae (Acarina). In *Advances in acarology* (ed. J. A. Naegele), Vol. 2, pp. 95–133, 13 pls. Comstock, Ithaca, New York.
(1965) **Baker, E. W.**

The genus *Lorryia*. *Ann. ent. Soc. Am.* **61**: 986–1008, 55 figs. (1968) **Baker, E. W.**

The genus *Tydeus*: subgenera and species groups with descriptions of new species. (Acarina: Tydeidae). *Ann. ent. Soc. Am.* **63**: 163–77, 53 figs.
(1970) **Baker, E. W.**

Les Ereynetidae de la collection Berlese à Florence; designation d'une espèce type pour le genre *Ereynetes* Berlese. *Redia* **49**: 87–111, 5 figs. (1965) **Fain, A.**
Descriptions, illustrations, list of known species of *Ereynetes*.

Notes sur les speleognathines parasites nasicoles des mammifères (Ereynetidae: Trombidiformes). *Acarologia* **12**: 509–21, 21 figs.
(1970) **Fain, A.**
Key to species, host list.

Clé et liste des espèces du genre *Boydaia* Wormersley (Ereynetidae: Trombidiformes). *Acarologia* **13**: 98–112. (1971) **Fain, A.**

Acarina. Tydeidae, Ereynetidae. *Terreich* **60**: 1–84, 102 figs. (1933) **Thor, S.**

New and redescribed species of Tydeidae (Acari) from moorland soils in Britain. *Acarologia* **7**: 663–72, 5 figs. (1965) **Wood, T. G.**

A generic revision of the family Tydeidae (Acari: Actinedida).
1. Introduction, paradigms and general classification. *Annals Soc. r. zool. Belg.* **108**(1979): 189–208, illustrated.
2. Organotaxy of the idiosoma and gnathosoma. *Acarologia* **22**(1981): 31–46, illustrated.
3. Organotaxy of the legs. *Acarologia* **22**(1981): 165–78. Illustrated.

   4. Generic descriptions, keys and conclusions. *Bull. Annls Soc. r. ent. Belg.* **116**(1980): 103–30, illustrated; 139–68, illustrated.   (1979) **André, H. M.**

PYCNOGONIDA

Pycnogonides. *Faune Fr.* **7**: 1–71, 61 figs.   (1923) **Bouvier, E.-L.**
Note sur les Pycnogonides de la Belgique. *Bull. Annls Soc. r. ent. Belg.* **68**: 193–229, 13 figs.   (1928) **Giltay, L.**
Quelques pycnogonides des environs de Banyuls (France). *Bull. Annls Soc. r. ent. Belg.* **69**: 172–6, 2 figs.   (1929) **Giltay, L.**
Remarques sur le genre *Ammothea* Leach, et description d'une espèce nouvelle de la Mer d'Irlande. *Bull. Mus. r. Hist. nat. Belg.* **10**(18): 1–6, 3 figs.
                    (1934) **Giltay, L.**
The Pycnogonida of the western North Atlantic and the Caribbean. *Proc. US natn. Mus.* **97**: 157–342, 339 figs, 3 charts.   (1948) **Hedgpeth, J. W.**
British sea spiders. *Synopses Br. Fauna* (N.S.) **5**: 68 pp, 28 figs, 5 maps.
                    (1974) **King, P. E.**
Pycnogonida. *Norske Norhavs-Exped.* **20**: 1–163, 15 pls, 1 map.
                    (1891) **Sars, G. O.**
Die Pantopoden der deutschen Küsten. *Wiss. Meeresunters.* **18**: 1–10, 8 figs.
                    (1932) **Schlottke, E.**
Revision of the European representatives of the genus *Callipallene* Flynn, 1929. *Beaufortia* **13**: 1–15, 27 figs.   (1952) **Stock, J. H.**

PENTASTOMIDA

Pentastomids. In *Parasites of laboratory animals* (ed. R. J. Flynn), pp. 493–503. Iowa State University Press, Ames, Iowa.   (1973) **Fain, A.**
Pentastomida. *Bronn's Kl. Ordn. Tierreichs* **5**(4): 1, vi + 268 pp, 148 figs.
                    (1935) **Heymons, R.**
Phylogénèse at systématique. Le phylum des Pentastomida. *Annls Parasit. hum. comp.* **38**: 483–516.   (1963) **Nicoli, R.-M.**
A synopsis of the family Linguatulidae. *J. trop. Med. Hyg.* **25**: 188–206, 391–428, 47 figs.   (1922) **Sambon, L. W.**
Pentastomida (Tongue worms). In *Insects and other arthropods of medical importance* (ed. K. G. V. Smith), pp. 479–81, 1 fig. British Museum (Natural History), London.   (1973) **Sheals, J. G.**

CRUSTACEA

BRANCHIOPODA

ANOSTRACA

Monographie systematique des phyllopodes Anostraces. *Ann. sci. nat. Zool. Paris* (9) **11**: 91–489.   (1910) **Daday de Dees, E.**

NOTOSTRACA

A review of the Notostraca. *Bull. Br. Mus. nat. Hist.* (Zool.) **3**: 1–57.
(1955) **Longhurst, A. R.**

CLADOCERA

Cladocera Sueciae. *Nova Acta R. Soc. Scient. upsal.* (3) **19**: 1–700.
(1900) **Lilljeborg, W.**

The British species of the genus *Daphnia* (Crustacea, Cladocera). *Proc. zool. Soc. Lond.* **122**: 435–62.   (1952) **Johnson, D. S.**

A key to the British species of freshwater Cladocera with notes on their ecology. *Scient. Publs Freshwat. biol. Ass.* **5**: 1–55.
(1958) **Scourfield, D. J. and Harding, J. P.**

Cladocera. *Fich. Ident. Zooplancton* **143**: 4 pp, 8 figs.   (1974) **Della Croce, N.**

OSTRACODA

Ostracoda. *An account of the Crustacea of Norway with short descriptions and figures of all the species.* Bergen Museum, Bergen. Vol. 9, 277 pp, many pls.
(1922–28) **Sars, G. O.**

Die Ostracoden des Golfes von Neapel und der angrenzenden Meeres Abschnitte. *Fauna Flora Golfo Napoli* **21**: 1–404.   (1894) **Muller, G. W.**

Ostracoda Muschelkrebse. *Tierwelt Dtl.* **34**(3): 1–230.   (1938) **Klie, W.**

Zur Kenntnis der mariner Ostracoden Schwedens mit besonderer Berücksichtigung des Skageraks. *Zool. Bidr. Uppsala* **19**: 215–534, 52 figs, 42 maps. English translation (1969) by Israel Program for scientific translations, Jerusalem, no. 5113.   (1941) **Elofson, O.**

A revision of the genus *Paradoxostoma* Fischer (Crustacea: Ostracoda) in British waters. *Zool. J. Linn. Soc.* **85**: 131–203.
(1985) **Horne, D. J. and Whittaker, J. E.**

British coastal Ostracoda (Benthic Podocopina). *Synopses Br. Fauna.*
(In press) **Athersuch, J., Horne, D. J., and Whittaker, J. E.**

MYSTACOCARIDA

Revision des mystacocarides du genre *Derocheilocaris*. *Vie Milieu* **4**: 321–80.
(1953) **Delamare-Deboutteville, C.**

BRANCHIURA

The parasitic Copepoda and Branchiura of British freshwater fishes. A handbook and key. *Scient. Publs Freshwat. biol. Ass.* **46**: 1–87.
(1982) **Fryer, G.**

COPEPODA

GENERAL

Copepoda. *An account of the Crustacea of Norway.* Bergen Museum, Bergen. Vol. 4 (1901–03), xiii + 171 pp, pls. Vol. 5 (1903–11) xiv + 499 pp, pls. Vol. 6 (1913–18), ix + 225 pp, pls. Vol. 7 (1919–21), 121 pp, pls. Vol. 8 (1921), 91 pp, pls.   (1901–21) **Sars, G. O.**

Crustacea Copepoda 2. Cyclopoida Gnathostoma. *Tierreich* **53**: 1–102.
                              (1929) **Kiefer, F.**

Monographie der Harpacticiden. Hakan Ohlssons Boktrycheri, Lund. 2 vols, 1683 pp, 610 figs, 378 maps.   (1948) **Lang, K.**

MARINE

Catalogue des nouveaux Copepodes harpacticoides marins. Université de Bretagne Occidentale, Brest. 228 pp.   (1979) **Bodin, P.**

The planktonic copepods of the northeastern Atlantic Ocean: Harpacticoida, Siphonostomatoida, and Mormonilloida. *Bull. Br. Mus. nat. Hist.* (Zool.) **35**: 201–64.   (1979) **Boxshall, G. A.**

Copepoda, *Oncaea. Fich. Ident. Zooplancton.* **169–171**: 11 pp. (1983) **Malt, S. J.**

Taxonomy and distribution of the family Oithonidae (Copepoda, Cyclopoida) in the Pacific and Indian Oceans. *Bull. Ocean res. Inst.* **20**: 1–167.
                              (1985) **Nishida, S.**

Copepodes pelagiques. *Faune Fr.* **26**: 1–374, 456 figs.   (1933) **Rose, O.**

Keys to aid in the identification of marine harpacticoid copepods. University of Aberdeen, Aberdeen. 215 pp.   (1976) **Wells, J. B. J.**

Keys to aid in the identification of marine harpacticoid copepods. *Zool. Publ. Victoria Univ. Wellington* (Amendment Bulletin No. 2.) **73**: 1–8.
                              (1979) **Wells, J. B. J.**

FRESHWATER

British freshwater Copepoda. Ray Society, London. Vol. 1, lii + 238 pp, figs; Vol. 2, ix + 336 pp, figs; Vol. 3, xxix + 384 pp. figs.   (1931–33) **Gurney, R.**

A key to the British freshwater cyclopid and calanoid copepods. *Scient. Publs Freshwat. biol. Ass.* **18**: 1–54, 16 figs.
                              (1960) **Harding, J. P. and Smith, W. A.**

Les copepodes des eaux continentales d'Europe occidentale. Calanoides et Harpacticolides. N. Boubée, Paris. 500 pp, text illustrations.   (1967) **Dussart, B.**

Les Copepodes des eaux continentales d'Europe occidentale. II. *Cyclopoides et Biologie.* N. Boubée, Paris. 292 pp, figs, tables.   (1969) **Dussart, B.**

Sur la biologie et l'ecologie des copepodes cyclopides hypoges (Crustaces). *Annls Spéléol.* **3**: 429–502; **4**: 581–674.   (1973) **Lescher-Moutoue, F.**

ASSOCIATED AND PARASITIC

Die Asterocheriden. *Fauna Flora Golfo Napoli* **25**: 1–217.
(1899) **Giesbrecht, W.**

Copepoda parasitica. *Tierwelt N.-u. Ostsee* **10**(c): 73–198, 192 figs.
(1936) **Oorde de Lint, G. M. van and Schuurmans Stekhoven, J. H.**

A key to the ascidicolous copepods of British waters with distributional notes. *Ann. Mag. nat. Hist.* (13) **3**: 211–29, 30 figs.   (1960) **Gotto, R. V.**

A revision of the family Lichomolgidae Kossmann, 1877, cyclopoid copepods mainly associated with marine invertebrates. *Smithsonian Contr. Zool.* **127**, i–v + 1–368, 190 figs.   (1973) **Humes, A. G. and Stock, J. H.**

Cyclopoida Siphonostoma (Crustacea) von Banyuls (Frankreich, Pyrenees Orientales) mit besonderer Berücksichtigung des Gast-Wirtverhältnisses. *Bijdr. Dierk.* **43**(1): 64–92, 12 figs.   (1973) **Schirl, K.**

Parasitic Copepoda of British fishes. Ray Society, London. 468 pp.
(1979) **Kabata, Z.**

The parasitic Copepoda and Branchiura of British freshwater fishes. A handbook and key. *Scient. Publs Freshwat. biol. Ass.* **46**: 1–87.
(1982) **Fryer, G.**

Some new parasitic copepods (Siphonostomatoida: Nicothoidae) from deep-sea asellote isopods. *J. nat. Hist.* **17**: 891–900.
(1983) **Boxshall, G. A. and Lincoln, R. J.**

CIRRIPEDIA

A monograph of the sub-class Cirripedia. Ray Society, London. Vol. 1 (1851) The Lepadidae; or, pedunculated cirripedes, 400 pp; Vol. 2 (1854) The Balanidae, the Verrucidae, 672 pp.   (1851, 1854) **Darwin, C.**

Crustacea Rhizocephala. *Marine invertebrates of Scandinavia.* Norwegian University Press, Oslo. Vol. 6, 92 pp.   (1985) **Hoeg, J. and Lutzen, J.**

A revision of the balanomorph barnacles; including a catalog of the species. *Mem. S. Diego Soc. nat. Hist.* **9**: 108 pp, 17 figs.
(1976) **Newman, W. A. and Ross, A.**

Cirripedia Thoracica and Acrothoracica. *Marine invertebrates of Scandinavia.* Norwegian University Press, Oslo. Vol. 5, 133 pp.
(1978) **Nilsson-Cantell, C. A.**

On the taxonomic status and distribution of *Chthamalus stellatus* (Crustacea, Cirripedia) in the north-east Atlantic region; with a key to the common intertidal barnacles of Britain. *J. mar. biol. Ass. UK* **56**(4): 1007–28, 6 figs, 2 pls.   (1976) **Southward, A. J.**

Barnacles of European waters. Catalogue of main marine fouling organisms found on ships coming into European waters [Introduced by D. L. Ray]. Organisation for Economic Cooperation and Development, Paris. 46 pp, figs.
(1963) Southward, A. J. and Crisp, D. J.

MALACOSTRACA

A revised key to the British species of Crustacea: Malacostraca occurring in fresh water with notes on their ecology and distribution. *Scient. Publs Freshwat. biol. Ass.* **32**: 1–72, 48 figs.
(1976) Gledhill, T., Sutcliffe, D. W., and Williams, W. D.

BATHYNELLACEA

Bathynellacea. In A revised key to the British species of Crustacea: Malacostraca occurring in fresh water with notes on their ecology and distribution. *Scient. Publs Freshwat. biol. Ass.* **32**: 7–11.
(1976) Gledhill, T., Sutcliffe, D. W., and Williams, W. D.

MYSIDACEA

The British Mysidacea. Ray Society, London. viii + 460 pp, 118 figs.
(1951) Tattersall, W. M. and Tattersall, O. S.

A guide to the British coastal Mysidacea. *Fld Stud.* **4**: 575–95, 2 figs.
(1977) Makings, P.

CUMACEA

Cumacea. *An account of the Crustacea of Norway.* Bergen Museum, Bergen. Vol. 3, 115 pp, 72 pls.
(1900) Sars, G. O.

The marine fauna of the west coast of Ireland. Cumacea. *Scient. Invest. Fish Brch. Ire.* **1**: 1–52, 5 pls.
(1905) Calman, W. T.

Cumacea. *Fich. Ident. Zooplancton* **71–76**: 46 pp, 248 figs. (1958) **Jones, N. S.**

Résultats scientifiques de la campagne du N.O. 'Jean Charcot' en Méditerranée occidentale, Mai–Juin–Juillet 1970. Cumaces. *Crustaceana* (Suppl.) **3**: 362–77, 7 figs.
(1972) Reyss, D.

British Cumacea. *Synopses Br. Fauna* (N.S.) **7**: 66 pp, 20 figs.
(1976) Jones, N. S.

The family Nannastacidae (Crustacea: Cumacea) from the deep Atlantic. *Bull. Br. Mus. nat. Hist.* (Zool.) **46**(3): 207–89, 50 figs. (1984) **Jones, N. S.**

ISOPODA and TANAIDACEA

MARINE AND FRESHWATER

Isopoda. *An account of the Crustacea of Norway.* Bergen Museum, Bergen. Vol. 2, 270 pp, 104 pls.
(1899) Sars, G. O.

Isopoda en Tanaidacea (KV). *Fauna Ned.* **16**: 1–280, 89 figs. [Dutch].
(1956) **Holthuis, I. B.**

The Bopyrid parasites of the Anomura from British and Irish waters. *J. Linn. Soc.* (Zool.) **42**: 219–37, 1 fig, 5 pls. (1953) **Pike, R. B.**

Isopoda, Valvifera, Asellota (Vol. 77); Isopoda, Flabellifera (Vol. 78). *Fich. Ident. Zooplancton.* 8 pp, 32 figs. (1957) **Naylor, E.**

Revision of the European marine forms of the Cirolanidae, a subfamily of Crustacea Isopoda. *J. Linn. Soc.* (Zool.) **29**: 337–73, pls 33–5.
(1905) **Hansen, H. J.**

Les Gnathiidae; essay monographique. *Mem. Soc. Sci. nat. Phys. Maroc.* **13**: 1–668, 277 figs, 1 pl. (1926) **Monod, T.**

Notes on British Sphaeromatidae. (Crustacea Isopoda). *Rep. Dove mar. Lab.* **2**(3): 22–58, 6 pls. (1934) **Omer-Cooper, J. and Rawson, J. H.**

The comparative external morphology and revised taxonomy of the British species of *Idotea.J. mar. biol. Ass. UK* **34**: 467–93, 11 figs. (1955) **Naylor, E.**

The occurrence of *Microcharon* in the Plymouth offshore bottom fauna, with description of new species. *J. mar. biol. Ass. UK* **38**: 57–63, 2 figs.
(1959) **Spooner, G. M.**

Étude des représentants du genre *Munna* Kröyer sur les Côtes françaises de la Manche. *Bull. Soc. linn. Normandie* **10**(2): 222–42, 8 pls. (1961) **Carton, Y.**

Observations on the British species of *Jaera* (Isopoda: Asellota). *J. mar. biol. Ass. UK* **41**: 817–28, 4 figs.
(1961) **Naylor, E., Slinn, D. J., and Spooner, G. M.**

Tanaidacea from Trondheimsfjorden. *K. norske Vidensk. Selsk. Forh.* **38**: 140–3.
(1966) **Greve, L.**

The distribution of *Eurydice* (Crustacea: Isopoda) in British waters, including *E. affinis* new to Britain. *J. mar. biol. Ass. UK* **47**: 373–82, figs, tables.
(1967) **Jones, D. A. and Naylor, E.**

A systematic revision of the genus *Dynamene* (Crustacea: Isopoda) with a description of three new species. *Pubbl. Staz. zool. Napoli* **36**: 401–26, figs.
(1968) **Holdich, D. M.**

Tanaidacea from Hardangerfjorden, western Norway. *Sarsia* **36**: 77–84, 3 figs.
(1968) **Greve, L.**

Some new records of Tanaidacea from Norway. *Sarsia* **48**: 33–8, 1 fig.
(1972) **Greve, L.**

British marine isopods. *Synopses Br. Fauna* (N.S.) **3**: 1–86, 24 figs, 1 table.
(1972) **Naylor, E.**

Les Asellides d'Europe et Pays Limitrophes. *Archs Zool. exp. gén.* (Notes et Revue) **86**: 78–94, figs. (1949) **Chappius, P. A.**

A re-examination of certain records for the genus *Asellus* (Isopoda) in the British Isles. *Proc. zool. Soc. Lond.* **123**: 411–17. (1953) **Moon, H. P.**

Observations on Epicaridea obtained from hermit-crabs in British waters, with notes on the longevity of the host-species. *Ann. Mag. nat. Hist.* (13)**4**: 225–40, 7 figs.   (1961) **Pike, R. B.**

Les Bopyridae des Mers Européenes. *Mémoires Mus. natn. Hist. nat. Paris* (Serie A, Zoologie) **50** (2 et dernier): 77–424, 190 figs.   (1968) **Bourdon, R.**

Keys to suborders and families of Tanaidacea (Crustacea). *Proc. biol. Soc. Wash.* **91**(4): 840–6, 3 figs.   (1978) **Sieg, J. and Winn, R.**

Nomenclatura e geonemia di *Sphaeroma monodi* Arcangeli, 1934, del Mediterraneo e del Mar Nero (Crustacea, Isopoda Flabellifera). *Boll. Mus. civ. Stor. nat. Verona* **7**: 227–34, 1 fig.   (1980) **Argano, R. and Ponticelli, A.**

A new contribution to the systematics and phylogeny of the suborder Monokonophora (Crustacea, Tanaidacea). *Trav. Mus. Hist. nat. 'Gr. Antipa'* **23**: 81–108.   (1981) **Gutu, M.**
Key to genera of *Monokonophora*.

Tanaids. *Synopses Br. Fauna* (NS) **27**: 98 pp, 32 figs.
                                        (1983) **Holdich, D. M. and Jones, J. A.**

Cirolana cranchi Leach, 1818 (Crustacea, Isopoda: Cirolanidae) redescribed, with notes on its distribution. *Bull. Br. Mus. nat. Hist.* (Zool.) **44**(2): 75–84, 4 figs.   (1983) **Bruce, N. L. and Ellis, J. P.**

Deep-sea asellote isopods of the north-east Atlantic: the family Dendrotionidae and some new ectoparasitic copepods. *Zool. J. Linn. Soc.* **79**: 297–318, 11 figs.   (1983) **Lincoln, R. J. and Boxshall, G. A.**

Deep-sea asellote isopods of the north-east Atlantic: the family Haploniscidae. *J. nat. Hist.* **19**: 655–95, 21 figs.   (1985) **Lincoln, R. J.**

TERRESTRIAL

Isopodes terrestres. *Faune Fr.* **64**: 1–416, 205 figs; **66**: 417–931, 409 figs.
                                                        (1960–62) **Vandel, A.**

British woodlice. *Synopses Br. Fauna* **9**: 55 pp, 33 figs.   (1954) **Edney, E. B.**

Krebstiere oder Crustacea. 5. Isopoda 2 Liefeerung. *Tierwelt Dtl.* **53**: 151–380, 262 figs.   (1966) **Gruner, H.-E.**

Woodlice (Facsimile reprint of 1st edn). Pergamon Press, Oxford. 144 pp, 37 figs, 8 pls.   (1980) **Sutton, S. L.**

Woodlice in Britain and Ireland: distribution and habitat. Institute of Terrestrial Ecology, Huntingdon. 151 pp, 14 figs + maps.
                                        (1985) **Harding, P. T. and Sutton, S. L.**

AMPHIPODA

Amphipoda. *An account of the Crustacea of Norway.* Museum, Bergen. Vol. 1, 711 pp, 248 pls.   (1890–95) **Sars, G. O.**

Amphipodes. *Faune Fr.* **9**: 1–488, 438 figs. (1925) **Chevreux, E. and Fage, L.**

The Amphipoda of N. Norway and Spitsbergen with adjacent waters. *Tromso Mus. Skr.* **3**(1): 1–140, 19 figs, 2 maps; **3**(2): 141–278; **3**(3): 279–362; **3**(4): 363–526, figs.   (1935–42) **Stephensen, K.**

Krebstiere oder Crustacea. IV: Flohkrebse oder Amphipoda. *Tierwelt Dtl.* **40**(4): 1–252, 204 figs.   (1942) **Schellenberg, A.**

Caprellidae (Amphipoda, Crustacea). *Synopses Br. Fauna* **2**: 1–27, figs.
                                                    (1944) **Harrison, R. J.**

The occurrence of *Ingolfiella* in the Eddystone shell gravel, with description of a new species. *J. mar. biol. Ass. UK* **39**: 319–29, 5 figs. (1960) **Spooner, G. M.**

A revision of the European species of the *Gammarus locusta* – group (Crustacea: Amphipoda). *Zool. Verh. Leiden* **90**: 1–56, figs.   (1967) **Stock, J. H.**

A revision of the amphipod genus *Microdeutopus* Costa (Gammaridea: Aoridae). *Bull. Br. Mus. nat. Hist.* (Zool.) **17**(4): 93–148, 22 figs, 1 pl.
                                                    (1968) **Myers, A. A.**

The families and genera of marine gammaridean Amphipoda. *Bull. US natn. Mus.* **271**: 1–535, 173 figs.   (1969) **Barnard, J. L.**

Scutocyamus parvus, a new genus and species of whale-louse (Amphipoda: Cyamidae) ectoparasitic on the North Atlantic white-beaked dolphin. *Bull. Br. Mus. nat. Hist.* (Zool.) **26**: 59–64, 2 figs, 1 pl.
                                        (1974) **Lincoln, R. J. and Hurley, D. E.**

A revision of the European species of the *Echinogammarus pungens* – group (Crustacea, Amphipoda). *Beaufortia* **16**: 13–78, 35 figs. (1968) **Stock, J. H.**

On members of the *Gammarus pulex* – group (Crustacea, Amphipoda) from western Europe. *Bijdr. Dierk.* **42**: 164–91, 7 figs, 3 maps. (1972) **Pinkster, S.**

The *Echinogammarus berilloni* – group, a number of predominantly Iberian Amphipod species (Crustacea). *Bijdr. Dierk.* **43**(1): 1–38, 16 figs, 3 maps.
                                                    (1973) **Pinkster, S.**

The *Niphargus kochianus* – group in north-western Europe. *Crustaceana* (Suppl.) **4**: 212–43, 1 pl, 16 figs.   (1977) **Stock, J. H. and Gledhill, T.**

Amphipoda. In *Limnofauna Europaea*, 2nd edn (ed. J. Illies), pp. 244–53. G. Fischer, Stuttgart and New York and Swets and Zeitlinger b.v., Amsterdam.   (1978) **Pinkster, S.**
Check-list and bibliography.

British Marine Amphipoda: Gammaridea. British Museum (Natural History), London. vi + 658 pp, 1 pl., 280 figs.   (1979) **Lincoln, R. J.**

A new species of *Stenothoe* Dana (Amphipoda, Gammaridea) from maerl deposits in Kilkieran Bay. *J. Life Sci. R. Dublin Soc.* **2**: 15–18, 2 figs.
                                        (1980) **Myers, A. A. and McGrath, D.**

Taxonomic studies on British and Irish Amphipoda. The genus *Photis* with the re-establishment of *P. pollex* (=*P. macrocoxa*). *J. mar. biol. Ass. UK* **61**: 759–68, 5 figs.   (1981) **Myers, A. A. and McGrath, D.**

The Amphipoda of the Mediterranean. Part 1 Gammaridea (Acan-
thonotozomatidae to Gammaridae). *Mem. Inst. oceanogr. Monaco* **13**: xiii +
364 pp, 243 figs.   (1982) **Ruffo, S.**, ed.

Taxonomic studies on British and Irish Amphipoda. Re-establishment of
Leucothoe procera. J. mar. biol. Ass. UK **62**: 693–8, 3 figs.
                                        (1982) **Myers, A. A. and McGrath, D.**

Taxonomic studies on British and Irish Amphipoda. The genus *Gammaropsis.*
J. mar. biol. Ass. UK **62**: 93–100, 3 figs.
                                        (1982) **Myers, A. A. and McGrath, D.**

The genus *Listriella* (Crustacea: Amphipoda) in British and Irish waters, with
the description of a new species. *J. mar. biol. Ass. UK* **63**: 347–53, 3 figs.
                                        (1983) **Myers, A. A. and McGrath, D.**

Freshwater amphipoda of the world. Hayfield Associates, Mt. Vernon, Virginia. 2
vols, vii + 830 pp, maps.   (1983) **Barnard, J. L. and Barnard, C. M.**

The amphipod *Monoculodes gibbosus* (Crustacea) in British waters. *J. mar. biol.
Soc. UK* **64**: 271–8, 4 figs.   (1984) **Moore, P. G.**

A revision of the north-east Atlantic species of *Ericthonius* (Crustacea:
Amphipoda). *J. mar. biol. Ass. UK* **64**: 379–400, 14 figs.
                                        (1984) **Myers, A. A. and McGrath, D.**

The amphipod genus *Aora* in British and Irish waters. *J. mar. biol. Ass. UK* **64**:
279–83, 2 figs.   (1984) **Myers, A. A. and Costello, M. J.**

DECAPODA

The Decapoda Natantia of the coasts of Ireland. *Scient. Invest. Fish. Brch Ire.* **1**:
1–190, 23 pls.   (1910) **Kemp, S.**

The Decapoda Reptantia of the coasts of Ireland. Pt. 1. Palinura, Astacura
and Anomura (except Paguridea). *Scient. Invest. Fish. Brch Ire.* **1**: 1–116, 15
pls.   (1914) **Selbie, C. M.**

The Decapoda Reptantia of the coasts of Ireland. Pt. 2. Paguridea. *Scient.
Invest. Fish. Brch Ire.* **1**: 1–68, 9 pls.   (1921) **Selbie, C. M.**

A revision of the genus *Portunus* (A. Milne Edwards, Bell, etc.) *J. mar. biol. Ass.
UK* **14**: 877–908, 9 figs.   (1927) **Palmer, R.**

The larval stages of the Plymouth Brachyura. *Proc. zool. Soc. Lond.* **1928**(2):
473–560, 5 figs, 16 pls.   (1928) **Lebour, M. V.**

Notes on the British species of the genus *Galathea* Fab. *Rep. Dove mar. Lab.* **4**:
38–52, 6 pls.   (1937) **Bull, H. O.**

Bibliography of the larvae of decapod Crustacea. Ray Society, London. i–vi + 1–123
pp.   (1939) **Gurney, H. O.**

Larvae of decapod Crustacea. Ray Society, London. viii + 306 pp, 122 figs.
                                        (1942) **Gurney, R.**

The larvae of the genus *Porcellana* (Crustacea Decapoda) and related forms. *J.
mar. biol. Ass. UK* **25**: 721–37, 12 figs.   (1943) **Lebour, M. V.**

The larval stages of *Portunus* (Crustacea Brachyura) with notes on some other genera. *J. mar. biol. Ass. UK* **26**: 7–15, 5 figs.   (1944) **Lebour, M. V.**

Decapoda (K IX) A. Natantia, Macrura Reptantia, Anomura en Stomatopoda (K X). *Fauna Ned.* **15**: 1–166, figs.   (1950) **Holthuis, L. B.**

Description et repartition des *Xantho* des mers d'Europe. *Archs Zool. exp. gen.* **90**: 1–36, 24 figs.   (1953) **Drach, P. and Forest, J.**

Un *Hippolyte* (Crust. Decap. Nat.) méconnu, nouveau pour les côtes de France et commensal de la comatule, *Antedon bifida. Archs Zool. exp. gen.* **90**: 71–86, 41 figs.   (1953) **Nouvel, H.**

The names of the european species of the genus *Xantho* Leach, 1814 (Crustacea Decapoda Brachyura). *Proc. K. Akad. Wet.* **57C**: 103–7.
                              (1954) **Holthuis, L. B.**

On a pelagic Penaeid prawn, *Funchalia woodwardi* Johnson, new to the British Fauna. *J. mar. biol. Ass. UK* **35**: 475–81, 1 fig.
                              (1956) **Gordon, I. and Ingle, R. W.**

Brachynotus sexdentatus (Risso), a Grapsoid crab new to Britain. *Ann. Mag. nat. Hist.* (12) **10**: 521–3, 1 fig.   (1957) **Naylor, E.**

Les Processidae (Crustacea Decapoda Natantia) des eaux européennes. *Zool. Verh. Leiden* **32**: 1–53, 220 figs.   (1957) **Nouvel, H. and Holthuis, L. B.**

Larvae of the British species of *Diogenes, Pagurus, Anapagurus* and *Lithodes* (Crustacea, Decapoda). *Proc. zool. Soc. Lond.* **128**: 209–57, 11 figs.
                   (1957) **Macdonald, J. M., Pike, R. B., and Williamson, D. I.**

The Crustacea, Decapoda of Belfast Loch. *Ann. Mag. nat. Hist.* (12) **10**: 656–60.   (1958) **Macdonald, R.**

Crustacea Decapoda: Larvae XI. Paguridea, Coenobitidea, Dromiidea and Homolidea. (Revised 1959). *Fich. Ident. Zooplancton* **81**: 1–9, 66 figs.
                              (1958) **Pike, R. B. and Williamson, D. L.**

Sur deux espèces de crevettes nouvelles pour la faune marine des côtes de Bretagne: *Periclimenes amethysteus* (Risso) et *Hippolyte leptocerus* (Heller) (Decapoda Natantia). *Bull. Lab. marit. Dinard* **44**: 4–6. (1959) **Sollaud, E.**

A new decapod larva referred to *Calocarides coronatus* (Trybon). *Univ. Bergen Arb.* **7**: 1–10, 20 figs.   (1959) **Elofsson, R.**

Sobre a variabilidade e a posicao sistematica do *Xantho incisus* Leach (=*X. floridus* (Montagu)) da zona intercotidal do litoral Portugues. 1. Populacoes do Sul do Cabo da Roca. *Rev. Fac. Cienc. Lisboa* **2C**: 233–52, 1 fig. [Portuguese].   (1959) **Almaça, C.**

A re-examination of the spider crab *Eurynome* Leach from British waters. *Crustaceana* **2**: 171–82, 7 figs.   (1961) **Hartnoll, R. G.**

A hermit crab new to Britain. *Nature, Lond.* **190**: 931.
                              (1961) **Carlisle, D. B. and Tregenza, N.**

The biology of Manx spider crabs. *Proc. zool. Soc. Lond.* **141**: 423–96, 30 figs.
                              (1963) **Hartnoll, R. G.**

Le genre *Macropodia* Leach en Mediterranee. I. *Bull. Mus. nat. Hist. Paris* (2) **36**: 222–34, 16 figs. (1964) **Forest, J. and Zariquiey Alvarez**

The Fauna of the Clyde Sea area. Crustacea: Euphausiacea and Decapoda, with an illustrated key to the British species. Scottish Marine Biological Association, Millport. 116 pp. (1967) **Allen, J. A.**

Marine invertebrates of Scandinavia 2. Crustacea Decapoda Brachyura. *Mar. Invert. Scand.* **2**: 1–143, figs, maps. (1969) **Christiansen, M. E.**

A new species of Processidae (Crustacea, Decapoda, Caridea) and the larvae of the north European species. *J. nat. Hist.* **13**: 11–33. (1979) **Williamson, D. I. and Rochanaburanon, T.**

Larval keys and diagnoses for the subfamily Palaemoninae (Crustacea: Decapoda: Palaemonidae) in the north-east Atlantic and aspects of functional morphology. *J. nat. Hist.* **20**: 203–24. (1986) **Fincham, A. A. and Figueras, A. J.**

Crustacea: Decapoda: Larvae VI. Caridea, Families Palaemonidae and Processidae. *Fich. Ident. Zooplancton* **159/160**: 1–8. (1978) **Fincham, A. A. and Williamson, D. I.**

Crustacea: Decapoda: Larvae IV. Caridea, Families Pandalidae and Alpheidae. *Fich. Ident. Zooplancton* **109**: 1–5. (1967) **Williamson, D. I.**

Crustacea: Decapoda: Larvae III. Caridea, Families Oprlophoridae, Nematocarcinidae and Pasiphaeidae. *Fich. Ident. Zooplancton* **92**: 1–5. (1962) **Williamson, D. I.**

Crustacea: Decapoda: Larvae VII. Caridea, Family Crangonidae. Stenopodidea. *Fich. Ident. Zooplancton* **90**: 1–5. (1960) **Williamson, D. I.**

Crustacea: Decapoda: Larvae V. Caridea, Family Hippolytidae. *Fich. Ident. Zooplancton* **68**: 1–5. (1957) **Williamson, D. I.**

Crustacea: Decapoda: Larvae I. General. *Fich. Ident. Zooplancton* **67**: 1–7. (1957) **Williamson, D. I.**

The recent genera of the caridean and stenopodidean shrimps (Class Crustacea: Order Decapoda; Supersection Natantia) with keys for their determination. *Zool. Verh. Leiden* **26**: 1–157. (1955) **Holthuis, L. B.**

British coastal shrimps and prawns. *Synopsis Br. Fauna* (N.S. **15**: 1–126. (1979) **Smaldon, G.**

The juvenile stages of eight swimming crab species (Crustacea: Brachyura: Portunidae); a comparative study. *Bull. Br. Mus. nat. Hist.* (Zool.) **46**(4): 345–54, 5 figs. (1984) **Ingle, R. W. and Rice, A. L.**

Northeastern Atlantic and Mediterranean hermit crabs (Crustacea: Anomura: Paguroidea: Paguridae). I. the genus *Pagurus* Fabricius, 1775. *J. nat. Hist.* **19**: 745–69, 77 figs. (1985) **Ingle, R. W.**

British crabs. British Museum (Natural History), London and Oxford University Press, Oxford. 222 pp, 114 figs, 34 pls. (1980) **Ingle, R. W.**

A key to the crabs and crab-like animals of British inshore waters. *Fld Stud.* **5**: 753–806, 53 figs.   (1983) **Crothers, J. H. and Crothers, M.**

Shallow-water crabs. *Synopses Br. Fauna* (NS) **25**: 206 pp, 54 figs.
                    (1983) **Ingle, R. W.**

MYRIAPODA

DIPLOPODA

Millipedes. *Synopses Br. Fauna* (N.S.) **35**: 240 pp, 74 figs. (1986) **Blower, J. G.**

Les milles-pattes (Myriapodes). Société Nouvelle des Editions Boubée, Paris. 284 pp, 309 figs, 4 pls.   (1981) **Demange, J.-M.**
General text and keys to the species (centipedes included) of France.

Myriapodes Diplopodes (Chilognathes I). *Faune Fr.* **29**: 369 pp, 748 figs.
                    (1935) **Brölemann, H. W.**
Orders Nematophora and Colobognatha only.

Tausenfüssler oder Myriapoda I: Diplopoda. *Tierwelt Dtl.* **28**: vii + 318 pp, 480 figs.   (1934) **Schubart, O.**

CHILOPODA

British species about 47.

Centipedes of the British Isles. Warne, London. x + 294 pp, 495 figs, 5 pls.
                    (1964) **Eason, E. H.**

Eléments d'une faune des Myriapodes de France. Chilopodes. Toulousaine, Toulouse. xx + 450 pp, 481 figs.   (1930) **Brölemann, H. W.**

SYMPHYLA AND PAUROPODA

The Pauropoda and Symphyla of the Geneva Museum II. A review of the Swiss Paurpoda (Myriapoda). *Revue suisse Zool.* **83**: 3–37, 25 figs.
                    (1976) **Scheller, U.**
Keys to and illustrations of 26 species, most of which are likely to occur throughout NW Europe.

Quelques stations de Symphyles et de Pauropodes dans les Îles Britanniques. *Ann. Mag. nat. Hist.* (12) **9**: 287–8.   (1956) **Remy, P. A.**

A revision of the British Symphyla. *Proc. zool. Soc. Lond.* **132**: 403–39, 13 figs (compound).   (1959) **Edwards, C. A.**

Sur la microfaune du sol de Grande-Bretagne. *Ann. Mag. nat. Hist.* (13)**4**: 149–54, 3 figs.   (1961) **Remy, P. A.**

Pauropoda from arable soil in Great Britain. *Symp. zool. Soc. Lond.* **32**: 405–10, 1 fig. (1947) **Scheller, U.**

Some Symphyla and Pauropoda from south-western Germany. *Mitt. bad. Landesver. Naturk. N.F.* **8**: 261–5, 1 fig. (1962) **Scheller, U.**

INSECTA (GENERAL)

Imms' general textbook of entomology, 10th edn. Chapman and Hall, London. viii + 1354 pp in 2 vols, 591 figs. (1977) **Richards, O. W. and Davies, R. G.** The standard British textbook. Vol. 2 (Classification and biology) covers most families of insects but keys to families have been omitted in this edition.

Handbooks for the identification of British insects. Royal Entomological Society of London, London. 11 vols, 1949 onwards, still being issued under numerous authors.
Keys to species and genera with numerous figures by specialists. The Parts cover individual families or larger groups where appropriate. No vols are complete except for Vol. 11 which is the revised Kloet and Hincks *Check List of British Insects.*

Classification of insects, 2nd edn. Harvard University Press, Cambridge, Massachusetts. v + 917 pp.
(1954) **Brues, C. T., Melander, A. L., and Carpenter, F. M.**
Keys to families and some subfamilies, including fossils.

Bibliographie der Bestimmungstabellen europäischer Insekten (1964–1973). *Beitr. Ent.* **26**: 49–66. (1976) **Gaedike, R.**
Cites publications listing works published in earlier years.

Bibliographie der Bestimmungstabellen europäischer Insekten (1974–1978). *Beitr. Ent.* **31**: 235–304. (1981) **Gaedike, R.**

A field guide to the insects of Britain and northern Europe. Collins, London. 352 pp, many line drawings, 60 col. pls. (1973) **Chinnery, M.**
Keys to families or superfamilies; illustrations in colour of several hundred of the commoner British species.

Fauna von Deutschland, 12th edn. Quelle and Meyer, Heidelberg. viii + 580 pp, 1836 figs, 10 pls. (1974) **Brohmer, P.**

Svenska Insekter. Nordstedt and Söner, Stockholm. 812 pp, 642 figs, 9 col. pls. [Swedish]. (1920–22) **Tullgren, A. and Wahlgren, E.**
Keys to families, genera; notes on species. Out of print.

[Keys to the insects of the European USSR. 1. Apterygota, Palaeoptera, Hemimetabola.] *Opred. Faune SSSR* **84**. [Russian]. English translation (1967) by Israel program for scientific translations, Jerusalem. 1214 pp.
(1964) **Bei-Bienko, G. Y.**, ed.

Other parts cited below under different orders; it is likely that yet other parts will be translated.

Insects and other arthropods of medical importance. British Museum (Natural History), London. xiv + 561 pp, 217 figs, 12 pls. (1963) **Smith, K. G. V.**, ed.
Keys to families, larval Diptera and sometimes genera of medical importance, includes a larval key.

Insects and hygiene, 3rd edn. Chapman and Hall, London. viii + 568 pp, many figs (compound).   (1980) **Busvine, J. R.**
Useful background information on insects, mites, and ticks of public health importance in Britain, Europe, and N. America; includes some keys to species.

A manual of forensic entomology. British Museum (Natural History), London. 256 pp, 402 figs.   (1986) **Smith, K. G. V.**
Useful for the identification of insects found on carrion with keys to adult and larval Diptera. Includes figures of Coleoptera and Lepidoptera (adults and larvae), with some carrion-frequenting species of other orders.

Forstinsektenkunde, 4th edn. Parey, Berlin. xvi + 625 pp, 480 figs.
(1927) **Nüsslin, O. and Rhumbler, L.**

Sveriges Skogsinsekter, 2nd edn. Geber, Stockholm. 509 pp, 570 figs. [Swedish].
(1939) **Trägardh, I.**

How to know the immature insects. W. C. Brown, Dubuque, Iowa. 234 pp, 631 figs.
(1949) **Chu, H. F.**

A practical key to the orders and suborders of soil and litter inhabiting animals. In *Soil zoology* (ed. D. K. McE. Kevan), pp. 452–88. Butterworth, London.   (1955) **Kevan, D. K. McE.**

Biology of insect eggs. Pergamon Press, Oxford. 3 vols, 1125 pp (vol III bibliogr. + index only), 296 figs (compound), 155 pls (SEM photos of surface structure etc.).   (1982) **Hinton, H. E.**
Useful for approximate identification of eggs, includes keys to a few orders, especially Ephemeroptera, Orthoptera, Hemiptera, and Coleoptera.

[*Determination of the larvae of insects found in the soil.*] Akad. Nauk SSSR, Moscow. 919 pp, 567 figs. [Russian].   (1964) **Gilyarov, M. S.**, ed.

Handledning för Insektsamlare. Entomological Society, Stockholm. 67 pp, well-illustrated.   (1974) **Lindroth, C. H.**
Instructions (in Swedish) for collecting, rearing, and mounting with guide to literature.

[*The identification of freshwater invertebrates of European USSR i.e. USSR west of the Urals (plankton and benthos).*] USSR Academy of Sciences, Zoological Institute, Leningrad. 510 pp [Russian].
(1977) **Kutikova, L. A. and Starabogatov, I.**, eds.

British plant-galls: a classified textbook of cecidology (Methuen, London. xv + 287 pp, illustrated (partly in colour).   (1912) **Swanton, E. W.**

Bestimmungstabellen der Gallen an Pflanzen Mittel- und Nordeuropas. G. Fischer, Jena. 2 vols, xvi + 1572 pp, 25 pls.   (1964–65) **Buhr, H.**

Bestimmungstabellen der Blattminen von Europa. W. Junk, The Hague. Vols. 1–2,
1185 pp; Vol. 3, 221 pp.   (1957) **Hering, E. M.**

The pocket encyclopaedia of plant galls in colour. Blandford, London. 191 pp, 3 black
and white, 293 col. figs.   (1968) **Darlington, A.**
A useful pocket guide to the commoner British galls.

A guide to freshwater invertebrate animals. Longmans Green, London. x + 118 pp.
                                                               (1959) **Macan, T. T.**

Collecting, preserving and studying insects, 2nd edn. Hutchinson, London. 336 pp,
135 figs, 10 pls.   (1970) **Oldroyd, H.**

Insects—instructions for collectors, revised edn. British Museum (Natural
History), London. vi + 169 pp. (1973) **Cogan, B. H. and Smith, K. G. V.**

THYSANURA and DIPLURA

British species 22.

Thysanura and Diplura. *Handbk Ident. Br. Insects* **1**(2): 1–7, 15 figs.
                                                               (1954) **Delany, M. J.**
Thysanura. Family Nicolettidae. *Genera Insectorum* **216e**: 1–58, 36 figs.
                                                               (1963) **Palct, J.**
Thysanura. Families Lepidotrichidae, Maindroniidae, Lepismatidae. *Genera
Insectorum* **218e**: 1–86, 21 figs.   (1967) **Palct, J.**

PROTURA

British species about 12.

The Protura. A revision of the species of the world with keys for determination. Hermann,
Paris. 360 pp, 567 figs.   (1964) **Tuxen, S. L.**
*The European Protura, their taxonomy, ecology, and distribution with keys for determina-
tion.* Muséum d'Histoire naturelle, Geneva. 345 pp, 111 figs.
                                                               (1973) **Nosek, J.**

COLLEMBOLA

British species about 300.

Collembolenfauna Europas. Muséum d'Histoire naturelle, Geneva. 360 pp, 554
figs.   (1960) **Gisin, H.**

The Collembola of North America north of the Rio Grande. Grinnell College, Grinnell. 4 vols, 1322 pp, 1043 figs.   (1980) **Christiansen, K. and Bellinger, P.**
Valuable generic descriptions with many British and European species covered.

Identification keys to Norwegian Collembola. Norwegian Entomological Society, Oslo. 152 pp, 416 figs.   (1980) **Fjellberg, A.**
Contains keys covering many British species.

A key for the identification of the families of Collembola recorded from the British Isles. *Entomologist's mon. Mag.* **113**: 193–7, 18 figs.
                                                                      (1979) **Gough, H. J.**

EPHEMEROPTERA

British species 47.

A key to the adults of the British Ephemeroptera. *Scient. Publs Freshwat. biol. Ass.* **47**: 1–101, 44 figs.   (1983) **Elliott, J. M. and Humpesch, U. H.**

A key to the nymphs of British Ephemeroptera, 3rd edn. *Scient. Publs Freshwat. biol. Ass.* **20**: 1–80, 27 figs.   (1979) **Macan, T. T.**

6. Ordnung: Eintagsfliegen, Ephemeroptera (Agnatha). *Tierwelt Mitteleur.* **4** (1, 3): 1–43, 150 figs.   (1929) **Ulmer, G.**

Revision der europäischen Arten der Gattung *Baetis* Leach, 1815 (Insecta, Ephemeroptera). *Gewäss.* Abwäss. **48/49**: 1–214, 155 figs (some compound).   (1969) **Müller-Liebenau, I.**

Ephemeroptera. In *Limnofauna Europa*, 2nd edn (ed. J. Illies), pp. 256–63. G. Fischer, Stuttgart.   (1978) **Puthz, V.**
Distribution of all the European species, no keys or figures.

Die europäischen Arten der Gattung *Caenis* Stephens (Insecta: Ephemeroptera). *Stuttg. Beitr. Naturk.* (A) **373**: 1–48, 26 pls of grouped figures.
                                                                      (1984) **Malzacher, P.**

ODONATA

British species 44.

Les Odonates de l'Europe, du Nord de l'Afrique et des Îles Atlantiques. Masson et Cie, Paris. 258 pp, 116 figs, 4 pls.   (1968) **Aguesse, P.**
Out of print but contains useful keys to species.

Guide des Libellules d'Europe et d'Afrique du Nord. Delachaux et Niestlé, Neuchâtel. 341 pp, 77 figs, 6 pls line drawings, 40 col. pls, maps.
                                    (1985) **d'Aguilar, J. and Dommanget, J.-L.**
English translation published by Collins 1986.

The dragonflies of Great Britain and Ireland. Curwen Press, London. 115 pp, 23 figs (compound), 20 pls, 44 maps. (1977) **Hammond, C. O.**
Includes Gardner's key to the nymphs.

Sveriges Trollsländor (Odonata). Fältbiologerna, Sollentuna. 151 pp, 46 figs (compound) + maps. (1985) **Sahlén, G.**
Keys to nymphs and adults of the Scandinavian species.

PLECOPTERA

British species 34.

A key to the adults and nymphs of British stoneflies (Plecoptera), 3rd edn. *Scient. Publs Freshwat. biol. Ass.* **17**: 1–92, 47 figs. + maps.
(1977) **Hynes, H. B. N.**

7. Ordnung: Steinfliegen, Uferfliegen, Plecoptera. Neubearbeitung. *Tierwelt Mitteleur.* **4** (2, 5): 1–19, 15 pls. (1963) **Illies, J.**

Bäckslándor. Plecoptera. *Svensk Insektfauna* **15**: 1–126, 76 figs. [Swedish]..
(1952) **Brinck, P.**

Die europäischen Arten der Plecopterengattung *Isoperla* Banks (=*Chloroperla* Pictet). *Beitr. Ent.* **2**: 369–424, 19 figs (some compound). (1952) **Illies, J.**

Revision der Gattung *Chloroperla* Newman (Plecoptera). *Mitt. schweiz. ent. Ges.* **40**: 1–26, 20 figs. (1967) **Zwick, P.**

A revision of the stonefly family Nemouridae (Plecoptera): a study of the world fauna at the generic level. *Smithson. Contr. Zool.* **211**: 1–74, 186 figs.
(1975) **Baumann, R. W.**

ORTHOPTERA, PHASMIDA, DERMAPTERA, and DICTYOPTERA

British species 30, 3, 7, and 8, respectively.

Die Orthopteren Europas I. Junk, The Hague. 749 pp, 2360 figs. (1969) **Harz, K.**
Keys, in German and English, to adults of all European Ensifera (bush-crickets and crickets).

Die Orthopteren Europas II. Junk, The Hague. 939 pp, 3519 figs. (1975) **Harz, K.**
Keys, in German and English, to adults of all European Caelifera (grasshoppers and groundhoppers).

Die Orthopteren Europas III. Junk, The Hague. 434 pp, 1185 figs.
(1976) **Harz, K. and Kaltenbach, A.**
Keys, in German and English, to adults of all European Phasmida, Dermaptera, Mantodea, and Blattodea.

Dermaptera and Orthoptera. *Handbk Ident. Br. Insects* **1** (5, revised). 24 pp, 74 figs. (1956) **Hincks, W. D.**

Grasshoppers, crickets and cockroaches of the British Isles (with companion gramophone record of the songs). Warne, London. xii + 299 pp, 130 figs, 22 col. pls. (1956) **Ragge, D. R.**
Keys to adults. Life histories, habits, songs, distribution maps. Colour plates of every species. Includes the three species of stick insect established in small colonies in Britain.

The British Orthoptera: a supplement. *Entomologist's Gaz.* **24**: 227–45.
(1973) **Ragge, D. R.**
Supplement to Ragge (1956) above.

Crickets and grasshoppers of the British Isles. British Naturalists' Association, London. 15 pp, 3 pls. (1973) **Haes, E. C. M.**
Introductory account, well suited to beginners.

Grasshoppers. Cambridge University Press, Naturalist's Handbooks, Vol. 2, 65 pp. (1983) **Brown, V. K.**
Keys to families of British Orthoptera and to adult Tettigoniidae and Acrididae.

Studies on the biology and population dynamics of British grasshoppers. *Anti-locust Bull.* **17**, 182 pp, 67 figs. (1954) **Richards, O. W. and Waloff, N.**
Keys to nymphs of British Acrididae in every instar.

The British Orthoptera since 1800. In *The changing flora and fauna of Britain* (ed. D. L. Hawksworth), pp. 307–22. Academic Press, London.
(1974) **Marshall, J. A.**
An account of recent trends in the British Orthopteran fauna.

A comparative study of the stridulatory files of the British Gomphocerinae (Orthoptera: Acrididae). *J. nat. Hist.* **10**, 17–28, 27 figs. (1976) **Pitkin, L. M.**

Provisional atlas of the insects of the British Isles. Part 6. Orthoptera, 2nd edn. Biological Records Centre, Huntingdon. 40 pp. (1979) **Haes, E. C. M.**, ed.

A key to the nymphal instars of the British species of *Ectobius* Stephens (Dictyoptera: Blattidae). *Entomologist* **100**, 202–9, 4 figs.
(1973) **Brown, V. K.**

Notes and key to the oothecae of the British *Ectobius* (Dictyoptera: Blattidae). *Entomologist's mon. Mag.* **116** (1980): 151–4. (1981) **Brown, V. K.**

British earwigs (Dermaptera). *Entomologist's Gaz.* **28**: 29–37, 11 figs.
(1977) **Brindle, A.**
Key to adults.

Bestimmungstabellen für die Larven mittel-deutscher Orthopteren. *Dt. ent. Z.* (N.F.) **16**, 227–91, 2 figs. (1969) **Oschmann, M.**
Keys to nymphs of selected German Orthoptera and Dermaptera in every instar.

Heuschrecken: beobachten, bestimmen. Neumann-Neudamm, Melsungen. 210 pp, 80 col. pls. (1985) **Bellman, H.**
Keys in German to adult German Saltatoria (grasshoppers, bushcrickets, and crickets) and to their songs, with colour photographs and song diagrams.

Die Stimmen der heimischen Heuschrecken. JNN-Naturführer. Sound cassette— companion to Bellman 1985 (above). (1985) **Bellman, H.**

ISOPTERA

No British species. Termites are not indigenous to the fauna of the area. The timber-infesting species *Reticulitermes flavipes* (Kollar) was described from Vienna, from where it has been eradicated. It is a native of the eastern USA and N. Mexico and in 1937 successfully infested a centrally heated system of buildings in Hamburg. Certain dry-wood termites have been maintained in culture in the UK.

PSOCOPTERA

British species 90.

Psocoptera. *Handbk Ident. Br. Insects* **1**(7): 1–102, 350 figs. (1974) **New, T. R.**
An introduction to the natural history of the British Psocoptera. *Entomologist* **104**: 59–97. (1971) **New, T. R.**
The early stages and life histories of some British foliage-frequenting Psocoptera, with notes on the overwintering stages of British arboreal Psocoptera. *Trans. R. ent. Soc. Lond.* **121**: 59–77, 7 figs. (1969) **New, T. R.**
Includes keys to the immature stages of British foliage-frequenting species.
Some features of the nymphs of common species of British Trogiidae (Psocoptera). *Entomologist's mon. Mag.* **105**: 55–9, 2 figs. (1969) **Fahy, E. D.**
Staubläuse. Psocoptera. *Tierwelt Dtl.* **61**: 1–314, illustrated. (1974) **Gunther, K. K.**

PHTHIRAPTERA (TRUE LICE)

British species about 425. As these have a predominantly host rather than geographical distribution, published keys deal with genera on a world basis.

AMBLYCERA AND ISCHNOCERA (BITING AND CHEWING LICE)

British species about 400.

A key to the genera of the Menoponidae (Amblycera: Mallophaga: Insecta). *Bull. Br. Mus. nat. Hist.* (Ent.) **24**: 1–26, 29 figs, 7 pls.   (1969) **Clay, T.**

The Amblycera (Phthiraptera: Insecta). *Bull. Br. Mus. nat. Hist.* (Ent.) **25**: 75–98, 9 figs, 5 pls.   (1970) **Clay, T.**

Zeitgenössische Mallophagen-Literatur (I). *Angew. Parasit.* **10**: 53–60, 104–24.   (1969) **Eichler, W. and Zlotorycka, J.**

A check list of the genera and species of Mallophaga. British Museum (Natural History), London. 362 pp.   (1952) **Hopkins, G. H. E. and Clay, T.**
For additions and corrections see *Ann. Mag. nat. Hist.* (12) **6**: 434–48 and **8**: 177–90.

Bibliographie der Mallophagen. *Mitt. zool. Mus. Berl.* **36**: 146–403.
(1960) **Keler, S. von**

[16. Order Mallophaga. (Biting lice).] In [Keys to the insects of the European USSR (ed. G. Ya. Bei Bienko).] *Opred. Fauna SSSR* **84**: 309–23 [Russian]. English translation (1967) (keys to families and genera only) by Israel program for scientific translations, Jerusalem, pp. 385–403.
(1964) **Blagoveshchenskii, D. J.**

ANOPLURA (SUCKING LICE)

British species about 25.

The sucking lice. *Mem. Pacif. Cst ent. Soc.* **1**: ix + 320 pp, 124 pls.
(1961) **Ferris, G. F.**
Keys to Anoplura of the world.

The host-association of the lice of mammals. *Proc. zool. Soc. Lond.* **B119**: 387–604.   (1949) **Hopkins, G. H. E.**

A cladistic analysis and classification of trichodectid mammal lice (Phthiraptera: Ischnocera). *Bull. Br. Mus. nat. Hist.* (Ent.) **51**: 187–346.
(1985) **Lyal, C. H. C.**
Keys to subfamilies, genera, and subgenera on a world-wide basis.

HEMIPTERA

British species approximately 1709.

HOMOPTERA—AUCHENORHYNCHA

British species approximately 351.

Palaearctic Auchenorrhyncha (Homoptera). An annotated check list. Polish Academy of Sciences Institute of Zoology, Warsaw. 550 pp. (1972) **Nast, J.**

Palaearctic Auchenorrhyncha (Homoptera). Part 2, Bibliography, addenda and corrigenda. Part 3, New taxa and replacement names introduced till 1980. *Annls zool. Warsaw*: **34**: 481–99 and **36**: 290–362.

(1979 and 1982) **Nast, J.**

Hemiptera, Fulgoromorpha. *Handbk Ident. Br. Insects* **2**(3): 1–68, 382 figs.

(1960) **Le Quesne, W. J.**

Hemiptera, Cicadomorpha (excluding Deltocephalinae and Typhlocybinae). *Handbk Ident. Br. Insects* **2**(2a): 1–64, 323 figs. (1965) **Le Quesne, W. J.**

Hemiptera, Cicadomorpha, Deltocephalinae. *Handbk Ident. Br. Insects* **2**(2b): 65–148, figs 324–830. (1969) **Le Quesne, W. J.**

Cicadellidae (Typhlocybinae) with a check list of the British Auchenorhyncha (Hemiptera, Homoptera). *Handbk Ident. Br. Insects* **2**(2c): 95 pp, illustrated.

(1981) **Le Quesne, W. J. and Payne, K. R.**

Homoptères Auchénorhynques. I. Typhlocybidae. *Faune Fr.* **31**: 1–228, 629 figs. (1936) **Ribaut, H.**

Homoptères Auchénorhynques. II. Jassidae. *Faune Fr.* **57**: 1–474, 1212 figs.

(1952) **Ribaut, H.**

The Auchenorrhyncha (Homoptera) of Fennoscandia and Denmark. Part 1: Introduction, infraorder Fulgoromorpha. Part 2: The families Cicadidae, Cercopidae, Membracidae and Cicadellidae (excl. Deltocephalinae). *Fauna ent. Scand.* **7**(1 and 2): 1–593. (1981) **Ossiannilsson, F.**

The Auchenorrhyncha (Homoptera) of Fennoscandia and Denmark. Part 3: The family Cicadellidae: Deltocephalinae, catalogue, literature and index. *Fauna ent. Scand.* **7**(3): 594–979. (1983) **Ossiannilson, F.**

HOMOPTERA—STERNORHYNCHA

PSYLLOIDEA

British species 80.

Psylloidea. [Keys to the insects of the European USSR. 1. Apterygota, Palaeoptera, Hemimetabola (ed. G. Ya. Bei-Bienko).] *Opred. Faune SSSR* **84**: 551–608 [Russian] English translation (1967) by Israel program for scientific translations, Jerusalem. (1964) **Loginova, M. M.**

Beiträge zur Systematik und Faunistik der schweizerischen Psyllodea (Sternorrhyncha). *Ent. Basiliensia* **8**: 43–83. (1983) **Burckhardt, D.**
Many northern European species treated; gives most up-to-date synonymy and generic placements.

Homoptera Psylloidea. *Handbk Ident. Br. Insects* **2**(5a): iv + 1–98, 321 figs.

(1979) **Hodkinson, I. D. and White, I. M.**

Psylloidea (Nymphal stages). Hemiptera, Homoptera. *Handbk Ident. Br. Insects*
2(5b): 1–50, many figs.   (1982) **White, I. M. and Hodkinson, I. D.**

ALEYROIDEA

British species 17.

A revision of the British Aleyrodidae (Hemiptera Homoptera). *Bull. Br. Mus.
nat. Hist.* (Ent.) **17**(9): 397–428, 28 figs.   (1966) **Mound, L. A.**

Homoptera Aleyrodoidea, Subfamilia Aleyrodinae. *Fauna Repub. pop. rom.
Insecta* **8**(5): 1–152 [Romanian]. (1969) **Dobreanu, E. and Manolache, C.**
Most of the northern European fauna is covered, with excellent figures.

APHIDOIDEA

British species 552.

The plant lice or Aphididae of Great Britain. Headley, Ashford and London. 1147
pp in 3 vols, 591 figs.   (1926–29) **Theobald, F. V.**

Recent additions to the British aphid fauna [and similar titles]. *Trans. R. ent.
Soc. Lond.* **101**: 89–124; **106**: 283–340; **109**: 311–59; **116**: 29–72; **124**: 37–79;
also *Zool. J. Linn. Soc.* **65**: 1–54.   (1950–79) **Stroyan, H. L. G.**

The identification of aphids of economic importance. *Plant Path.* **1**: 9–15, 42–8,
92–9, 123–99, many figs.   (1952) **Stroyan, H. L. G.**

Homoptera Aphidoidea (Part). Chaitophoridae and Callaphididae. *Handbk
Ident. Br. Insects* **2**(4a): viii + 130, many figs.   (1977) **Stroyan, H. L. G.**

Aphids—Pterocommatinae and Aphidinae (Aphidini). *Handbk Ident. Br.
Insects* **2**(6), 232 pp, many figs.   (1984) **Stroyan, H. L. G.**

Conifer woolly aphids (Adelgidae) in Britain. *Bull. For. Comm.* **42**, 51 pp, 32
figs.   (1971) **Carter, C. I.**

Conifer Lachnids. *Bull. For. Comm.* **58**: iv + 75 pp.
          (1982) **Carter, C. I. and Maslen, N. R.**

Contributions to a monograph of the Aphididae of Europe. *Temminckia* **3**: 1–44
+ 4 pls; **4**: 1–234 + 6 pls; **7**: 179–320 + 7 pls; **8**: 182–323 + 6 pls; **9**: 1–176 +
6 pls.   (1938–53) **Hille Ris Lambers, D.**

The Aphidoidea (Hemiptera) of Fennoscandia and Denmark. I. General pt.
The families Mindaridae, Hormaphididae, Thelaxidae, Anoeciidae and
Pemphigidae. *Fauna ent. Scand.* **9**: 1–236.   (1980) **Heie, O. E.**

The Aphidoidea (Hemiptera) of Fennoscandia and Denmark. II. The family
Drepansiphidae. *Fauna Ent. Scand.* **11**: 1–176.   (1982) **Heie, O. E.**

Europae centralis Aphides. *Mitt. Thüring. bot. Ges. Beih.* **3**: 1–488.
          (1952) **Börner, C.**

Aphidoidea. *Handbk Pfl Krank*, 5th edn. P. Parey, Berlin. Vol. 5, Part 4, pp. 1–402, 194 figs. (1957) **Börner, C. and Heinze, K.**

Pflanzenschädliche Blttlausarten der Familien Lachnidae, Adelgidae und Phylloxeridae, eine systematisch-faunistische Studie. *Deuts. ent. Z.* (N.F.) **9**: 143–227, 41 figs. (1962) **Heinze, K.**

Aphids. Ginn, London. 175 pp, 37 figs, 8 pls. (1975) **Blackman, R.**

Aphidoidea. In [Keys to the insects of the European USSR. 1. Apterygota, Palaeoptera, Hemitabola (ed. G. Ya. Bei Bienko).] *Opred. Faune SSSR* **84**: 616–799, 273 figs. [Russian] English translation (1967) by Israel program for scientific translations, Jerusalem. (1964) **Shaposhnikov, R. K.**

HETEROPTERA

British species approximately 533.

Land and water bugs of the British Isles. Warne, London. 436 pp, 153 figs, 63 pls. (1959) **Southwood, T. R. E. and Leston, D.**

A list of the Irish Hemiptera (Heteroptera and Cicadina). *Proc. R. Ir. Acad.* **42B**(8): 211–307. (1935) **Halbert, J. N.**

Illustrierte Bestimmungstabellen der Wanzen. II. Europa (Hemiptera-Heteroptera Europae). Hersdorf, Berlin. Over 2000 pp, profusely illustrated, in 4 vols plus index. (1955–62) **Stichel, W.**

Hémiptères Tingidae euro-méditerranéens. *Faune Fr.* **69**: x + 620 pp, 250 figs. (1983) **Péricart, J.**

Wanzen oder Heteropteren. II. Cimicomorpha. *Tierwelt Dtl.* **55**: 1–179, 114 figs. (1967) **Wagner, E.**

Hémiptères Anthocoridae, Cimicidae et Microphysidae de l'ouest paléarctique. *Faune de l'Europe et du Bassin Méditerranéen* **7**: 1–402, 198 figs. (1972) **Péricart, J.**

[Insecta Rhynchota. Vol. XIII, pt 2. Hemiptera of the family Nabidae.] *Fauna SSSR* 326 pp, 513 figs. [Russian]. (1981) **Kerzhner, I. M.**

Blindwanzen oder Miriden. *Tierwelt Dtl.* **41**: 1–218, 125 figs. (1952) **Wagner, E.**

Die Miridae Hahn, 1831, des Mittelmeerraumes und der Makaronesischen Inseln (Hemiptera, Heteroptera). In three parts and supplement. *Ent. Abh. Mus. Tierk. Dresden* **37, 39, 40, 42** (suppls): about 1500 pp, many illustrations. (1971–78) **Wagner, E.**
Includes keys to all European species of most genera.

Blomstertaeger [Miridae]. *Danmarks Fauna* **81**: 279 pp, 127 figs. (1974) **Gaun, S.**

Wanzen oder Heteropteren. I. Pentatomorpha. *Tierwelt Dtl.* **54**: 1–235, 149 figs.   (1952) **Wagner, E.**

Hémiptères Berytidae euro-méditerranéens. *Faune Fr.* **70**: iv + 172 pp,
illustrated.   (1984) **Péricart, J.**

Revision of Palaearctic Piesmatidae (Heteroptera). *Mitt. Münch. ent. Ges.* **73**:
61–171, 45 figs.   (1983) **Heiss, E. and Péricart, J.**

A revised key to the British water bugs (Hemiptera-Heteroptera) with notes
on their ecology. *Scient. Publs Freshwat. biol. Ass.* **16**: 1–77, 48 fgs.
                                                     (1965) **Macan, T. T.**

THYSANOPTERA

British species 158.

Thysanoptera. *Handbk Ident. Br. Insects* **1**(11): 1–79, 327 figs.
                                                     (1976) **Mound, L. A.** *et al.*

The hosts and distribution of British thrips. *Ecol. Ent.* **1**: 41–7.
                                                     (1976) **Pitkin, B. R.**

Thysanoptera, Fransenflügler. *Tierwelt Dtl.* **66**: 477 pp.
                                          (1979) **Schliephake, G. and Klimt, K.**

Ordnung Thysanoptera. *Bodenfauna Europas* (Berlin) **2**: 242 pp, 212 figs, 7 pls.
                                                     (1964) **Priesner, H.**

NEUROPTERA, MEGALOPTERA, RAPHIDIOPTERA, and MECOPTERA

British species 57, 3, 4, 4 respectively.

Die Neuropteren Europas. Goecke and Evers, Krefeld. 2 vols, 495 and 355 pp, 913
figs, 26 col. pls, 221 maps, 259 wing photographs.
                              (1980) **Aspöck, H., Aspöck, U., and Hölzel, H.**
Monograph of the European Neuroptera, Megaloptera, and Raphidioptera.

Mecoptera, Megaloptera, Neuroptera. *Handbk. Ident. Br. Insects* **1**(12–13): 1–
40, 15 figs (compound).   (1959) **Fraser, F. C.**

A key to British freshwater Megaloptera and Neuroptera. *Scient. Publs Freshwat. biol. Ass.* **35**: 1–52, 18 figs, 1 pl.   (1977) **Elliott, J. M.**

The Neuroptera and Mecoptera of eastern Fennoscandia. *Fauna fenn.* **13**: 1–96,
127 figs, 39 maps.   (1962) **Meinander, M.**

COLEOPTERA

British species about 3850.

General

The Coleoptera of the British Islands. Reeve, London. 5 vols, 120 pls (in large paper
  edn).   (1886–91) **Fowler, W. W.**

The Coleoptera of the British Islands. Reeve, London. **6** (Suppl), 20 pls.
                               (1913) **Fowler, W. W. and Donisthorpe, H. St J. K.**
  Keys, descriptions of adults; frequency and distribution; some bionomic data.

A practical handbook of British beetles. Witherby, London. 2 vols, xxvii + 622 pp,
  169 pls, 1 map.   (1932) **Joy, N. H.**
  Reprinted (1976) by Classey, Faringdon. Keys to species; the only comprehensive
  work since Fowler.

Atlas des Coléoptères de France, Belgique, Suisse. Nouvel atlas d'Entomologie, 2nd edn.
  Boubée, Paris. 2 vols, 522 pp, 112 figs, 60 pls (36 col.).   (1960) **Auber, L.**
  Keys to major groups; brief descriptions of each species; excellent illustrations.

Field book of beetles. Brown, London and Hull. xxv + 197 p, 14 pls.
                                                          (1948) **Dibb, J. R.**
  Primary division by bionomic characters.

Coleoptera: introduction and key to families. *Handbk Ident. Br. Insects* **4**(1): 1–
  59, 118 figs.   (1956) **Crowson, R. A.**

Einführung in die Käferkunde. *Die Käfer Mitteleuropas.* Goecke and Evers,
  Krefeld. Vol. 1, 214 pp, 204 figs, 7 col. pls.
                               (1965) **Freude, H., Harde, K. W., and Lohse, G. A.**
  General introduction and key to families.

Bibliographie der Bestimmungstabellen europäischer Insekten (1880–1963).
  Teil III: Coleoptera und Strepsiptera. *Dt. ent. Zt.* **17**: 33–118, 433–76; **18**: 1–
  84, 287–360.   (1970–71) **Göllner-Scheiding, U.**
  With index to taxa giving cross-references to authors.

A check list of British insects. 2nd edn Coleoptera. Revised by R. D. Pope.
  Handbk Ident. Br. Insects **11**(3): xiv + 105 pp.
                                      (1977) **Kloet, G. S. and Hincks, W. D.**

Catalogus coleopterorum Fennoscandiae et Daniae. Entomological Society, Lund.
  478 pp, 1 map. Additions and corrections (1970). Strand, Norway. *Norsk ent.
  Tidsskr.* **17**: 125–45.   (1970) **Lindroth, C. H.**, ed.
  Distribution by provinces in tabular form.

A list of the beetles of Ireland. *Proc. R. Ir. Acad.* (3)**6**: 535–827.
                                  (1902) **Johnson, W. F. and Halbert, J. N.**

The biology of the Coleoptera. Academic Press, London. xi + 802 pp.
                                                          (1981) **Crowson, R. A.**

The most useful general work concerning beetle biology, including an extensive bibliography.

A key to the families of British beetles. *FSC Publication S15*. (Reprinted from *Field Studies* **6**(4).) (1984) **Unwin, D. M.**
An AIDGAP guide designed for the non-specialist. Simple and practical illustrated keys but with some inaccuracies.

A coleopterist's handbook, 2nd edn. Amateur Entomologist's Society, London. 142 pp, 62 figs, 18 pls. (1975) **Anon.**
A useful compendium of collecting methods, field work techniques, etc.

Changes in the British coleopterous fauna. In *The changing flora and fauna of Britain* (ed. D. L. Hawksworth), pp. 323–69. Academic Press, London.
(1974) **Hammond, P. M.**
Summarizes changes in British beetle fauna over the last 10 000 years.

Coleoptera I. Synopsis. *Zoology Iceland* **3**(46a): 1–218.
(1959) **Larsson, S. G. and Gigja, A.**
Full annotated list of Icelandic species; no keys.

Faunistik der Mitteleuropäischen Käfer. Several volumes; various publishers.
(1941 on) **Horion, A.**
The only source of data concerning distribution, habitats, and general biology of European Coleoptera, which covers a wide range of families in depth.

Balfour-Browne Club Newsletter. Founded in 1976 to inform those interested in aquatic British beetles—new literature, noteworthy finds, etc. Available from Dr G. N. Foster, 20 Angus Avenue, Prestwick, Ayrshire KA9 2HZ.

The Coleopterist's Newsletter. A semiannual newsletter aimed at informing British coleopterists of new literature, recent noteworthy discoveries, etc. Available from J. Cooter, 222 Whittern Way, Hereford, HR1 1QP.

A field guide in colour to beetles. Octopus Books, London. 334 pp, numerous figs and col. pls. English edn edited and with additional introductory material by P. M. Hammond. Original German title (1981): *Die Kosmos-Käferführer*. Kosmos-Verlag, Stuttgart. (1984) **Harde, K. W.**
1080 species illustrated in colour, including most of the larger species found in N.W. Europe.

Larves et nymphes des Coléoptères aquatiques du globe. Paris. 804 pp, 561 figs.
(1972) **Bertrand, H. P. I.**
Keys to genera.

Larvae of British beetles. III. Keys to families. *Entomologist's mon. Mag.* **78**: 206–26, 253–72, 54 figs. (1942) **Emden, F. I. van**

[Morphology of larvae of beetles predaceous on and associated with bark beetles in the north-west Caucasus.] In [*Evolutionary morphology of the larvae of insects* (ed. B. M. Mamayev), pp. 175–201, 22 figs (compound).] Science Publishers, Moscow. [Russian]. (1976) **Nikitskiy, N. B.**

Larvas de coleopteros. *Boln Serv. Plagas for.* **5**: 87–91; **6**: 103–21, 40 figs.
(Reprinted 1964, *Graellsia* **20**: 245–75.) [Spanish].
(1962–63) Viedma, M. G. de
Keys to 31 families affecting forestry.

Ordnung Coleoptera (Larven). Bestimmungs bücher zur Bodenfauna Europas. Junk,
The Hague. 378 pp, 1000 figs.   (1978) **Klausnitzer, B.**
Keys to families and genera of soil and surface-substrate groups.

The woodworm problem. Hutchinson, London. 123 pp, 80 figs, 1 col. pl.
(1963) Hickin, N. E.
Covers all species affecting converted timber.

The insect factor in wood decay. Hutchinson, London. 336 pp, 263 figs, 1 col. pl.
(1963) Hickin, N. E.

Coléoptères. *Entomologie appliqués à l'agriculture.* Masson et Cie, Paris. (in 2 pts)
1391 pp.   (1962–63) **Balachowsky, A. S.**

Die Bestimmung von Insektenschäden imm Walde. P. Parey, Hamburg and Berlin.
196 pp.   (1955) **Schimitschek, E.**
Includes keys based on damage caused.

A monograph of the beetles associated with stored products. British Museum (Natural
History), London. Vol. 1, 443 pp, 505 figs, 55 tables (reprinted 1963,
Johnson Reprint).   (1945) **Hinton, H. E.**
Keys to families, species (adults + larvae). Covers: Carabidae, Staphylinidae,
Nitidulidae, Lathridiidae, Mycetophagidae, Colydiidae, Murmidiidae
(Cerylonidae), Endomychidae, Erotylidae, Anthicidae, Crytophagidae, and
Dermestidae.

Systematic

CARABOIDEA

Coleoptera, Carabidae. *Handbk Ident. Br. Insects* **4**(2): 1–148, 97 figs.
(1974) Lindroth, C. H.
See corrections to key in *Antenna* **1**: 25.

Coléoptères carabiques. *Faune Fr.* **39**: 1–571, 213 figs; **40**: 1–601, 368 figs; **51**:
1–51, 12 figs, 20 pls.   (1941, 42, and 49) **Jeannel, R.**

Catalogue des Coléoptères carabiques de France. *Nouv. Revue Ent.* (suppl) **1**: 1–
177.   (1971) **Bonadona, P.**

Carabidae. *Die Käfer Mitteleuropas.* Goecke and Evers, Krefeld. Vol. 2, 302 pp,
many figs.   (1976) **Freude, H., Harde, K. W., and Lohse, G. A.**

9. Skalbaggar, Coleoptera. Sandjagare och jordlopare. Fam. Carabidae, 2nd
edn. *Svensk Insektfauna* **35**: 1–209, 135 figs. [Swedish].
(1961) Lindroth, C. H.

Die *Asaphidion*-Arten aus der Verwandtschaft des *A. flavipes* L. *Ent. Bl. Biol.*

Syst. Käfer **79**: 33–6.   (1983) **Lohse, G. A.**
Key to the three British species of this group.

British water beetles. Ray Society, London. Vol. 1 (1940 Haliplidae, Hygrobiidae and Dytiscidae-part) xxi + 375 pp, 89 figs, 72 maps, 5 pls; Vol. 2 (1950 Dytiscidae-part and Gyrinidae) 394 pp, 90 figs, 56 maps. (Facsimile reprints 1964.)   (1940, 1950) **Balfour-Browne, F.**

Hygrobiidae, Haliplidae, Dytiscidae, Gyrinidae, Rhysodidae. *Die Käfer Mitteleuropas*. Goecke and Evers, Krefeld. Vol. 3, 7–95, many figs.
                    (1971) **Freude, H., Harde, K. W., and Lohse, G. A.**

Zur Kenntnis der europäischen *Gyrinus*-Arten. *Ent. Bl. Biol. Syst. Käfer* **63**: 174–86; (Corrections) **64**: 64.   (1967) **Ochs, G.**

A key to the genera of larval Carabidae. *Trans. R. ent. Soc. Lond.* **92**: 1–99, 100 figs.   (1942) **Emden, F. I. van**

Larvae of British beetles. IV. Various small families. Cicindelidae, Hygrobiidae. *Entomologist's mon. Mag.* **79**: 209–13, 6 figs.
                                            (1943) **Emden, F. I. van**

Løbebillernes larver. *Danm. Fauna* **76**: 282–433, 76 figs. [Danish].
                                            (1968) **Larsson, S. G.**

The larvae of British Carabidae (Coleoptera). I. Carabini and Cychrini. *Entomologist* **102**: 245–63, 75 figs.   (1969) **Luff, M. L.**

The larvae of the British Carabidae (Coleoptera). II. Nebriini. *Entomologist* **105**: 161–79, 61 figs.   (1972) **Luff, M. L.**

The larvae of the British Carabidae (Coleoptera). III. Patrobini. *Entomologist's Gaz.* **26**: 59–64, 15 figs.   (1975) **Houston, W. W. K. and Luff, M. L.**

The larvae of the British Carabidae (Coleoptera). IV. Notiophilini and Elaphrini. *Entomologist's Gaz.* **27**: 51–67, 47 figs.   (1976) **Luff, M. L.**

The larvae of the British Carabidae (Coleoptera). V. Omophronini, Loricerini, Scaritini and Broscini. *Entomologist's Gaz.* **29**: 265–87.
                                            (1978) **Luff, M. L.**
Illustrated key to all genera and 13 of the British species.

Coleoptera. 3b. Carabidae: Bembidiinae, Trechinae. *Fauna Pol.* **19**: 1–94.
                                            (1974) **Pawlowski, J.**

Some notes on the generic characters of the larvae of the subfamily Colymbetinae (Dytiscidae, Coleoptera) with a key for the identification of the European genera. *Polskie Pismo ent.* **43**: 215–24, 48 figs. (1973) **Galewski, K.**

Diagnostic characters of larvae of European species of *Graphoderus* Dejean (Coleoptera, Dytiscidae) with an identification key and some notes on their biology. *Bull. Acad. pol. Sci. Sér. Biol.* **22**: 485–94, 29 figs.
                                            (1974) **Galewski, K.**

The Fennoscandian species of the genus *Hydaticus* Leach (Coleoptera: Dytiscidae). *Entomologica scand.* **12**: 103–8.   (1981) **Nilsson, A. N.**

Key to the British species of *Hydroporus*. *Balfour-Browne Club Newsletter* **33**: 1–19.
(1985) **Foster, G. N. and Angus, R. B.**
Excellent illustrated key to all British species with notes on other species from NW Europe.

The British species of *Gyrinus*. *Balfour-Browne Club Newsletter* **20**: 3–7.
(1981) **Foster, G. N.**
Well illustrated.

Colymbetes schildknechti, a new water beetle from Sardinia with a key to European species of *Colymbetes* (Coleoptera, Dytiscidae). *Aquatic Insects* **5**: 39–44.
(1983) **Dettner, K.**

Generic characters of the larvae of the subfamily Dytiscinae (Dytiscidae) with a key to the central European genera. *Polskie Pismo ent.* **43**: 491–8, 24 figs.
(1973) **Galewski, K.**

HYDROPHILOIDEA (including HYDRAENIDAE)

British water beetles. Ray Society, London. Vol. 3, liii + 210 pp, 87 figs, 67 maps.
(1958) **Balfour-Browne, F.**

Hydraenidae, Spercheidae, Hydrophilidae. *Die Käfer Mitteleuropas*. Goecke and Evers, Krefeld. Vol. 3, pp. 95–156, many figs.
(1971) **Lohse, G. A. and Vogt, H.**

Hydrophilidae Europae (Coleoptera, Palpicornia). Arnaldo Forni, Bologna, 199 pp.
(1959) **Chiese, A.**
Keys to genera and species.

The British species of *Helophorus*. *Balfour-Browne Club Newsletter* **11**: 2–15.
(1978) **Angus, R. B.**
Well illustrated. Essential for identification of British members of this genus.

The habitats, life-histories and immature stages of *Helophorus* F. (Coleoptera: Hydrophilidae). *Trans. R. ent. Soc. Lond.* **125**: 1–26, 7 figs, 4 pls.
(1973) **Angus, R. B.**
Key to larvae.

A re-evaluation of the taxonomy and distribution of some European species of *Hydrochus* Leach (Coleoptera: Hydrophilidae). *Entomologist's mon. Mag.* **112**: 177–201.
(1977) **Angus, R. B.**
Illustrated. Key includes the six British recorded species.

Revisional notes on some European *Helochares* Muls. (Coleoptera, Hydrophilidae). *Entom. Scand.* **13**: 201–11.
(1982) **Hansen, M.**
English summary with key to the three British species in *Balfour-Browne Club Newsletter* **25**: 1–3 (1982).

Limnebius crinifer Rey new to Britain, with a revised key to the British *Limnebius* species (Coleoptera: Hydraenidae). *Entomologist's Gaz.* **35**: 99–102, illustrated.
(1984) **Carr, R.**
Good key to British species.

Cercyon-Studien. *Ent. Bl. Biol. Syst. Käfer* **64**: 172–91.
(1968) **Vogt, H.**

The key to the central European species of *Cercyon* is translated into English in *Balfour-Browne Club Newsletter* **7**: 4–12 (1978).

Revisione dei *Laccobius* paleartici (Coleoptera, Hydrophilidae). *Mem. Soc. ent. ital.* **54**: 5–187.   (1975) **Gentili, E. and Chiesa, A.**
Figures of male genitalia and distribution maps for most species; English key pp. 179–84. Key reproduced in *Balfour-Browne Club Newsletter* **4**: 8–10 (1977). Addenda by Gentili in *Boll. Soc. ent. ital.* **111**: 43–50 (1979).

Enochrus (Methydrus) isotae sp.n.—eine neue Hydrophiliden-Art aus Jugoslawien. *Ent. Bl. Biol. Syst. Käfer* **77**: 137–9.   (1981) **Hebauer, F.**
Summary in English with figures for separation of the three British species of the *E. coarctatus* group in *Balfour-Browne Club Newsletter* **24**: 1–2 (1982).

HISTEROIDEA

Coleoptera Histeroidea. Sphaeritidae and Histeridae. *Handbk Ident. Br. Insects* **4**(10): 1–16, 38 figs.   (1963) **Halstead, D. G. H.**

The Histeridae associated with stored products. *Bull. ent. Res.* **35**: 309–40, 56 figs.   (1945) **Hinton, H. E.**
Keys to adults and larvae.

Histeridae, Sphaeritidae. *Die Käfer Mitteleuropas.* Goecke and Evers, Krefeld. Vol. 3, pp. 156–89, many figs.   (1971) **Witzgall, K.**

STAPHYLINOIDEA (including SPHAERIIDAE)

Staphylinidae

Staphylinidae (Piestinae to Euaesthetinae). *Handbk Ident. Br. Insects* **4**(8a): 1–79, 196 figs.   (1954) **Tottenham, C. E.**

Coléoptères Staphylinidae de la région paléarctique occidentale. 1. Généralités, sous-familles Xantholininae et Leptotyphlinae. *Nouv. Revue Ent.* (Suppl.) **2**: 1–651, 219 figs.   (1972) **Coiffait, H.**

Staphylinidae I (Micropeplinae bis Tachyporinae); II (Hypocyphtinae und Aleocharinae). *Die Käfer Mitteleuropas.* Goecke and Evers, Krefeld. Vol. 4, pp. 1–264; Vol. 5, pp. 1–304, many figs. (1964 and 1974) **Lohse, G. A.** *et al.*

Kortvingar, fam. Staphylinidae. *Svensk Insektfauna* **38**, **48–53**: 985 pp, many figs. [Swedish].   (1948–72) **Palm, T.**

Notes on British Staphylinidae 2. On the British species of *Platystethus* Mannerheim, with one species new to Britain. *Entomologist's mon. Mag.* **107**: 93–111, 14 figs, 7 maps.   (1971) **Hammond, P. M.**

Notes on British Staphylinidae 3. The British species of *Sepedophilus* Gistel (*Conosomus* auct.). *Entomologist's mon. Mag.* **108**: 130–65, 32 figs, 8 maps.
                                                    (1973) **Hammond, P. M.**

Philonthus mannerheimi Fauvel (Col., Staphylinidae) and related species. *Entomologist's mon. Mag.* **109**: 85–8, 12 figs.   (1974) **Last, H.**

The British species of *Staphylinus* subgenus *Ocypus* Steph (Col., Staphylinidae).
Entomologist's mon. Mag. **84**: 271–5, 16 figs.   (1948) **Steel, W. O.**

Notes on the Omaliinae (Col., Staphylinidae). *Entomologist's mon. Mag.* **98**:
157–64, 30 figs.   (1957) **Steel, W. O.**
Key to *Acrolocha*.

A revision of the British species of *Arphirus* Tottenham (subgenus of *Quedius*
Stephens) (Col., Staphilinidae). *Entomologist's mon. Mag.* **84**: 241–58, 27 figs.
                    (1948) **Tottenham, C. E.**

Identification of the *Aleochara diversa* (J. Sahlberg) group (Col., Staphylinidae),
including a species new to Britain. *Entomologist* **102**: 231–4, 12 figs.
                    (1969) **Welch, R. C.**

The British species of the genus *Amischa* Thomson (Col., Staphylinidae),
including *A. soror* Krastz, an addition to the list. *Entomologist's mon. Mag.*
105: 38–42, 6 figs.   (1969) **Williams, S. A.**

Notes on the genus *Oligota* Mannerheim (Col., Staphylinidae) and key the
British species. *Entomologist's mon. Mag.* **106**: 54–62, 15 figs.
                    (1970) **Williams, S. A.**

Kritische Faunistik der bisher aus Mitteleuropa bekannten *Stenus*-Arten nebst
systematischen Bemerkungen und Neubeschreibungen (Coleoptera,
Staphylinidae). *Ent. Bl. Biol. Syst. Käfer* **67**: 74–121, 37 figs, 3 maps.
                    (1971) **Puthz, V.**

Notes on some British Staphylinidae (Col.). 1. The genus *Scopaeus* Er., with the
addition of *S. laevigatus* Gyll. to our list. *Entomologist's mon. Mag.* **104**: 198–
207, 22 figs.   (1969) **Allen, A. A.**

Notes on some British Staphylinidae (Col.). 2. Three additions to our species
of *Philonthus* Curt. *Entomologist's mon. Mag.* **106**: 157–61, 14 figs.
                    (1971) **Allen, A. A.**

Studien an den paläarktischen Arten der Gattung *Myrmecopora* Saulcy (Col.,
Staphylinidae). *Koleopt. Rdsch.* **50**: 93–109.   (1972) **Scheerpeltz, O.**

Studi sui Paederinae—III. I *Medon* Steph. paleartici con descrizione di nuove
specie mediterranee (Col. Staphylinidae). *Boll. Lab. Ent. agr. Filippo Silvestri*
37: 73–125.   (1980) **Bordoni, A.**

Coléoptères Staphylinides de la Region Paléarctique occidentale. III. Sous
famille Staphylininae, Tribu Quediini; sous famille Paederinae, Tribu
Pinophilini. *Nouv. Revue Ent. (Suppl.)* **8**(4): 1–364.   (1978) **Coiffait, H.**

Coléoptères Staphylinidae de la region paléarctique occidentale. 2. Sous
famille Staphylininae. Tribus Philonthini et Staphylinini. *Nouv. Revue Ent.*
(Suppl.) **4**(4): 1–593.   (1974) **Coiffait, H.**

Monographie der Gattung Tachinus *Gravenhorst (Coleoptera: Staphylinidae), mit
Bermerkungen zur Phylogenie und Verbreitung der Arten* Kiel (privately
published). 365 pp.   (1975) **Ullrich, W. G.**

Gnypeta ripicola (Kiesenw.) (Col., Staphylinidae) new to Britain. *Entomologist's*

mon. Mag. **116**: 37–9.   (1980) **Williams, S. A.**
Key to the five British species.

The subfamilies of the larvae of Staphylinidae (Coleoptera) with keys to the larvae of British genera of Steninae and Proteininae. *Trans. R. ent. Soc. Lond.* **118**: 261–83, 92 figs.   (1966) **Kasule, F. K.**

The larval characters of some subfamilies of British Staphylinidae with keys to the known genera. *Trans. R. ent. Soc. Lond.* **120**: 115–38, 116 figs.
                                                            (1968) **Kasule, F. K.**

The larvae of Paederinae and Staphylininae (Coleoptera: Staphylinidae) with keys to the known British genera. *Trans. R. ent. Soc. Lond.* **122**: 49–80, 144 figs.   (1970) **Kasule, F. K.**

A revision of the Staphylinid subfamily Proteininae (Coleoptera) I. *Trans. R. ent. Soc. Lond.* **118**: 285–311, 120 figs.   (1966) **Steel, W. O.**
Keys to genera of larvae.

The larvae of the genera of the Omaliinae (Coleoptera: Staphylinidae) with particular reference to the British fauna. *Trans. R. ent. Soc. Lond.* **122**: 1–47, 191 figs.   (1970) **Steel, W. O.**

A provisional key to the British species of *Tachyporus* (Col. Staph.) based on elytral chaetotaxy. *Circaea* **2**: 15–19.   (1984) **Booth, R. G.**

The larvae of some British species of *Gyrophaena* Mannerheim (Coleoptera: Staphylinidae) with notes on the taxonomy and biology of the genus. *Zool. J. Linn. Soc.* **60**: 297–317.   (1977) **White, I. M.**
Key to adults of 19 British species as appendix.

Other families

Silphidae, Leptinidae, Catopidae, Colonidae, Liodidae, Clambidae, Scydmaenidae, Orthoperidae (=Corylophidae), Sphaeriidae, Ptiliidae, Scaphidiidae. *Die Käfer Mitteleuropas.* Goecke and Evers, Krefeld. Vol. 3, pp. 190–347, many figs. (1971) **Freude, H., Karde, K. W., and Lohse, G. A.**

Five species of Ptiliidae (C.) new to Britain, and corrections to the British list of the family. *Entomologist's Gaz.* **26**: 211–23, 13 figs.   (1975) **Johnson, C.**

Revision of the Fenno-Scandian species of the genus *Acrotrichis* Motsch., 1848. *Norsk ent. Tidsskr.* **10**: 241–77, 6 figs (see also pp. 173–80). (1958) **Sundt, E.**

The aedeagi of the British species of the genus *Catops* Pk. (Col., Cholevidae). *Entomologist's mon. Mag.* **81**: 69–72, 13 figs (see also pp. 121–5).
                                                            (1945) **Kevan, D. K.**

The sexual characters of the British species of the genus *Choleva*, including *C. cisteloides*, new to the British list (Col., Cholevidae). *Entomologist's mon. Mag.* **82**: 122–30, 24 figs.   (1946) **Kevan, D. K.**

A revision of the British species of the genus *Colon* (Col., Cholevidae). *Entomologist's mon. Mag.* **83**: 249–67, 50 figs.   (1947) **Kevan, D. K.**

Zwei neue Catopiden-Gattungen aus Europa (Auflösung der *nigrita*-Gruppe in der Gattung *Catops*). *Ent. Bl. Biol. Syst. Käfer* **64**: 1–16, 17 figs.
(1968) **Zwick, P.**

Pselaphidae. *Handbk Ident. Br. Insects* **4**(9): 1–32, 41 figs. (1957) **Pearce, E. J.**

A revised annotated list of the British Pselaphidae (Coleoptera). *Entomologist's mon. Mag.* **110**: 13–26.
(1974) **Pearce, E. J.**

Coléoptères Pselaphides. *Faune Fr.* **53**: 1–421, 169 figs.   (1950) **Jeannel, R.**

Pselaphidae. In *Die Käfer Mitteleuropas* (ed. H. Freude, K. W. Harde, and G. A. Lohse). Goecke and Evers, Krefeld. Vol. 5, pp. 305–62, many figs.
(1974) **Besuchet, C.**

Revision der paläarktischen Arten der Gattungen *Scaphisoma* Leach und *Caryoscapha* Ganglbauer der Tribus Scaphisomini (Col. Scaphidiidae). *Rev. suisse Zool.* **77**: 727–99.
(1970) **Löbl, I.**

Observations on scydmaenid (Col.) larvae with a tentative key to the main British genera. *Entomologist's mon. Mag.* **115**: 49–59.
(1980) **Brown, C. and Crowson, R. A.**
Good illustrations; key to seven of eight British genera.

Revision der paläarktischen Arten der Tribus Leiodini Leach (Col., Leiodidae). *Folia ent. hung.* **44**: 9–163.   (1983) **Daffner, H.**
Keys to genera and species; good figures of male genitalia, etc.

The British species of *Agathidium* Panzer (Col., Leiodidae). *Entomologist's mon. Mag.* **113**: 125–35.   (1978) **Cooter, J.**
Well illustrated; good practical key.

SCARABAEOIDEA

Lamellicornia. In *Die Käfer Mitteleuropas* (ed. H. Freude, K. W. Harde, and G. A. Lohse). Goecke and Evers, Krefeld. Vol. 8, pp. 265–371.
(1969) **Machatschke, J. W.**

Scarabaeoidea (Lucanidae, Trogidae, Geotrupidae, Scarabaeidae). *Handbk Ident. Br. Insects* **5**(11): 1–29, 68 figs.   (1956) **Britton, E. B.**

Coléoptères Scarabaeoidea. Faune de l'Europe occidentale, Belgique. France. Grande-Bretagne. Italie. Peninsule Iberique. *Nouv. Revue Ent.* **7**(3) (Suppl): 1–352.   (1977) **Baraud, J.**

Encyclopédie Entomologique. Faune des Coleopteres de France II. Lucanidae et Scarabaeoidea. Lechevalier, Paris.   (1982) **Paulian, R. and Baraud, J.**

Larvae of British beetles. II. A key to the British Lamellicornia larvae. *Entomologist's mon. Mag.* **77**: 117–27; 181–92, 19 figs.
(1941) **Emden, F. I. van**

EUCINETOIDEA (CLAMBIDAE and HELODIDAE)

Clambidae. *Handbk Ident. Br. Insects* **4**(6a): 1–13, 22 figs. (1966) **Johnson, C.**

Clambidae. *Die Käfer Mitteleuropas*. Goecke and Evers, Krefeld. Vol. 3, pp. 266–70, 27 figs.   (1971) **Endrödi-Younga, S.**

The British species of the genus *Cyphon* Paykull (Col., Helodidae), including three new to the British list. *Entomologist's mon. Mag.* **98**: 114–21, 11 figs.
(1962) **Kevan, D. K.**

Die nordeuropäischen Arten der Gattung *Cyphon* Paykull (Coleoptera). Taxonomie, Biologie, Ökologie und Verbreitung. *Entomologica scand.* (Suppl.) **3**: 1–100, 25 figs, 8 maps.   (1972) **Nyholm, T.**

Zur Kenntnis der Larven der mitteleuropäischen Helodidae. *Dt ent. Z.* **22**: 61–5, 14 figs.   (1975) **Klausnitzer, B.**

Fossipedes [families Helodidae and Eucinetidae]. In *Die Käfer Mitteleuropas* (ed. H. Freude, K. W. Harde, and G. A. Lohse), Vol. 6, pp. 25–264. Goecke and Evers, Krefeld.   (1979) **Lohse, G. A.**

BYRRHOIDEA

Byrrhidae. In *Die Käfer Mitteleuropas* (ed. H. Freude, K. W. Harde, and G. A. Lohse), Vol. 6, pp. 328–50. Goecke and Evers, Krefeld.
(1979) **Paulus, H. F.**

The British species of the genus *Byrrhus* L., including *B. arietinus* Steffahny (Col., Byrrhidae) new to the British list. *Entomologist's mon. Mag.* **101**: 111–15, 10 figs.   (1966) **Johnson, C.**

Taxonomic notes on British Coleoptera. No. 4 *Simplocaria maculosa* Erichson (Byrrhidae). *Entomologist* **99**: 155–6, 3 figs.   (1966) **Johnson, C.**

Les *Byrrhus (sensu lato)* de France (Col., Byrrhidae). *Entomologiste* **31**: 193–209, 20 figs.   (1975) **Bonadona, P.**

Revision der Familie Byrrhidae I. Zur Systematik und Faunistik der westpaläarktischen Vertreter der Gattung *Curimopsis* Ganglebauer 1902 (Col., Byrrhidae, Sincalyptinae). *Senckenbergiana biol.* **54**: 353–67, 23 figs.
(1973) **Paulus, H. F.**

Notes on Byrrhidae (Col.); with special reference to a species new to the British fauna. *Entomologist's Rec. J. Var.* **90**: 141–7.   (1978) **Johnson, C.**

Über die Larvenmerkmale einige deutscher Byrrhidengattungen. *Mitt. dt. ent. Ges.* **17**: 39–40.   (1958) **Emden, F. I. van**

DRYOPOIDEA

Monographie der palaearktischen Dryopidae, mit Berücksichtigung der eventuell transgredierenden Arten (Col.). *Mitt münch. ent. Ges.* **28**: 147–87, 319–71; **29**: 109–45; **30**: 24–71; **31**: 1–88, 391 figs, 1 pl. (1938–41) **Bollow, H.** Includes Psephenidae and Elminthidae (=Elmidae).

Dryopidae. In *Die Käfer Mitteleuropas* (ed. H. Freude, K. W. Harde, and G. A. Lohse), Vol. 8, pp. 265–94. Goecke and Evers, Krefeld.

(1979) **Steffan, A. W.**

Includes the family Elmidae.

Heteroceridae. *Handbk Ident. Br. Insects* **5**(2c): 1–15, 14 figs.

(1973) **Clarke, R. O. S.**

Heteroceridae. In *Die Käfer Mitteleuropas* (ed. H. Freude, K. W. Harde, and G. A. Lohse), Vol. 8, pp. 296–304. Goecke and Evers, Krefeld.

(1979) **Drechsel, U.**

La larve d'*Heterocerus aragonicus* Kiesw. et son milieu biologique (Col., Heteroceridae). *Revue fr. Ent.* **12**: 166–74, 31 figs.   (1945) **Pierre, F.**

Key to larvae of *Heterocerus*.

A key to the larvae, pupae and adults of the British species of Elminthidae. *Scient. Publs Freshwat. biol. Ass.* **26**: 1–58, 29 figs (some compound), 11 maps.

(1972) **Holland, D. G.**

Elmidae de la region paléarctique occidental: systématique et repartition (Coleoptera Dryopidae). *Annls Limnologie* **15**: 1–102.

(1979) **Berthélemy, C.**

BUPRESTOIDEA

Les buprestides de France. Tableaux analytiques des Coléoptères de la faune franco-rhénane (France, Rhénanie, Belgique, Holland, Valais, Corse). Le Moult, Paris. 511 pp, 25 pls.   (1949) **Schaefer, L.**

Buprestidae. *Handbk Ident. Br. Insects* **5**(5a): 1–8, 15 figs.   (1977) **Levey, B.**

Zur Kenntnis der früheren Entwicklungsstadien schwedischer Käfer. 2. Buprestiden-Larven, die in Bäumen leben. *Opusc. ent.* **27**: 65–78, 4 figs (compound).   (1962) **Palm, T.**

The Buprestidae (Coleoptera) of Fennoscandia and Denmark. *Fauna ent. scand.* **10**.   (1982) **Bily, S.**

Buprestidae (Prachtkäfer). In *Die Käfer Mitteleuropas* (ed. H. Freude, K. W. Harde, and G. A. Lohse), Vol. 8, pp. 204–48. Goecke and Evers, Krefeld.

(1979) **Harde, K. W.** *et al.*

ELATEROIDEA

Coléoptères Elateridae de la faune de France continentale et de Corse. *Bull. mens. Soc. linn. Lyon* **41** (Suppl.): 1–379, 384 figs.   (1972) **Leseigneur, L.**

Sternoxia [families Elateridae, Cerophytidae, Eucnemidae, and Throscidae]. In *Die Käfer Mitteleuropas* (ed. H. Freude, K. W. Harde, and G. A. Lohse), Vol. 6, pp. 101–203 Goecke and Evers, Krefeld.   (1979) **Lohse, G. A.**

Beiträge zur Systematik der europäischen und nordwestafrikanischen *Agriotes* s.str. (Elateridae). *Ent. Bl. Biol. Syst. Käfer* **63**: 65–86, 22 figs.

(1967) **Franz, H.**

Zur Kenntnis der mitteleuropäischen *Hypnoidus*-Arten aus dem Subgenus *Zorochrus* Thoms. *Ent. Bl. Biol. Syst. Käfer* **63**: 32–7, 12 figs. (1967) **Franz, H.**

Analytische Tabelle zur Bestimmung der europäischen *Throscus*-Arten. *Wien. ent. Ztg* **8**: 35–7 (see also pp. 37–9).   (1889) **Reitter, E.**

Notes on some British serricorn Coleoptera, with adjustments to the list. 1. Sternoxia. *Entomologist's mon. Mag.* **104**: 208–16, 8 figs. (1969) **Allen, A. A.** Keys to British Eucnemidae.

Essai d'un genera [sic!] des Eucnemididae paléarctiques. *Revue fr. Ent.* **2**: 1–18.
                                                  (1935) **Fleutiaux, E.**

Les *Hypocoelus* (Eucnemidae) de la faune de France. Systématique et distribution. *Entomologiste* **34**: 105–23.   (1978) **Leseigneur, L.**

Bestimmungstabelle der bekanntesten deutschen Elateridenlarven (Coleoptera, Elateridae). *Arb. morph. taxon. Ent. Berl.* **8**: 217–30, 3 pls.
                                                  (1941) **Korschefsky, R.**

Zur Kenntnis der Entwicklungsstadien und der Lebensweise nord und mitteleuropäischer Eucnemiden (Coleoptera). *Ent. Bl. Biol. Syst. Käfer* **72**: 10–50.   (1976) **Leiler, T. E. von**

Die skandinavischen Elateriden-Larven (Coleoptera). *Entomologica scand.* (Suppl.) **2**: 1–63, 31 figs (compound).   (1972) **Palm, T.**

Larvae of British beetles. V. Elateridae. *Entomologist's mon. Mag.* **81**: 13–38, 54 figs.   (1945) **Emden, F. I. van**

Morphology and identification of the British larvae of the genus *Elater* (Col., Elateridae). *Entomologist's mon. Mag.* **92**: 167–88, 39 figs.
                                                  (1956) **Emden, F. I. van**

Larvae of the British beetles. IV. Various small families. Eucnemidae. *Entomologist's mon. Mag.* **79**: 218–19, 1 fig.   (1943) **Emden, F. I. van**

Zur Kenntnis der früheren Entwicklungsstadien schwedischer Käfer. I. Bisher bekannte Eucnemiden-Larven. *Opusc. ent.* **25**: 157–69, 11 figs (compound).   (1960) **Palm, T.**

CANTHAROIDEA

Malacodermata (families Lycidae, Cantharidae, Lampyridae, and Drilidae). In *Die Käfer Mitteleuropas* (ed. H. Freude, K. W. Harde, and G. A. Lohse), Vol. 8, pp. 9–53. Goecke and Evers, Krefeld. (1979) **Geisthardt, M.** *et al.*

Beiträge zur Kenntnis der Gattung *Rhagonycha* (Col., Cantharidae). *Ent. Bl. Biol. Syst. Käfer* **64**: 93–124, 16 figs (compound).   (1968) **Dahlgren, G.**

The larvae of the British genera of Cantharidae (Coleoptera). *J. Ent.* (B) **44**: 243–54, 34 figs.   (1975) **Fitton, M. G.** Illustrated key to six genera.

Zur Kenntnis der Canthariden-Larven. 2. Beitrag. *Arch. Naturgesch.* **89**(1): 110–37.   (1923) **Verhoeff, K. W.**

Bestimmungstabelle der bekanntesten deutschen Lyciden-, Lampyriden- und Driliden larven (Coleoptera). *Beitr. Ent.* **1**: 60–4, 3 figs, 1 pl.
(1951) **Korschefsky, R.**

DERMESTOIDEA

Dermestidae of Poland. *Fauna Pol.* **4**: 1–163.   (1975) **Mroczkowski, M.**
The most comprehensive recent treatment of European Dermestidae.

Révision des Dermestes (Col., Dermestidae). *Annls Soc. ent. Fr.* **115**: 37–68, 42 figs.
(1949) **Lepesme, P.**

Dermestidae. In *Die Käfer Mitteleuropas* (ed. H. Freude, K. W. Harde, and G. A. Lohse), Vol. 8, pp. 304–27. Goecke and Evers, Krefeld.
(1979) **Lohse, G. A.**

Dermestes peruvianus Cast., *D. haemorrhoidalis* Küst. and other *Dermestes* spp. (Col., Dermestidae). *Entomologist's mon. Mag.* **111**: 1–14, 33 figs, 1 pl.
(1976) **Peacock, E. R.**

The species of *Anthrenus* that have been found in Britain, with a description of a recently introduced species. (Coleoptera, Dermestidae). *Entomologist* **78**: 6–9, 6 figs.
(1945) **Hinton, H. E.**

Attagenus smirnovi Zhantiev (Coleoptera: Dermestidae) a species new to Britain, with keys to the adults and larvae of British *Attagenus*. *Entomologist's Gaz.* **30**: 131–6.
(1979) **Peacock, E. R.**

The 'gin traps' of some beetle pupae; a protective device which appears to be unknown. *Trans. R. ent. Soc. Lond.* **97**: 473–96, 27 figs. (1946) **Hinton, H. E.**

BOSTRYCHOIDEA (TEREDILIA)

Lyctidae, Bostrychidae, Anobiidae, Ptinidae. In *Die Käfer Mitteleuropas*, Vol. 8, pp. 7–74. Goecke and Evers, Krefeld.
(1969) **Freude, H., Harde, K. W., and Lohse, G. A.**, eds

The Fennoscandian, Danish and British species of the genus *Ernobius* Thomson (Col., Anobiidae). *Opusc. ent. Lund* **31**: 81–92, 30 figs.
(1966) **Johnson, C.**

The Ptinidae of economic importance. *Bull. ent. Res.* **31**: 331–81, 59 figs.
(1941) **Hinton, H. E.**

Mature larvae of the beetle-family Anobiidae. *Biol. Meddr* **22**(2): 1–298, 50 pls.
(1954) **Böving, A. F.**

The larvae of some wood-boring Anobiidae (Coleoptera). *Bull. ent. Res.* **24**: 33–68, 15 figs (compound).
(1933) **Parkin, E. A.**

A revised key to the larvae of the Ptinidae associated with stored products. *Bull. ent. Res.* **44**: 85–96, 92 figs.   (1953) **Hall, D. W. and Howe, R. W.**

Larvae of British beetles. IV. Various small families. Lyctidae, Bostrychidae. *Entomologist's mon. Mag.* **79**: 265–9, 6 figs.   (1943) **Emden, F. I. van**

CLEROIDEA

Malacodermata [families Malachiidae, Melyridae and Cleridae]. In *Die Käfer Mitteleuropas* (ed. H. Freude, K. W. Harde, and G. A. Lohse), Vol. 8, pp. 53–98. Goecke and Evers, Krefeld. (1979) **Evers, A. M. J.** *et al.*
Includes the family Phloiophilidae.

Ostomidae [=Trogossitidae]. In *Die Käfer Mitteleuropas* (ed. H. Freude, K. W. Harde, and G. A. Lohse), Vol. 7, pp. 14–18, 8 figs. Goecke and Evers, Krefeld. (1979) **Vogt, H.**

Die Buntkäfer (Cleridae). *Neue Brehm Büch.* **281**: 1–108, 82 figs, 2 pls.
(1961) **Winkler, J. R.**

Étude des *Divales* et *Dasytes* de France et de Corse (Col., Dasytidae). *Revue fr. Ent.* **13**: 19–27. (1946) **Fagniez, C.**

Über die paläarktischen Arten der Gattungen *Sphinginus* Rey und *Fortunatius* nov. gen. [Col., Melyridae]. *Ent. Bl. Biol. Syst. Käfer* **67**: 21–41, 4 figs.
(1971) **Evers, A. M. J.**

Aufteilung der paläarktischen Arten des Gattungskomplexes *Malachius* F. *Ent. Bl. Biol. Syst. Käfer* **81**: 1–40. (1985) **Evers, A. M. J.**
Key to males of all Palaearctic species of *Malachius* and closely related genera. No figures.

Species of the genus *Aplocnemus* of Central Europe. *Dt. ent. Z.* **29**: 421–45.
(1982) **Majer, K.**
Key includes all British species.

Larvae of British beetles. IV. Various small families. Trogositidae, Cleridae. *Entomologist's mon. Mag.* **79**: 214–18, 12 figs. (1943) **Emden, F. I. van**

LYMEXYLOIDEA

Lymexylonidae. In *Die Käfer Mitteleuropas* (ed. H. Freude, K. W. Harde, and G. A. Lohse), Vol. 8, pp. 100–1. Goecke and Evers, Krefeld.
(1979) **Lohse, G. A.**

Larvae of British beetles. IV. Various small families. Lymexylidae. *Entomologist's mon. Mag.* **79**: 259, 261, 2 figs. (1943) **Emden, F. I. van**

CUCUJOIDEA [CLAVICORNIA (*PARS* + CORYLOPHIDAE)]

Clavicornia. In *Die Käfer Mitteleuropas* (ed. H. Freude, K. W. Harde, and G. A. Lohse), Vol. 7, pp. 19–309, many figs. Goecke and Evers, Krefeld.
(1979) **Vogt, H.** *et al.*

Coccinellidae and Sphindidae. *Handbk Ident. Br. Insects* **5**(7): 1–12, 23 figs.
(1953) **Pope, R. D.**

Systématique de la tribu des Scymnini (Coccinellidae) [de France]. *Annls Zool. Ecol. anim.* (hors Séries): 1–221, 43 pls.   (1974) **Gourreau, J. M.**

Revision einiger europäischer *Scymnus (s. str.)* Arten (Col., Coccinellidae). *Mitt. Abt. Zool. Bot. Landesmus. Joanneum* **28**: 207–59, 125 figs.
          (1967) **Fürsch, H. and Kreissl, E.**

The species of *Scymnus (s. str.)*, *Scymnus (Pullus)* and *Nephus* (Col., Coccinellidae) occurring in the British Isles. *Entomologist's mon. Mag.* **109**: 3–39, 30 figs, 9 maps.   (1973) **Pope, R. D.**

Larvae of British beetles. VII. Coccinellidae. *Entomologist's mon. Mag.* **85**: 265–83, 61 figs.   (1949) **Emden, F. I. van**

Zur Larvalsystematik der mitteleuropäischen Coccinellidae (Coleoptera). *Ent. Abh. Mus. Tierk. Dresden* **38**: 55–10, 14 figs (compound).
          (1970) **Klausnitzer, B.**
Keys to and descriptions of 58 species.

Bestimmungstabelle für mitteleuropäische Coccinelliden-Larven nach leicht sichtbaren Merkmalen. *Beitr. Ent.* **23**: 93–8, 26 figs. (1973) **Klausnitzer, B.**
Keys to 41 species by 'hand lens' characters.

Morphology and taxonomy of the larvae with keys for their identification. In *Biology of Coccinellidae.* J. Hodek, Prague, pp. 36–55, 160 figs.
          (1973) **Savoïskaya, G. I. and Klausnitzer, B.**

Phalacridae. *Handbk Ident. Br. Insects* **5**(5b): 1–17, 47 figs.
          (1958) **Thompson, R. T.**

A revision of the British species of the genus *Cryptophagus* (Herbst) (Coleoptera, Cryptophagidae). *Trans. R. ent. Soc. Lond.* **106**: 237–82, 117 figs.
          (1955) **Coombs, C. W. and Woodroffe, G. E.**

Additions and corrections to the British list of *Atomaria s. str.* (Col., Cryptophagidae), including a species new to science. *Entomologist* **100**: 39–47, 6 figs.
          (1967) **Johnson, C.**

The *Atomaria gibbula* group of species. (Coleoptera, Cryptophagidae). *Reichenbachia* **14**: 125–41, 9 figs.   (1973) **Johnson, C.**

The Lathridiidae of economic importance. *Bull. ent. Res.* **32**: 191–247, 67 figs.
          (1941) **Hinton, H. E.**

On the British species of *Lathridius* Herbst (Col., Lathridiidae). *Entomologist's mon. Mag.* **108**: 193–9, 18 figs, 1 pl.   (1973) **Tozer, E. R.**

A revision of the genus *Silvanus* Latreille (Coleoptera, Silvanidae). *Bull. Brit. Mus. nat. Hist.* (Ent.) **29**: 39–112, 179 figs.   (1973) **Halstead, D. G. H.**

Cisiden Studien IV. *Rhopalodontus perforatus* und seine Verwandten. *Ent. Bl. Biol. Syst. Käfer* **65**: 48–52, 2 figs (compound).   (1969) **Lohse, G. A.**

Rhizophagidae. *Handbk Ident. Br. Insects* **5**(5a): 1–19, 54 figs.
          (1977) **Peacock, E. R.**

Coléoptères: Colydiidae et Anommatidae paléarctiques. *Faune de l'Europe et du Bassin Méditerranéen.* Masson, Paris, Vol. 8, 280 pp.   (1977) **Dajoz, R.**

Epuraea adumbrata Mannerheim and *E. biguttata* (Thunberg) (Col.,
Nitidulidae) new to Britain. *Entomologist* **105**: 126–9, 10 figs.
(1972) **Johnson, C.**

Die westpaläarktischen Arten der *umbrosus*-Gruppe der Gattung *Meligethes*
Steph. (Coleoptera, Nitidulidae). *Mitt. münch. ent. Ges.* **68**: 1–11.
(1979) **Jelinek, J. and Spronraft, K.**

On the apparent conspecificity of *Cis pygmaeus* (Marsh) and *C. rhododactylis*
(Marsh) and on other closely allied species (Col., Ciidae). *Entomologist's mon.
Mag.* **102**: 138–44, 5 figs. (1967) **Kevan, D. K.**

Revisional notes on the British species of *Orthoperus* Steph. (Col., Coryl-
ophidae). *Entomologist's Rec. J. Var.* **82**: 112–20. (1970) **Allen, A. A.**

Bestimmungstabelle der Merophysiidae des Mittelmeeräumes und der
angrenzenden Gebeite (Coleoptera: Merophysiidae). *Ent. Bl. Biol. Syst.
Käfer* **75**: 141–54. (1980) **Rücker, H. W.**

HETEROMERA

Heteromera. In *Die Käfer Mitteleuropas* (ed. H. Freude, K. W. Harde, and G. A.
Lohse), Vol. 8, pp. 75–138, 160–264. Goecke and Evers, Krefeld.
(1969) **Kaszab, Z.**

Lagriidae, Alleculidae, Tetratomidae, Melandryidae, Salpingidae, Pythidae,
Mycteridae, Oedemeridae, Mordellidae, Scraptiidae, Pyrochroidae,
Rhipiphoridae, Anthicidae, Aderidae and Meloidae. *Handbk Ident. Br. Insects*
5(9): 1–30, 63 figs. (1954) **Buck, F. D.**

Tenebrionidae. *Handbk Ident. Br. Insects* **5**(10): 1–22, 36 figs.
(1975) **Brendell, M. J. D.**

A revision of the genus *Palorus (sens. lat.)* (Coleoptera, Tenebrionidae). *Bull.
Br. Mus. nat. Hist.* (Ent.) **19**: 61–148, 56 figs (compound), 2 maps.
(1967) **Halstead, D. G. H.**

Les Notoxinae de France (Col., Anthicidae). *Entomologiste* **27**: 132–48, 15 figs.
(1971) **Bonadona, P.**

La classification de Anthicidae de la faune de France (Coleoptera). *Entomologiste* **30**: 101–11, 23 figs. (1974) **Bonadona, P.**

Coleoptera—Anthicidae. *Fauna Ital.* **16**: 1–240. (1980) **Bucciarelli, I.**

Two species of *Anaspis* (Col., Mordellidae) new to Britain, with a consideration of the status of *A. hudsoni* Donis., etc. *Entomologist's Rec. J. Var.* **87**: 269–
74. (1975) **Allen, A. A.**

Beiträge zur Insektenfauna der DDR: Coleoptera—Oedemeridae. *Beitr. Ent.*
29: 249–66. (1979) **Liebenow, K.**

Larvae of British beetles. IV. Various small families. Meloidae,
Rhipiphoridae, Lagriidae, Synchroidae, Pyrochroidae. *Entomologist's mon.
Mag.* **79**: 219–23; 259–65,7 figs. (1943) **Emden, F. I. van**

Larvae of British beetles. VI. Tenebrionidae. *Entomologist's mon. Mag.* **83**: 154–71, 46 figs (see also **84**: 10).   (1947) **Emden, F. I. van**

Notes on the British species of *Pyrochroa* (Col., Pyrochroidae) with a key to their first-stage larvae. *Entomologist's mon. Mag.* **82**: 92–3, 1 fig.
(1946) **Duffy, E. A. J.**

Contribución al conocimiento de las larvas de Melandryidae de Europa (Coleoptera). *Eos Madr.* **41**: 483–506, 52 figs (see also pp. 507–13). [Spanish].   (1966) **Viedma, M. G. de**

CHRYSOMELOIDEA
Cerambycidae

Cerambycidae. In *Die Käfer Mitteleuropas* (ed. H. Freude, K. W. Harde, and G. A. Lohse), Vol. 9, pp. 7–94, many figs. Goecke and Evers, Krefeld.
(1966) **Harde, K. W.**

Cerambycidae. *Handbk Ident. Br. Insects* **5**(12): 1–18, 32 figs.
(1952) **Duffy, E. A. J.**

Coléoptères Cerambycidae. *Faune Fr.* **20**: vii + 166, 71 figs. (1929) **Picard, F.**

Une nouvelle nomenclature des Lepturines de France (Col., Cerambycidae). *Entomologiste* **30**: 207–17, 34 figs.   (1974) **Villiers, A.**
Keys to genera.

A monograph of the immature stages of British and imported timber beetles (Cerambycidae). British Museum (Natural History), London. viii + 350 pp, 292 figs, 8 pls.   (1953) **Duffy, E. A. J.**

Bockkäfer oder Cerambycidae. 1. Biologie mitteleuropäischer Bockkäfer (Col., Cerambycidae) unter besonderer Berücksichtigung der Larven. *Tierwelt Dtl.* **52**(2): vii + 115, 97 figs, 9 pls.   (1966) **Demelt, C. von**

Bruchidae

Bruchidae (Samenkäfer). In *Die Käfer Mitteleuropas* (ed. H. Freude, K. W. Harde, and G. A. Lohse), Vol. 10, pp. 7–21. Goecke and Evers, Krefeld.
(1981) **Brandl, P.**

[Zhukizernovski] (Bruchidae). *Fauna SSSR* **24**(1): 1–209, 209 figs. [Russian]. English translation (1971): Graham, P. A. J., National Lending Library for Science and Technology, Boston Spa, UK.
(1957) **Lukyanovich, F. K. and Ter-Minasyan, M. E.**

Coléoptères Bruchides et Anthribides. *Faunes Fr.* **44**: 1–184, 382 figs. + illustrations of adults.   (1945) **Hoffmann, A.**

Chrysomelidae

Chrysomelidae. In *Die Käfer Mitteleuropas* (ed. H. Freude, K. W. Harde, and G. A. Lohse), Vol. 9, pp. 95–280, many figs. Goecke and Evers, Krefeld.
(1966) **Mohr, K. H.**

Galerucinae de la faune française. *Annls Soc. ent. Fr.* **103**: 1–108, 54 figs.
(1934) **Laboissière, V.**

The British species of the genera *Pyrrhalta* Joannis and *Galerucella* Crotch (Col., Chrysomelidae). *J. Soc. Br. Ent.* **3**: 150–6, 9 figs. (1950) **Hincks, W. D.**

The British species of the genus *Haltica* Geoffroy (Col., Chrysomelidae). *Entomologist's mon. Mag.* **98**: 189–96, 43 figs. (1963) **Kevan, D. K.**

Hannens genitalorgan hos de nordiske *Longitarsus*-arter (Col., Chrysomelidae). *Norsk ent. Tiddsskr.* **12**: 25–6, 3 pls. [Norwegian].
(1962) **Strand, A.**
Photographs of aedeagi of most British species.

The British species of the genus *Longitarsus* Latreille (Col., Chrysomelidae). *Entomologist's mon. Mag.* **103**: 83–110, 43 figs. (1967) **Kevan, D. K.**

Longitarsus jacobaeae Waterhouse: identity and distribution (Col., Chrysomelidae). *Entomologist's mon. Mag.* **111**: 33–9, 16 figs.
(1976) **Shute, S. L.**

A note on the specific status of *Psylliodes luridipennis* Kutschera (Col., Chrysomelidae). *Entomologist's mon. Mag.* **111**: 123–7, 24 figs.
(1976) **Shute, S. L.**

Les *Phyllodecta* de la faune française (Col., Chrysomelidae). *Entomologiste* **31**: 154–8, 4 figs. (1975) **Lambelet, J.**

Larver. Chrysomelidae. *Danm. Fauna* **31**: 290–376, 21 figs. [Danish].
(1927) **Henriksen, K. L.**

Larvae of the British genera of chrysomeline beetles (Coleoptera, Chrysomelidae). *Syst. Ent.* **7**: 297–310. (1982) **Cox, M. L.**
Key to 11 genera.

The bionomics and comparative morphology of the early stages of certain Chrysomelidae (Coleoptera, Phytophaga). *Proc. zool. Soc. Lond.* **1931**: 879–949, 30 figs (compound), 3 pls. (1931) **Paterson, N. F.**

The larvae of the British species of *Chrysolina* (Chrysomelidae). *Syst. Ent.* **4**: 409–17. (1979) **Marshall, J. E.**
Key to 15 species. Good illustrations.

Notes on the biology of *Orsodacne* Latreille with a subfamily key to the larvae of the British Chrysomelidae (Coleoptera). *Ent. Gaz.* **32**: 123–35.
(1981) **Cox, M. L.**

Key to species of British Cassidinae larvae (Col., Chrysomelidae). *Entomologist's mon. Mag.* **98**: 33–6, 11 figs. (1962) **Emden, F. I. van**

A key to some larvae of the British Galerucinae and Halticinae (Coleoptera: Chrysomelidae). *Entomologist's Gaz.* **31**: 275–83. (1980) **Marshall, J. E.**

CURCULIONOIDEA

Curculio: a twice yearly newsletter dealing with Curculionoidea on a world basis, providing information on current research, literature, etc. Available from Dept. of Entomology, Texas A & M University, College Station, Texas 77843, USA. **Burke, H. R. and Clark, W. E.**, eds.

Coléoptères Curculionides. *Faune Fr.* **52**: 1–486, 225 figs; **59**: 487–1208, 438 figs; **62**: 1209–839, 642 figs. (1950, 1954, 1958) **Hoffmann, A.**

Coléoptères Bruchides et Anthribides. *Faune Fr.* **44**: 1–184, 434 figs. (1945) **Hoffmann, A.**

Die Anthribiden der Westpaläarktis einschliesslich der Arten der UdSSR (Col., Anthribidae). *Mitt. Münch. Ent. Ges.* **71**: 33–107. (1981) **Frieser, R. von**

Anthribidae (Breitmaulrüssler). In *Die Käfer Mitteleuropas* (ed. H. Freude, K. W. Harde, and G. A. Lohse), Vol. 10, pp. 22–4. Goecke and Evers, Krefeld. (1981) **Frieser, R. von**

Curculionidae (Rüsselkafer). In *Die Käfer Mitteleuropas* (ed. H. Freude, K. W. Harde, and G. A. Lohse), Vol. 10, pp. 102–279. Goecke and Evers, Krefeld. (1981) **Lohse, G. A.** *et al.* Keys to subfamilies, genera, and species of 10 subfamilies (Rhinomacerinae to Leptopinae).

Familienreihe: Rhynchophora (Schluss). In *Die Käfer Mitteleuropas* (ed. H. Freude, K. W. Harde, and G. A. Lohse), Vol. 11, pp. 1–342. Goecke and Evers, Krefeld. (1983) **Lohse, G. A.** *et al.*

Beiträge zur Insektenfauna der DDR: Coleoptera–Curculionidae: Ceutorhynchinae. *Beitr. Ent.* **22**: 3–128, 141 figs. (1972) **Dieckmann, L.**

On the taxonomy of Rhynchophora larvae (Coleoptera). *Trans. R. ent. Soc. Lond.* **87**: 1–37, 108 figs. (1938) **Emden, F. I. van**

On the taxonomy of Rhynchophora larvae: Adelognatha and Alophinae (Insecta, Coleoptera). *Proc. zool. Soc. Lond.* **122**: 651–759, 153 figs. (1952) **Emden, F. I. van**

Die Entwicklungsstadien der mitteleuropäischen Curculioniden (Morphologie, Bionomie, Ökologie). *Abh. senckenb. naturforsch. Ges.* **506**: 1–335, 497 figs. (1964) **Scherf, H.** Key to species based on bionomic characters.

The identification of *Otiorrhynchus* larvae from blackcurrant roots with descriptions of the larvae of *O. clavipes* (Bonsd.) and *O. singularis* (L.) (Col., Curculionidae). *Entomologist's mon. Mag.* **99**: 210–12, 4 figs. (1964) **Fowler, V. W.**

Scolytidae and Platypodidae. *Handbk Ident. Br. Insects* **5**(15): 1–20, 40 figs. (1953) **Duffy, E. A. J.**

Scolytides. *Faune Fr.* **50**: 1–320, 300 figs. (1949) **Balachowsky, A.**

Scolytidae (Borken- und Ambrosiakäfer). Platypodidae (Kernkäfer). In *Die Käfer Mitteleuropas* (ed. H. Freude, K. W. Harde, and G. A. Lohse), Vol. 10, pp. 34–99, 100–1. Goecke and Evers, Krefeld. (1981) **Schedl, K.**

Revision of the palearctic species of the genus Scolytus Geoffroy (Coleoptera, Scolytidae).

Polish Academy of Sciences, Warsaw and Cracow. 214 pp, 49 pls.

(1973) **Michalski, J.**

The genus *Hylastes* Erichson (Col., Scolytidae). *The Coleopterist's Newsletter* **14**: 5–6.

(1983) **Nash, D. R.**

Practical key to British species of *Hylastes*.

Ernoporus caucasicus Lind. and *Leperesinus orni* Fuchs (Col., Scolytidae) in Britain. *Entomologist's mon. Mag.* **105**: 245–9.

(1970) **Allen, A. A.**

A contribution to the knowledge of the larvae of European bark beetles (Coleoptera, Scolytidae). *Acta ent. bohemoslovaka* **67**: 116–32.

(1970) **Kalina, V.**

Beiträge zur Insektenfauna der DDR: Coleoptera–Curculionidae (Rhinomacerinae, Rhynchitinae, Attelabinae, Apoderinae). *Beitr. Ent.* **24**: 5–54, 103 figs.

(1974) **Dieckmann, L.**

Beiträge zur Insektenfauna der DDR: Coleoptera–Curculionidae (Apionidae). *Beitr. Ent.* **27**: 7–143, figs 1–151.

(1977) **Dieckmann, L.**

Beiträge zur Insektenfauna der DDR: Coleoptera–Curculionidae (Brachycerinae, Otiorhynchinae, Brachyderinae). *Beitr. Ent.* **30**: 145–310.

(1980) **Dieckmann, L.**

Beiträge zur Insektenfauna der DDR: Coleoptera–Curculionidae (Erirhinae). *Beitr. Ent. Berlin* **36**: 119–81, 52 figs.

(1986) **Dieckmann, L.**

Keys to genera and species known from Germany.

On the synonymy of *Rhynchites sericeus* Hbst., *ophthalmicus* Steph. and *olivaceus* Gyll. (Col., Attelabidae). *Entomologist's mon. Mag.* **100**: 49–56.

(1964) **Allen, A. A.**

Taxonomic notes on British Coleoptera. No. 1. *Apion cerdo* Gerst. and its allies (Apionidae). *Entomologist* **98**: 80–2, 11 figs.

(1965) **Johnson, C.**

Revision der *Apion platalea*-Gruppe (Coleoptera, Curculionidae). *Ent. NachrBl., Wien* **20**: 117–28.

(1975) **Dieckmann, L.**

Le specie paleartiche occidentali della tribù Phyllobiini (Coleoptera Curculionidae). *Boll. Zool. agr. Bachic.*, Ser. II **15**: 49–230, illustrated. [Italian].

(1980) **Pesarini, C.**

Key to species of *Phyllobius*.

The British species of the genus *Sitona* Germar (Col., Curculionidae). *Entomologist's mon. Mag.* **95**: 251–61, figs (see also **100**: 91–3, 3 figs).

(1960) **Kevan, D. K.**

Les *Sitona* Germar 1817 du groupe de *Sitona humeralis* Stephens 1831. *Bull. Soc. ent. Fr.* **85**: 207–17, 2 figs.

(1980) **Roudier, A.**

The nordic species of the beetle genus *Bagous* (Coleoptera Curculionidae) with a key. *Ent. Tidskr.* **89**: 229–41, illustrated.

(1968) **Bruce, N.**

Die mitteleuropäischen Arten aus der Gattung *Bagous* Germ. *Ent. Bl. Biol. Syst. Käfer* **60**: 88–111, 51 figs.

(1964) **Dieckmann, L.**

The British species of *Anthonomus* Germar (Col., Curculionidae). *Entomologist's mon. Mag.* **112**: 19–40.   (1977) **Morris, M. G.**

Revision der westpaläarktischen Anthonomini (Coleoptera, Curculionidae). *Beitr. Ent.* **17**: 377–564, 114 figs, 4 maps (see also **19**: 679–82).
                                    (1968) **Dieckmann, L.**

Die palaearktischen Arten der Untergattung *Pseudorchestes* Bedel aus der Gattung *Rhynchaenus* Clairv. (Coleoptera, Curculionidae). *Ent. Abh. Mus. Tierk. Dresden* **29**: 275–327, 66 figs.   (1963) **Dieckmann, L.**

Die mitteleuropäischen Arten der Gattung *Neosirocalus* Ner. *et* Wagn. (mit Beschreibung von drei neuen Arten). *Ent. Bl. Biol. Syst. Käfer* **62**: 82–110, 28 figs.   (1966) **Dieckmann, L.**

Die westpaläarktischen *Thamiocolus*-Arten (Coleoptera, Curculionidae). *Beitr. Ent.* **23**: 245–73, 51 figs.   (1973) **Dieckmann, L.**

Ceuthorhynchus unguicularis C. G. Thomson (Col., Curculionidae) new to the British Isles, from the Suffolk Breckland and the Burren, Co. Clare. *Entomologist's mon. Mag.* **101**: 279–86, 6 figs.   (1966) **Morris, M. G.** Key to *Ceutorhynchus* s. str.

Bestimmungstabelle der paläarktischen Cossoninae (Coleoptera, Curculionidae). *Ent. Bl. Biol. Syst. Käfer* **69**: 65–180, 46 + 37 figs.
                                    (1973) **Folwaczny, B.**

Pentarthrum huttoni Woll. (Col., Curculionidae) and some imported Cossoninae. *Entomologist's mon. Mag.* **84**: 152–4, 5 figs.   (1948) **Buck, F. D.**

STYLOPOIDEA

Synopsis of British Strepsiptera of the genera *Stylops* and *Halictoxenus*. *Entomologist's mon. Mag.* **54**: 67–76 (see also pp. 107–8). (1918) **Perkins, R. C. L.**

Stylopidae, Fächerflügler (=Ornung: Strepsiptera). In *Die Käfer Mitteleuropas* (ed. H. Freude, K. W. Harde, and G. A. Lohse), Vol. 8, pp. 139–59, many figs. Goecke and Evers, Krefeld.   (1969) **Kinzelbach, R. K.**

Strepsiptera (Fächerflügler). *Handb. Zool.* **4**(2)2/24: 1–68, 57 figs.
                                    (1971) **Kinzelbach, R. K.**

Strepsiptera. *Tierwelt Dtl.* **65**: 116 pp.   (1978) **Kinzelbach, R. K.**

DIPTERA

British species about 6000.

General

Die Fliegen der Palaearktischen Region. E. Schweitzerbart'sche, Stuttgart.
                                    (1925 on) **Lindner, E.**, ed.

Volume I (by Lindner) contains a history of the study of Diptera, with sections on morphology, physiology, taxonomy, key to families, and short notes on each. The large number of subsequent parts cover nearly all families but parts are still unfinished. It should always be consulted when dealing with any particular family; it is only listed by name when it is a major reference for that family.

Faune de France. Lechevalier, Paris.
Many parts. Less comprehensive than Lindner, covers most British species. The most useful are part 35, Dolichopodidae by Parent and part 28, Muscidae, Acalyptratae and Scatophagidae by Séguy.

A check list of British insects, part 5: Diptera and Siphonaptera, 2nd edn (revised K. G. V. Smith *et al.*). *Handbk Ident. Br. Insects* **11**(5): 1–139.
(1975) **Kloet, G. S. and Hincks, W. D.**
Should be consulted for revised nomenclature for all families. It is kept updated in *Antenna*, the house journal of the Royal Entomological Society of London and in their Handbooks series.

Catalogue of Palaearctic Diptera. Elsevier, Amsterdam, Oxford, New York, Tokyo. (1984—still appearing) **Soos, A.**, ed.
Catalogues Palaearctic species with full synonymies and bibliography. To be completed in 14 vols of which Vol. 9 (Micropezidae—Agromyzidae, 460 pp) and Vol. 10 (Clusiidae—Chloropidae, 402 pp) have so far appeared.

Flies of the British Isles, 2nd edn. Warne, London. 384 pp, 52 pls.
(1968) **Colyer, C. N. and Hammond, C. O.**
Excellent general account with keys to families and unusually high standard of illustration.

Diptera Danica. Genera and species of flies hitherto found in Denmark. Gad, Wesley, Copenhagen and London. 7 vols. [in English]. (1907–12) **Lundbeck, W.**
Covers Brachycera, Aschiza, and Tachinidae. Keys to subfamilies, genera, and species; some biological data. Out of date but still useful for the full descriptions and comment.

Diptera I (exclusive of Ceratopogonidae and Chironomidae). *Zoology Iceland* **3**(48a): 1–189, 49 figs. (1954) **Nielsen, P., Ringdahl, O., and Tuxen, S. L.**
No keys but lists Icelandic fly fauna with ecological notes and distribution tables. Updated by Andersson, H. (1967) *Opusc. Ent.* **32**: 101–20.

Diptera (Zweiflügler). In *Handbuch der Zoologie* (ed. W. Kükenthal), Part 31, Vol. iv (2), pp. 1–337, 143 figs. W. de Gruyter, Berlin. (1973) **Hennig, W.**
Very good general account of Diptera with key to families on a world basis.

Diptera: Introduction and key to families, 2nd edn. *Handbk Ident. Br. Insects* **9**(1): 1–104, 150 figs. (1970) **Oldroyd, H.**

The natural history of flies. Weidenfeld and Nicolson, London. 324 pp, 40 figs, 32 pls. (1964) **Oldroyd, H.**
An informative and readable account.

The historical development of Diptera. University of Alberta Press. xv + 360 pp, 85 figs. (1974) **Rohdendorf, B.**

English translation of work originally published in Russian in 1964 by Nauka, Moscow. Includes account of and keys to fossil Diptera.

La biologie des Diptères. *Encycl. Ent.* Lechevalier, Paris. (A) Vol. 26, 609 pp, 225 figs.   (1950) **Séguy, E.**
Information elaborately classified under 998 headings with useful illustrations.

De Europese Diptera. Determineer-tabel, biologie en literatuuroverzicht van de families van de muggen en vleigen. *Wet. Meded. K. ned. natuurh. Veren.* **148**: 1–81. [Dutch].   (1981) **Oosterbroek, P.**

Manual of nearctic Diptera. Research Branch Agriculture Canada, Hull, Quebec. Vol. 1, 674 pp, illustrated.
                    (1981) **McAlpine, J. F., Peterson, B. V.** *et al.*, co-ordinators
An important work attempting to standardize terminology for morphology of adults and larvae, which provides an introduction to each family. Contains keys to adults and larvae of most families. Good illustrations. Main text of Vol. 1 covers Nematocerous and Brachycerous families with keys to genera. (Vol. 2 will cover family treatment for Cyclorrhapha.)

Flies of the nearctic region. E. Schweitzerbart'sche, Stuttgart.
                                         (1980 on) **Griffiths, G. C. D.**, ed.
This series plans to cover the nearctic fly fauna as Lindner's series does the palaearctic. Its coverage of species with a holarctic distribution is important for certain families, such as Anthomyiidae.

A key to families of British Diptera. AIDGAP (Aids to identification in difficult groups of animals and plants), 41 pp. (Originally published in 1981, *Field Studies* **5**: 513–33).   (1982) **Unwin, D. M.**
A simplified key tested by potential users before publication.

Les insectes nuisibles aux plantes cultivées. Busson, Paris. 2 vols, 1921 pp, 1349 figs.
                          (1935–6) **Balachowsky, A. S. and Mesnil, L.**
Very useful for adults and larvae of species of economic importance, as far as these were known in 1936.

Cyclorrhaphous flies breeding in seaweed. Series of papers in *Entomologist's mon. Mag.* **96** and *Entomologist* **93–94**, illustrated.
                                            (1960–61) **Egglishaw, H. J.**

A manual of forensic entomology. British Museum (Natural History), London. 256 pp, 402 figs.   (1986) **Smith, K. G. V.**
Keys to adults and larvae of carrion-frequenting Diptera occurring in Britain and Europe.

Myiasis in man and animals in the Old World. A textbook for physicians, veterinarians, and zoologists. Butterworths, London. 267 pp.   (1965) **Zumpt, F.**
Keys to genera and species, both adults and larvae, of Diptera causing myiasis.

British bloodsucking flies. British Museum (Natural History), London. 156 pp, 64 figs, 45 pls.   (1939) **Edwards, F. W., Oldroyd, H., and Smart, J.**
Gives keys and biological notes on the families that suck the blood of man and other animals in Britain (Culicidae, Ceratopogonidae, Simuliidae, Tabanidae, Muscidae

(Stomoxydinae), Hippoboscidae, Nycteribiidae). Out of print, good plates but nomenclature dated.

The immature stages of flies. In *A dipterist's handbook* (ed. A. E. Stubbs and P. J. Chandler). *Amat. Ent.* **15**: 38–64, 84 figs.

(1978) **Brindle, A. and Smith, K. G. V.**

Contains an illustrated key to final stage larvae.

An introduction to the immature stages of British flies (Diptera). *Handbk Ident. Br. Insects* **10** (in press).   (1988) **Smith, K. G. V.**

Final stage larvae keyed to family and to subfamilies and tribes for some families. Illustrations combined with biological data facilitate further identification to genus and sometimes species. Typical eggs and pupae/puparia illustrated where known. Full bibliography for further identification.

Die Larvenformen der Dipteren. Akademie-Verlag, Berlin. 3 vols, 1270 pp.

(1948–52) **Hennig, W.**

Covers world literature of immature stages up to about 1949. Some keys and figures facilitating identification to family level.

Terricole Dipterenlarven und Puppen terricoler Dipterenlarven. Musterschmidt, Göttingen. 2 vols, 355 pp, 171 figs.   (1954) **Brauns, A.**

Very useful guide to soil and leaf-litter larvae. Key to families.

Terrestrial Diptera larvae. *Entomologist's Rec. J. Var.* **75**: 47–62, 19 figs.

(1963) **Brindle, A.**

A summary with illustrated keys to families.

Taxonomic notes on the larvae of British Diptera. *Entomologist* **94–106.**

(1961 on) **Brindle, A.**

28 illustrated papers appeared in this series.

Myiasis in man and animals in the Old World. Butterworth, London. 267 pp, 346 figs.   (1964) **Zumpt, F.**

The flies that cause myiasis in man. *US Dept. Agric. Misc. Publ.* **631**: 175 pp, 96 figs.   (1947) **James, M. T.**

[*Classification key to Diptera larvae of arboricole insects.*] Izvestia Nauka, Moscow. 367 pp, 154 figs. [Russian].

(1967) **Krivosheina, N. P. and Mamaev, B. M.**

A preliminary classification of Diptera exclusive of Pupipara, based upon larval and pupal characters, with keys to imagines in certain families. Part I. *Bull. Illinois State Lab.* **12**: 161–407, 29 pls.   (1917) **Malloch, J. R.**

Eggs and larvae of flies. In *Insects and other arthropods of medical importance* (ed. K. G. V. Smith), pp. 289–323. British Museum (Natural History), London.

(1973) **Oldroyd, H. and Smith, K. G. V.**

Systematic

NEMATOCERA

British species over 200.

Diptera. 2. Nematocera: families Tipulidae to Chironomidae. *Handbk Ident. Br. Insects* **9**(2): 1–216, 199 figs (compound).
(1950) **Coe, R. L., Freeman, P., and Mattingly, P. F.**
Still useful but later works must be studied for some families.

Aquatic Diptera. Pt I. Nematocera, excluding Chironomidae and Ceratopogonidae. *Bull. Cornell Univ. agric. Exp. Sta.* **164**, 77 pp, 218 figs; Pt II. Orthorrhapha, Brachycera, and Cyclorrhapha, **177**, 62 pp, 132 figs; Pt III. Chironomidae, subfamilies Tanypodinae, Diamesinae, Orthocladiinae, **205**, 84 pp, 274 figs; Pt IV. Chironomidae, subfamily Chironominae and Pt V. Ceratopogonidae (L. C. Thomsen), **210**, 80 pp. (Reprinted (1969) by Classey, Faringdon.)
(1934–37) **Johanssen, O. A.**

Tipulidae and Trichoceridae

British short-palped crane-flies. *Trans. Soc. Br. Ent.* **5**: 1–168, 31 figs (compound), 5 pls.
(1938) **Edwards, F. W.**
Still useful in conjunction with later works.

[Tipulidae.] *Fauna SSSR* **2**(3): 1–486, 295 figs; **2**(4): 1–502, 380 figs; **2**(5): 1–281, 160 figs. [Russian].
(1961–73) **Savtshenko, E. N.**

European species of the genus *Dicranoptycha* Östen-Sacken (Diptera, Tipulidae). *Acta ent. bohemoslovaca* **69**: 401–16, 15 figs.
(1972) **Starý, J.**

A revision of palaearctic species of the subgenera *Protogonomya* Alexander, *Ellipteroides* Becker and *Ptilostenodes* Alexander (Diptera, Tipulidae: *Gonomyia* Meigen). *Acta ent. bohemoslovaca* **67**: 362–74, 28 figs.
(1970) **Starý, J. and Rozkõsný, R.**

Introduction to craneflies. *Bull. amat. Ent. Soc.* **31**: 46–54, 83–93; **32**: 14–23, 58–64, 101–7; **33**: 18–23, 142–5, numerous figs.
(1972 on) **Stubbs, A. E.**
Useful introduction to family.

A synopsis of the Swedish Tipulidae. 1. Subfam. Limoniinae: tribe Limoniini. *Opusc. Ent.* **23**: 133–69, 40 figs. 2. Subfam. Limoniinae: tribe Pediciini. *Opusc. Ent.* **24**: 1–9, 9 figs.
(1958) **Tjeder, B.**

The western palaearctic species of *Nephrotoma* Meigen, 1803 (Diptera, Tipulidae) Part 4, including a key to species. *Beaufortia* **29**: 129–97.
(1979) **Oosterbroek, P.**

The larvae and pupae of the British Cylindrotominae and Limoniinae (Diptera, Tipulidae). *Trans. Soc. Br. Ent.* **17**(7): 151–216, 173 figs.
(1967) **Brindle, A.**
Well illustrated key to immature stages.

The larvae and pupae of the British Tipulinae (Diptera: Tipulidae). *Trans. Soc. Br. Ent.* **14**: 63–114, 179 figs.
(1960) **Brindle, A.**

Die Entwicklungstadien der Tipuliden (Diptera, Nematocera), insbesondere der West-Palaearktischen Arten. *Tijdschr. Ent.* **100**: 195–308, 329 figs.
(1957) **Theowald, B.**

The British species of *Trichocera* (Diptera: Trichoceridae). *Proc. R. ent. Soc. Lond.* **A32**: 132–8, 2 figs. (1957) **Laurence, B. R.**

Notes on the taxonomy and distribution of Swedish Trichoceridae (Dipt., Nemat.). *Opusc. ent.* **31**: 93–118, 72 figs. (1966) **Dahl, C.**

A world catalogue of Trichoceridae Kertesz 1902 (Diptera). *Ent. Scand.* **7**: 7–18. (1976) **Dahl, C. and Alexander, C. P.**
Includes a key to world genera.

The early stages of the families Trichoceridae and Anisopodidae (=Rhyphidae) (Diptera: Nematocera). *Trans R. ent. Soc. Lond.* **90**: 39–62, 87 figs. (1940) **Keilin, D. and Tate, P.**

Trichoceridae (Dipt.) of Spitsbergen. *Ann. ent. fenn.* **39**(2): 49–59, 35 figs. (1973) **Dahl, C.**
Keys all known larvae of the family.

Psychodidae

Psychodidae. *Proc. R. ent. Soc. Lond.* **B22**:69–71; **25**: 147–56. (1950, 56) **Freeman, P.**

Some British Psychodidae (Diptera, Nematocera): descriptions of species and a discussion on the problems of species pairs. *Trans. R. ent. Soc. Lond.* **114**: 403–36, 17 figs (compound). (1962) **Duckhouse, D. A.**

Psychodidae (Diptera, Nematocera) of Southern Australia: subfamily Psychodinae. *Trans. R. ent. Soc. Lond.* **118**(6): 153–220, 235 figs. (1966) **Duckhouse, D. A.**
Includes some comments on the generic placing of British species.

Palaearctic moth-flies: a review of the Trichomyiinae (Psychodidae). *Syst. Ent.* **7**: 357–65. (1982) **Wagner, R.**

The larvae of the British species of *Psychoda* (Diptera: Psychodidae). *Parasitology* **38**: 51–69, 84 figs. (1947) **Satchell, G. H.**

The early stages of the British species of *Pericoma* Walker (Diptera, Psychodidae). *Trans. R. ent. Soc. Lond.* **100**: 411–47, 26 figs (compound). (1949) **Satchell, G. H.**

Epidemiologie des leischmanioses dans le sud de la France. *Paris Inst. nat. Santé Rech. méd., Monagr.* **37**: 1–223. (1969) **Rioux, J. A. and Golvan, Y. J.**
Includes key to French species and is the most up-to-date work covering the sandflies in Europe.

Ptychopteridae

[Key to larvae of Danish Ptychopteridae (Diptera, Nematocera), with notes on habitat preferences.] *Ent. Meddr.* **49**(2): 59–64. [Danish: English summary includes a key.] (1982) **Hansen, S. B.**

Dixidae

A key to British Dixidae. *Scient. Publs Freshwat. biol. Ass.* **31**: 1–78, 23 figs, 2 pls, maps. (1975) **Disney, R. H. L.**
Keys to mature larvae, pupae and adults.

Contributions to the knowledge of Danish Nematocera. I. Dixidae. *Vidensk. Meddr. Dansk. Naturh. Foren.* **101**: 119–24, 5 figs. (1937) **Nielsen, P.**

Über einige bisher nicht oder wenig bekannte *Dixa*—Arten de palaearctischen Fauna. *Arb. morph. taxon. Ent.* (Berlin-Dahlem) **1**: 195–204. (1934) **Peus, F.**

Chaoboridae

Das Zooplankton der Binnengewässer. 1. Teil VI. Chaoboridae. *Binnengewässer* **26**: 257–80. (1972) **Saether, O. A.**
Keys to all stages of genus *Chaoborus*; morphology of all stages with references.

Culicidae

The British mosquitoes. British Museum (Natural History), London. Reprinted 1966, Johnson Reprint, 341 pp, 172 figs, 20 pls. (1938) **Marshall, J. F.**
An elaborate monograph.

Contributions to the knowledge of the Danish and Fennoscandian mosquitoes. Culicini. *Norsk ent. Tidsskr.* (Suppl.) **1**: 1–567, 148 figs. (1948) **Natvig, L. R.**

A catalog of the mosquitoes of the world (Diptera: Culicidae), 2nd edn. Thomas Say Foundation, Vol. 6. Entomological Society of America, Washington DC. xi + 611 pp with supplement 107 pp (1978), second supplement (1984) *Mosquito Syst.* **16**: 227–70 and index to latter *Mosquito Syst.* **17**: 52–63 (1985). (1977) **Knight, K. L. and Stone, A.**

Keys to the British mosquitoes (Culicidae). *Scient. Publs Freshwat. biol. Ass.* (In press) **Cranston, P. S., Ramsdale, C. D., Snow, K. R., and White, G. B.**

The identification of larvae of British mosquitoes (Diptera, Culicidae). *Entomologist's mon. Mag.* **120**: 21–32, 23 figs. (1984) **Snow, K. R.**
An important paper which keys the fourth (final) instar of all British species.

6Thaumaleidae

A revision of the Thaumaleidae. *Zool. Anz.* **82**: 121–42, 46 figs. (1929) **Edwards, F. W.**

Ceratopogonidae

A bibliography and key word—in context index of the Ceratopogonidae (Diptera) from 1758 to 1973. Texas Technical Press, Lubbock, Texas. 300 pp. (1975) **Atchley, W. R., Wirth, W. W., and Gaskins, C. T.**

On the British biting midges. *Trans. ent. Soc. Lond.* **74**: 389–426, 3 figs (compound), 2 pls. (1926) **Edwards, F. W.**
Keys to genera and species.

A taxonomic review of the British species of '*Culicoides*' Latreille (Diptera: Ceratopogonidae). *Proc. R. Soc. Edinb. (B)* **67**: 181–302, 291 figs. (1960) **Campbell, J. A. and Pelham-Clinton, E. C.**

Notes sur les Cératopogonides (part). *Archs Inst. Pasteur Alger* **39**: 401–37, 20 figs (*Ceratopogon* and *Alluaudomyia*); **40**: 53–125, 32 figs (*Bezzia*); **40**: 225–8

(*Palpomyia*): **41**: 41–68, 17 figs (*Stilobezzia*).   (1961–63) **Clastrier, J.**

Descriptions of three species of *Culicoides* Latreille (Diptera: Ceratopogonidae) new to science, together with notes on, and a revised key to the British species of the *pulicaris* and *obsoletus* groups. *Proc. R. ent. Soc. Lond.* **B21**: 61–78, 8 figs.   (1952) **Downes, J. A. and Kettle, D. S.**

Descriptions of two species of *Culicoides* Latreille (Diptera: Ceratopogonidae) new to science. *Proc. R. ent. Soc. Lond.* **B24**: 37–47, 14 figs.
                    (1955) **Kettle, D. S.**
Revised couplets to existing keys to adults and larvae.

Contribution à l'étude de genre *Culicoides* Latreille particulièrement en France. *Encycl. ent.* (A) **39**: 3–299, 479 figs.   (1965) **Kremer, M.**

Synopsis of the genera of Ceratopogonidae (Diptera). *Annls Parasit. hum. comp.* **49**(5):595–613.
                    (1974) **Wirth, W. W., Ratanaworabhan, N. C., and Blanton, F. S.**
Key to world genera; outline of family classification. No illustrations.

On the life history and the anatomy of the early stages of *Forcipomyia* (Diptera, Ceratopogonidae). *Parasitology* **16**: 164–212, 26 figs, 3 pls.
                    (1924) **Saunders, L. G.**

The early stages of British biting midges *Culicoides* Latreille (Diptera: Ceratopogonidae) and allied genera. *Bull. ent. Res.* **43**: 421–67, 6 pls.
                    (1952) **Kettle, D. S. and Lawson, J. W. H.**

Chironomidae

Zur Systematik der Orthocladiinae (Dipt., Chironomidae). *Inst. Freshwat. Res. (Drottningholm)* **37**: 5–185, 137 figs.   (1956) **Brundin, L.**
Gives modern generic concepts.

Transantarctic relationships and their significance, as evidenced by Chironomid midges with a monograph of subfamilies Podonominae and Aphroteniinae and the Austral Heptagyiae. *K. svenska VetenskAkad. Handl.* **11**(1): 1–472, 638 figs, 30 pls.   (1966) **Brundin, L.**
Essential reading: keys to genera and all species of these subfamilies.

Die Tanypodinae (Diptera: Chironomidae). *Abh. Larvalsyst. Insekt.* **6**: 1–53, 409 figs.   (1962) **Fittkau, E. J.**
Keys to European adults and pupae with distribution.

Revision der Gattung *Cricotopus* v. d. Wulp und ihrer Verwandten (Diptera, Chironomidae). *Annls zool. fenn.* **10**: 1–363, 216 figs. (1973) **Hirvenoja, K.**
Especially useful for larvae and pupae.

A key to adult males of the British Chironomidae (Diptera) *Scient. Publs Freshwat. biol. Ass.* **37**: 1–169, 189 figs.   (1978) **Pinder, L. C. V.**

A key to the larvae of the British Orthocladiinae (Chironomidae). *Scient. Publs Freshwat. biol. Ass.* **45**: 1–152, 57 figs.   (1982) **Cranston, P. S.**

A key to pupal exuviae of British Chironomidae. 324 pp, 101 figs (compound). Privately published.   (1984) **Langton, P. H.**

Notes on the taxonomy of European species of *Chironomus* (Diptera: Chironomidae) *Entomologica Scand.* Suppl. **10**: 99–116.

(1981) **Lindeburg, B. and Wiederholm, T.**

Female genitalia in Chironomidae and other Nematocera: morphology, phylogenies, keys. *Bull. Fish. Res. Bd Can.* **197**: 1–209. (1977) **Saether, C. A.**
Keys to subfamilies, tribes, and genera of female Chironomidae.

Taxonomic studies on Chironomidae: *Nanocladius, Pseudochironomus* and the *Harnischia* complex. *Bull. Fish Res. Bd Can.* **196**: 1–143.

(1977) **Saether, O. A.**
Keys to the genera of larvae, pupae, males, and females of the *Harnischia* complex (Chironominae).

Chironomidae of the holarctic region. Keys and diagnoses. Pt 1. Larvae. *Entomologica scand.* Suppl. **19**: 1–457, illustrated.

(1983) **Wiederholm, T.**, ed.

Part 2 (Pupae) is in press.

A bibliography of the Chironomidae. *K. norske Vidensk. Selck. Mus. Gunneria* **26**: 1–177. (1976) **Fittkau, E. F., Reiss, F., and Hoffrichter, O.**
Bibliography to approximately 6400 papers on Chironomidae published up to the end of 1974, with a key to subjects including cytology, ecology, history, morphology, parasitology, physiology, and systematics.

A bibliography of the Chironomidae. Supplement 1. *K. norske Vidensk Selck. Mus. Gunneria* **37**: 1–68. (1981) **Hoffrichter, O. and Reiss, F.**
Adds some 1700 additional titles.

Simuliidae

Taxonomy of British black-flies (Diptera Simuliidae). *Trans. R. ent. Soc. Lond.* **118**: 413–508, 49 figs. (1966) **Davies, L.**

A key to the British species of Simuliidae (Diptera) in the larval, pupal and adult stages. *Scient. Publs Freshwat. biol. Ass.* **24**: 1–126, 49 figs, maps.

(1968) **Davies, L.**
For nomenclatural changes *see above* on p. 131 Crosskey in Kloet and Hincks (1975); the section on early stages is currently (1986) under revision.

Simuliidae (Diptera) from the Channel Islands: first records. *Entomologist's Gaz.* **21**: 125–32, 7 figs. (1970) **Crosskey, R. W.**
Keys to adults, larvae, and pupae of species occurring in Jersey.

A revision of the taxonomy and distribution of the Danish black-flies (Diptera: Simuliidae), with keys to the larval and pupal stages. *Natura Jutlandica* **21**: 69–116. (1984) **Jensen, F.**

The blackflies of Iceland (Diptera: Simuliidae) *Can. Ent.* **109**: 449–72, 30 figs.

(1977) **Peterson, B. V.**
Keys to adults, pupae, and larvae.

Studies on Scandinavian black flies (fam. Simuliidae Latr.). *Opusc. ent. suppl.* **21**: 1–280, 43 figs. (1962) **Carlsson, G.**

The identity and synonymy of *Simulium (Wilhelmia) pseudequinum* Séguy and the

occurrence of this species in England (Diptera: Simuliidae). *Entomologist's Gaz.* **32**: 137–48.   (1981) **Crosskey, R. W.**
Includes revised key to subgenus *Wilhelmia*.

Anisopodidae

The early stages of the families Trichoceridae and Anisopodidae (=Rhyphidae) (Diptera: Nematocera). *Trans. R. ent. Soc. Lond.* **90**: 39–62, 87 figs.   (1940) **Keilin, D. and Tate, P.**

Bibionidae and Scatopsidae

A contribution toward a monograph of the Scatopsidae (Diptera). Part VIII. The genus *Anapausis. Ann. Soc. ent. Am.* **58**: 7–18, 54 figs. (1965) **Cook, E. F.**
Includes a key to the European species.

A synopsis of the Scatopsidae of the Palaearctic. *J. nat. Hist.*: part I, Rhegmoclematini **3**: 393–407, 27 figs; part II, Swammerdamellini **6**: 625–34, 26 figs; part III, The Scatopsinae **8**: 61–100, 120 figs. (1969–74) **Cook, E. F.**
Deals comprehensively with the main subfamily Scatopsinae; later figures in part III reproduced from photographs and are sometimes difficult to interpret.

Bibionid and scatopsid flies. Diptera: Bibionidae and Scatopsidae. *Handbk Ident. Br. Insects.* **9**(7): 1–74, 189 figs. (1985) **Freeman, P. and Lane, R. P.**

The larval and pupal stages of the Bibionidae. *Bull. ent. Res.* **12**: 221–32, 17 figs; **13**: 189–95, 10 figs, 1 pl.   (1921–22) **Morris, R. M.**

Mycetophilidae and Sciaridae

British fungus-gnats. *Trans. ent. Soc. Lond.* **57** (1924): 507–670, 230 figs. (See also 1941 *Entomologist's mon. Mag.* **77**: 21–32, 67–82, 9 figs (compound).)
                                                      (1925) **Edwards, F. W.**
Dated but still useful.

Mycetophilidae (Bolitophilinae, Ditomyiinae, Diadocidiinae, Keroplatinae, Sciophilinae and Manotinae). *Handbk Ident. Br. Insects.* **9**(3): 1–111, 290 figs.
            (1980) **Hutson, A. M., Ackland, D. M., and Kidd, L. N.**

New species of the genus *Phronia* Winnertz (Diptera, Mycetophilidae) from Eastern Fennoscandia and notes on the synonymies in this genus. *Notul. ent.* **50**: 41–60, 82 figs.   (1970) **Hackmann, W.**

Zweiflügler oder Diptera VI: Pilzmücken oder Fungivoridae (Mycetophilidae). *Tierwelt Dtl.* **38**: 1–166, 337 figs.   (1940) **Landrock, K.**

Beitrag zur Kenntnis der europäischen *Fungivora*-Arten aus der Gruppe *vittipes* (Zett.) (Dipt., Fungivoridae). *Cas. csl. Spol. ent.* **60**: 312–27, 18 figs.
                                                      (1963) **Laštovka, P.**

Holarctic species of *Mycetophila ruficollis*-group (Diptera, Mycetophilidae). *Acta ent. bohemoslovaca* **69**: 275–94, 38 figs.   (1972) **Laštovka, P.**

Revision des *Diadocidia* Holarctiques (Dipt., Mycetophilidae). *Annls Soc. ent. Fr.* (N.S.) **8**: 205–23, 28 figs.   (1972) **Laštovka, P. and Matile, L.**

Review of the British and notes on other species of the *Mycetophila ruficollis*-group, with the description of a new species (Diptera, Mycetophilidae). *Entomologist's mon. Mag.* **110**: 203–14, 40 figs.
(1975) **Laštovka, P. and Kidd, L. N.**

Die Pilzmückengattung *Neoempheria* (Diptera: Fungivoridae). *Senckenberg biol.* **53**: 239–44, 7 figs. (1972) **Plassmann, E.**

Die Pilzmückengattung *Leia* (Diptera: Mycetophilidae). *Senckenberg biol.* **54**: 131–40, 23 figs. (1973) **Plassmann, E.**

Revision der europäischen Arten der Pilzmückengattung *Bolitophila* Meigen (Diptera, Mycetophilidae). *Entomologica scand.* **6**: 145–57, 29 figs.
(1975) **Plassmann, E.**

Revision der europäischen Arten der Pilzmückengattung *Phronia* (Diptera: Mycetophilidae). *Dt. ent. Z.* (NF) **24**(4–5): 305–44, 67 figs.
(1977) **Plassmann, E.**

Revision der palaearktischen Arten der Pilzmücken-Gattung *Ectrepesthoneura* (Diptera: Mycetophilidae). *Beitr. ent.* **30**: 1–7, 7 figs. (1980) **Plassmann, E.**

Notes on the Holarctic species of *Pseudexechia* Tuomikoski (Diptera: Mycetophilidae), with the description of a new British species. *Entomologist's Rec. J. Var.* **90**: 44–51, 10 figs. (1978) **Chandler, P. J.**
Keys to British species.

The European and eastern Nearctic fungus-gnats in the genus *Ectrepesthoneura* (Mycetophilidae). *Syst. ent.* **5**: 27–41, 27 figs. (1979) **Chandler, P. J.**
Keys to European species.

The European and North American species of *Epicypta* Winnertz (Diptera: Mycetophilidae). *Entomologica scand.* **12**: 199–212, 19 figs.
(1981) **Chandler, P. J.**
Keys to Holarctic species.

A monograph of *Trichonta* with a model for the distribution of holarctic Mycetophilidae (Diptera). *US Dept. Agr. tech. Bull.* **1638**: 1–64, illustrated.
(1981) **Gagné, R.**

Notes sur les Mycetophilidae (Diptera) de la faune de France. I. Le genre *Allodiopsis*. *Entomologiste* **27**(3): 64–70, 2 figs. (1971) **Matile, L.**
Keys to French species.

Révision des *Diadocidia* Holarctiques (Dipt., Mycetophilidae). *Annls Soc. ent. Fr.* (NS) **8**(1): 205–23, 28 figs. (1972) **Matile, L.**

Notes sur les Mycetophilidae (Diptera) de la faune de France. III. Le genre *Neuratelia*. *Entomologiste* **30**(1): 26–33, 9 figs. (1974) **Matile, L.**
Keys to French species.

Revision des *Asindulum* et des *Macrorrhyncha* de la région Paléarctique (Dipt., Mycetophilidae). *Annls Soc. ent. Fr.* (NS) **11**(3): 491–515, 26 figs.
(1975) **Matile, L.**

Notes sur les Mycetophilidae (Diptera) de la faune de France. IV. Le genre *Bolitophila*. 1: Sous-genre *Bolitophila* s. str. (1re partie). *Entomologiste* **32**(6):

235–44, 9 figs.   (1976) **Matile, L.**
Keys to French species.

Catalogue provisoire des Diptères Mycetophilidae de la faune de France. *Bull.*
Mus. natn Hist. nat. Paris (3ᵉ Ser) **456** (zool 319): 1–655. (1977) **Matile, L.**
Lists French species and taxonomic literature thereon.

Révision des Keroplatinae du genre *Antlemon* (Dipt., Mycetophilidae). *Annls*
Soc. ent. Fr. (NS) **13**(4): 639–49, 13 figs.   (1977) **Matile, L.**

Complément au Catalogue des Mycetophilidae de France. *Bull. Soc. ent. Fr.* **85**:
93–102.   (1980) **Matile, L.**
Lists French species and taxonomic literature thereon, supplements Matile (1977).

[Families 17–22, 24 (Bolitophilidae, Ditomyiidae, Ceroplatidae, Diado-
cidiidae, Macroceridae, Manotidae, Mycetophilidae).] In [Diptera, Fleas,
Part I. Keys to the insects of the European part of the USSR. V (ed. G. Ya.
Bei-Bienko).] *Opred. faune SSSR* **100**: 247–320. [Russian].
                          (1969) **Stackleberg, A. and Ostraverkhova, G. P.**

Generic taxonomy of the Exechiini (Dipt., Mycetophilidae). *Suom. hyönt.*
Aikak. **32**: 159–94.   (1966) **Tuomikoski, R.**
Keys to genera.

A review of the palaearctic Mycomyini (Diptera, Mycetophilidae). *Ann. Ent.*
Fenn. **48**: 37–42.   (1982) **Vaisanen, R.**

Morphologisch-taxonomische Untersuchungen an Fungivoridenlarven. *Dt.*
Ent. Z. **19**: 73–99, 15 figs (compound).   (1972) **Plassmann, E.**

Biology and morphology of the immature stages of Mycetophilidae. *Phil.*
Trans. R. Soc. **B227**: 1–110, 392 figs.   (1937) **Madwar, S.**

A study on the last instar larvae of some Czechoslovak *Mycetophila* (Diptera,
Mycetophilidae). *Acta Univ. Carol. Biol.* **1970**: 137–76, 75 figs.
                                                          (1971) **Laštovka, P.**

A contribution to the larval morphology of the genera *Platurocypta* and
Dynatosoma (Diptera, Mycetophilidae). *Entomologist* **105**: 59–76, 56 figs.
                                                          (1972) **Laštovka, P.**

Zur Kenntnis der Sciariden (Dipt.) Finnlands. *Annls zool. Soc. Vanamo* **21**(4):
1–164, 33 figs (compound).   (1960) **Tuomikoski, R.**
Essential for a study of the European Sciaridae; keys to genera and species.

Sciarid flies (Diptera, Sciaridae). *Handbk Ident. Br. Insects* **9**(6): 1–68, 162 figs.
                                                          (1983) **Freeman, P.**

Cecidomyiidae

Die Zoocecidien durch Tiere erzeugte Pflanzengallen Deutschlands und ihre
Bewohner. Die Cecidomyiden (Gallmücken) und ihre Cecidien. *Zoologica,*
Stuttg. **29**: 1–350.   (1926–39) **Rübsaamen, E. H. and Hedicke, H.**

The biology of some Cecidomyiidae (Diptera) galling the leaves of birch
(*Betula*), with special reference to their chalcidoid (Hymenoptera) parasites.

Trans. R. ent. Soc. Lond. **125**: 257–94, 17 figs.
(1974) **Askew, R. R. and Ruse, J. M.**
Illustrations of galls and comparative descriptions and their biology.

Gall midges of economic importance. Crosby Lockwood, London. 8 vols, illustrated, pls.
(1946–69) **Barnes, H. F.**
Vol. 8 is by W. Nijveldt and contains chapters on identification, including keys to tribes.

Gallenboek. Thieme, Zutphen. 332 pp, illustrated. [Dutch].
(1957) **Doctors van Leeuwen, W. M.**

On the British Lestremiinae, with notes on exotic species. Pts 1–7: *Proc. R. ent. Soc. Lond.* **B7**: 18–32, 102–8, 173–82, 199–210, 229–43, 253–65, 28 figs (compound).
(1938) **Edwards, F. W.**
Keys to tribes only: diagnoses of genera, subgenera, species, figs of genitalia; notes on mounting specimens.

Gall midge genera of economic importance (Dipt., Cecidomyiidae). Part I: introduction and subfamily Cecidomyiinae supertribe Cecidomyiidi. *Trans. R. ent. Soc. Lond.* **118**: 313–58, 119 figs.
(1966) **Harris, K. M.**

A systematic revision and biological review of the cecidomyiid predators (Diptera: Cecidomyiidae) on world Coccoidea (Hemiptera: Homoptera). *Trans. R. ent. Soc. Lond.* **119**: 401–94, 183 figs.
(1968) **Harris, K. M.**
Keys genera predacious on Coccids.

Aphidophagous Cecidomyiidae (Diptera): taxonomy, biology and assessments of field populations. *Bull. ent. Res.* **63**: 305–25, 17 figs, 5 tables.
(1973) **Harris, K. M.**

A revision of the European gall midges of the subfamily Porricondylinae (Diptera: Itoniidae). *Acta zool. fenn.* **113**: 1–157, 43 figs.
(1965) **Panelius, S.**
Illustrated keys to species.

[*Evolution of gall forming insects—gall midges.* (Translation from the Russian.) The British Library, Lending Division, Wetherby, Yorkshire. 317 pp, 79 figs.
(1975) **Mamaev, B. M.**
Essential primary reading for this family.

A new genus and species of Cecidomyiidae (Diptera) infesting mushrooms. *Proc. R. ent. Soc. Lond.* **B28**: 175–9, 11 figs.
(1959) **Wyatt, I. J.**

Pupal paedogenesis in the Cecidomyiidae (Diptera). *Proc. R. ent. Soc. Lond.* **A36**: 133–43, 14 figs; **38**: 136–44, 22 figs; *Trans. R. ent. Soc. Lond.* **119**: 71–98, 52 figs.
(1961–67) **Wyatt, I. J.**
Part 3 contains keys to genera of the tribe Heteropezini for adults, larvae, pupae, and hemipupae.

[*Larvae of gall midges (Diptera, Cecidomyiidae): comparative morphology, biology and identification tables.*] Izdatelstvo Nauka, Moscow. 278 pp, 94 figs (compound). [Russian].
(1965) **Mamaev, B. M. and Krivosheina, N. P.**

Beiträge zur Systematik der Larven der Itonididae (=Cecidomyiidae, Diptera) 1. Teil: Porricondylinae und Itonidinae Mitteleuropas. *Zoologica, Stuttg.* **38**: 1–247.   (1955) **Möhn, E.**

Immature stages of Lestremiinae (Diptera: Cecidomyidae) infesting cultured mushrooms. *Trans. R. ent. Soc. Lond.* **116**: 15–28, 36 figs. (1964) **Wyatt, I. J.**

Study on relationships between habits and external structures in Oligotrophidi larvae (Diptera, Cecidomyiidae). *Zoologica Scr.* **4**: 55–92.
                                                    (1975) **Sylven, E.**

Tabulated taxonomic data with illustrations.

BRACHYCERA

British species nearly 800.

Stratiomyiidae to Cyrtidae. *British flies*. Gurney and Jackson, London. Vol. 5, 780 + 34 pp, 407 figs.   (1909) **Verrall, G. H.**

Covers all British Brachycera except Empididae and Dolichopodidae. Keys to adults, very full descriptions, distribution, and biological data. Non-British families (Mydidae and Nemestrinidae) not covered.

Diptera Brachycera. *Handbk Ident. Br. Insects* **9**(4): 1–132, 339 figs.
                                                  (1969) **Oldroyd, H.**

Covers same families as Verrall (above).

Some distinctions between the larvae and pupae of the Empididae and Dolichopodidae (Diptera). *Proc. R. ent. Soc. Lond.* **A42**: 119–28, 11 figs.
                                                    (1967) **Dyte, C. E.**

Papers on larvae and pupae of British Empididae and Rhagionidae in *Entomologist's mon. Mag.* **97–98**, illustrated.
                          (1961–62) **Hobby, B. M. and Smith, K. G. V.**

The pupae of *Atherix ibis* (Fabricius) and *A. marginata* (Fabricius) (Diptera: Rhagionidae) and a key to the families of the aquatic Brachycera based on the pupae. *Entomologist's Gaz.* **31**: 157–61, 9 figs.
                            (1980) **Mackey, A. P. and Brown, H. M.**

Stratiomyiidae, Rhagionidae, and Tabanidae

A biosystematic study of the European Stratiomyidae (Diptera) in 2 vols. *Series Ent.* **25**: 1–40 + 431, 147 pls, 136 maps.   (1982–83) **Rozkosny, R.**

Keys and descriptions for adults and larvae, with biology and distribution. A superb monograph.

The Stratiomyidae (Diptera) of Fennoscandia and Denmark. *Fauna ent. scand.* **1**: 1–152, 456 figs.   (1973) **Rozkosny, R.**

Keys, maps, distribution, well illustrated.

The horse flies of Europe (Diptera: Tabanidae). Entomological Society of Copenhagen, Copenhagen. 499 pp, 164 figs, 8 pls.
                          (1972) **Chvála, M., Lyneborg, L., and Moucha, J.**

Révision systématique et biogéographique des Tabanidae (Diptera) Paléarctiques. Volume I. Pangoniinae et Chrysopinae. *Mém. Inst. r. Sci. nat. Belg.* (2^e
Sérié) **63**: 1–77.   (1960) **Leclerq, M.**

Révision systématique et biogéographique des Tabanidae (Diptera) Paléarctiques. Volume II. Tabaninae. *Mém. Inst. r. Sci. nat. Belg.* (2e Série) **80**: 1–
237.   (1966) **Leclerq, M.**

Asilidae, Scenopinidae, Acroceridae, Nemestrinidae, and Bombyliidae

Robber flies of the world, the genera of the family Asilidae. *Bull. US natn. Mus.*
224: 1–907, 2536 figs (in two vols).   (1962) **Hull, F. M.**
Keys to subfamilies, tribes, and genera of the world. Generic concepts somewhat
controversial—*see* Oldroyd (1974) below.

Some comments on the tribal classification of Asilidae (Diptera). *Israel J. Ent.*
9: 5–21.   (1974) **Oldroyd, H.**
A critique of Hull (1962).

Contributions to the biology, metamorphosis and distribution of the Swedish
Asilids. *Zool. Bidr. Uppsala* **8**: 1–316, 295 figs.   (1923) **Melin, D.**

A revision of the Scenopinidae (Diptera) of the World. *Bull. US natn. Mus.* **277**:
1–336.   (1969) **Kelsey, L. P.**
Keys to world genera and species.

Cyrtidae (Acroceridae). *Fliegen palaearkt. Reg.* **4**(21): 1–36.   (1936) **Sack, P.**
Keys to genera and species.

The genera of the family Nemestrinidae (Diptera: Brachycera). *Archos Zool. S.*
Paulo **24**: 208–318.   (1973) **Bernardi, N.**
Keys to genera, includes bibliography for all regions.

Bee flies of the world. The genera of the family Bombyliidae. Smithsonian Institution
Press, Washington. 687 pp, 1030 figs.   (1973) **Hull, F. M.**
Keys to subfamilies and genera, though use of genera somewhat controversial. A
partial critique of this work is given by Bowden (1985), *Entomologist's mon. Mag.* **121**:
99–107.

Empididae

Empididae. *British flies.* Cambridge University Press, Cambridge. Vol. 6, 782
pp, 317 figs.   (1961) **Collin, J. E.**
In series with Verrall's two volumes (on Brachycera on p. 143 and Syrphidae on
p. 146). In addition to keys and descriptions to all known British species, there is
critical comment on European species.

The Empidoidea (Diptera) of Fennoscandia and Denmark. II General part.
The families Hybotidae, Atelestidae and Microphoridae. *Fauna ent. scand.*
12: 1–281, 639 figs.   (1983) **Chvála, M.**
Keys, maps, distribution, excellent illustrations; covers all species occurring in
Britain. These families were formerly included in the Empididae. A final volume will
deal with the true Empididae as currently defined, i.e. subfamilies Oreogetoninae,
Empidinae, Brachystomatinae, Hemerodromiinae, and Clinocerinae.

Revision of the palaearctic species of *Tachydromia* Meig. (=*Tachista* Loew) (Diptera, Empididae). *Sb. ent. Odd. nar. Mus. Praze* **38**: 415–524, 111 figs.
(1969) **Chvála, M.**

The Tachydromiinae (Diptera) of Fennoscandia and Denmark. *Fauna ent. scand.* **3**: 1–336, 790 figs.
(1975) **Chvála, M.**
Keys, maps, distribution; updates this subfamily as given in Collin (1961) above. Includes all British species.

A revision of palaearctic Tachydromiinae genus *Dysaletria* Loew (Diptera: Empididae). *Vestn. csl. Spol. zool.* **39**: 167–72, 4 figs.
(1975) **Chvála, M.**

European species of the *Platypalpus albiseta*-group (Diptera, Empididae). *Acta ent. bohemoslovaca* **70**: 117–36, 19 figs.
(1973) **Chvála, M.**

Revision of the Palaearctic *Platypalpus nigritarsis*-group (Diptera, Empididae), with special reference to *P. excisus* Beck. *Acta ent. bohemoslovaca* **71**: 250–9, 8 figs.
(1974) **Chvála, M. and Kovalev, V. G.**

Chersodromia cursitans Zetterstedt (Dipt., Empididae) reinstated as a British species. *Entomologist's mon. Mag.* **99**: 127–8.
(1963) **Smith, K. G. V.**
Revised key to some *Chersodromia* species; see also Chvála (1977) below.

Revision of the genus *Chersodromia* Walker (Diptera, Empididae). *Acta ent. Mus. natn. Pragae* **39**: 55–138.
(1977) **Chvála, M.**

Some distinctions between the larvae and pupae of the Empididae and Dolichopodidae (Diptera). *Proc. R. ent. Soc. Lond.* (A) **42**: 119–28, 11 figs.
(1967) **Dyte, C. E.**
Reviews knowledge of immature stages.

Dolichopodidae

Diptera Orthorrhapha Brachycera Dolichopodidae. *Handbk Ident. Br. Insects* **9**(5): iv + 1–90, 220 figs.
(1978) **Fonseca, E. C. M. d'Assis**

Some distinctions between the larvae and pupae of the Empididae and Dolichopodidae (Diptera). *Proc. R. ent. Soc. Lond.* (A) **42**: 119–28, 11 figs.
(1967) **Dyte, C. E.**
Reviews knowledge of immature stages.

A revision of the palaearctic species of the genus *Medetera* Fischer (Dipt., Dolichopodidae). *Ann. ent. fenn.* **21**: 130–57, 50 figs. (See also *Entomologist's mon. Mag.* **97**: 100.)
(1955) **Thuneborg, E.**

Zur Kenntnis einiger palaearktischer Arten der Gattung *Asyndetus* Loew (Diptera, Dolichopodidae). *Beitr. Ent.* **23**: 157–67. (1973) **Negrobov, O. P.**

CYCLORRHAPHA—ASCHIZA

British species about 650.

Lonchopteridae

Lonchopteridae. *Handbk Ident. Br. Insects* **10**(2): 1–9, 14 figs.
(1969) **Smith, K. G. V.**

The Swedish species of *Lonchoptera* Meig. (Dipt., Lonchopteridae) with lectotype designations. *Opusc. Ent.* **31**: 77–80.   (1966) **Andersson, H.**
Lists species with their distribution in Sweden; no keys or descriptions.

Phoridae

A catalogue of the Phoridae of the world (Diptera, Phoridae). *Studia ent.* **11**: 1–367; **14**: 177–224.   (1938) **Borgmeier, T.**
Contains original citations of world species and world bibliography.

Scuttle flies. Diptera, Phoridae (except *Megaselia*). *Handbk Ident. Br. Insects* **10**(6): 1–81, 186 figs.   (1983) **Disney, R. H. L.**
A second part on *Megaselia* is nearing completion.

Platypezidae, Pipunculidae, Syrphidae, and Conopidae

Platypezidae, Pipunculidae, Syrphidae. *British flies*. Gurney and Jackson, London. Reprinted (1969) by Classey, Faringdon. Vol. 8, 683 pp, 458 figs.
(1901) **Verrall, G. H.**

The flat-footed flies (Diptera, Aschiza-Platypezidae) known to occur in Kent. With a key to the genera and species so far recorded from the British Isles. *Trans. Kent Fld Club* **5**(1): 15–44, 1 fig.   (1973) **Chandler, P. J.**

Additions and corrections to the British list of Platypezidae (Diptera), incorporating a revision of the palaearctic species of *Callomyia* Meigen. *Proc. Trans. Br. ent. nat. Hist. Soc.* **7**(1): 1–32, 34 figs.   (1974) **Chandler, P. J.**

Pipunculidae. *Handbk Ident. Br. Insects* **10**(26): 1–83, 192 figs, 1 pl.
(1966) **Coe, R. L.**

Scandinavian Pipunculidae. *Opusc. Ent.* **21**: 149–69, 19 figs.
(1956) **Collin, J. E.**

Syrphidae. *Handbk Ident. Br. Insects* **10**(1): 1–98, 46 figs (compound).
(1953) **Coe, R. L.**
See Kloet and Hincks (1975) on p. 131 for revised nomenclature.

British hoverflies: an illustrated identification guide. The British Entomological and Natural History Society, London. 253 pp, 12 col. pls, 8 figs (compound), keys illustrated.   (1983) **Stubbs, A. E. and Falk, S.**
Keys figured and superb colour illustrations.

De zweefvliegen van Noordwest-Europa en Europees Rusland, in het bijzonder van de Benelux [The Syrphidae of North-western Europe and European Russia, with particular reference to the Benelux countries.] Bibliotheek van de KNNV, no. 32, Amsterdam. 275 pp, 496 figs, 12 pls. [Dutch].
(1981) **Van der Goot, V. S.**
An excellent guide.

Danmarks dyreliv bind 1: Die Danske Svirrefluer (Diptera: Syrphidae). Kendetegn, levevis og udbredelse. Fauna Boger, Copenhagen. 300 pp, 381 figs, 4 pls.

[Danish].   (1984) **Torp, E.**
An excellent guide to the Danish Syrphidae.

Versuch zum Aufbau eines natürlichen Systems mitteleuropäischer Arten der Unterfamilie Syrphinae (Diptera). *Acta sci. nat. Brno* **1**: 349–90, illustrated.
(1967) **Dušek, J. and Láska, P.**

Révision du genre *Paragus* (Dipt., Syrphidae) de la région paléarctique occidentale. *Mitt. Schweiz. ent. Ges.* **49**: 79–108, 37 figs.
(1976) **Goeldlin de Tiefenau, P.**

A generic revision of the genus *Syrphus* and allied genera (Diptera, Syrphidae) in the palaearctic region, with descriptions of the male genitalia. *Acta ent. fenn.* **25**: 1–94, 331 figs.   (1968) **Hippa, H.**

Classification of the palaearctic species of the genera *Xylota* Meigen and *Xylotomima* Shannon (Dipt., Syrphidae). *Acta ent. fenn.* **34**(4): 179–97, 23 figs.
(1968) **Hippa, H.**

Classification of Xylotini (Diptera, Syrphidae). *Acta zool. fenn.* **156:** 1–153, 58 figs (compound).   (1978) **Hippa, H.**
Keys to genera and subgenera.

Diptères Syrphides de l'europe occidentale. *Mem. Mus. natn. Hist. nat. Paris* (A) **23**: 1–248, 48 figs.   (1961) **Séguy, E.**

Syrphidae. In [Classification key to the insects of the European part of USSR (ed. G. Ya Bei-Bienko).] Vol. 5(2) *Opred. Faune SSSR* **103**: 11–96. [Russian].
(1970) **Stackelberg, A. A.**
One of the more useful volumes of this extensive series.

Irish Syrphidae (Diptera); notes on the species and an account of their known distribution. *Proc. R. Ir. Acad.* **75B**(1): 1–80, 2 figs.
(1975) **Speight, M. C. D., Chandler, P. J., and Nash, R.**
Includes key to European genera of tribe Syrphina and key to *Parasyrphus* species.

A revision of the genera of the Syrphini (Diptera: Syrphidae). *Mem. ent. Soc. Can.* **6**: 1–176, 100 figs.   (1969) **Vockeroth, J. R.**

The problem of old names as illustrated by *Brachyopa conica* Panzer, with a synopsis of Palaearctic *Brachyopa* Meigen (Diptera: Syrphidae). *Entomologica scand.* **11**: 209–16, 11 figs.   (1980) **Thompson, F. C.**
Keys to Palaearctic species.

Key to and descriptions of the third instar larvae of some species of Syrphidae (Diptera) occurring in Britain. *Trans. R. ent. Soc. Lond.* **112**: 345–79, 8 figs (compound).   (1960) **Dixon, T. J.**

A taxonomic account of the larvae of some British Syrphidae. *Proc. zool. Soc. Lond.* **163**: 505–73, 117 figs.   (1961) **Hartley, J. C.**

Contribution à l'étude systématique et écologique des Syrphidae (Dipt.) de la Suisse occidentale. *Mitt. Schweiz. ent. Ges.* **47**: 151–252, 116 figs.
(1974) **Goeldlin de Tiefenau, P.**
Some 30 species of aphidophagous larvae described and illustrated.

A preliminary key to the eggs of some of the commonest aphidophagous Syrphidae (Diptera) occurring in Britain. *Trans. R. ent. Soc. Lond.* **120**: 199–218, 4 figs (compound).   (1968) **Chandler, A. E. F.**

Conopidae. *Fliegen palaearkt. Reg.* **4**(35): 1–48.   (1924–25) **Kröber, O.**

Conopidae. *Handbk Ident. Br. Insects* **10**(3a): 1–19, 34 figs.
                                                         (1969) **Smith, K. G. V.**

The identity of *Myopa polystigma* Rondani, and an additional British and continental species of the genus (Diptera: Conopidae). *Entomologist* **103**: 186–9, 1 pl.   (1970) **Smith, K. G. V.**
Revises couplets in Smith (1969) above.

Czechoslovak species of the subfamily Conopinae (Diptera: Conopidae). *Acta Univ. Carol. Biologica* **1961**(2): 103–45, 31 figs, maps, tables.
                                                         (1961) **Chvála, M.**

A review of the Conopid flies of the genus *Sicus* Scop. (Diptera, Conopidae). *Acta Univ. Carol. Biologica* **1963**(3): 275–82, 7 figs.   (1963) **Chvála, M.**

Czechoslovak species of the subfamilies Myopinae and Dalmanniinae (Diptera, Conopidae). *Acta Univ. Carol. Biologica* **1965**(2): 93–149, 33 figs, tables, maps.   (1965) **Chvála, M.**
The keys in the above three works cover central Europe but are of value for the more northern territories. See also Lyneborg (1962–64) in the following section (Acalyptratae).

Conopidae (Insecta: Diptera). *Fauna Pol.* **7**: 134. [Polish].
                                                     (1979) **Bańkowska, R.**

The larvae of *Thecophora occidensis* with comments upon the biology of Conopidae (Dipt.). *J. Zool. Lond.* **149**: 263–76, 6 figs, 1 pl.
                                                     (1966) **Smith, K. G. V.**
Keys, reviews, and biology of world species.

CYCLORRHAPHA—ACALYPTRATAE

British species over 1200. In this group of 40 or more families, many quite small, the family limits are often ill-defined and the literature scattered. The more general papers covering several or all of the families and papers with keys to the families are placed first and should be consulted where specialist works are not listed. These are followed by specialist papers under families with the latter arranged alphabetically. Modern family keys are to be found in Oldroyd (1970) and Hennig (1973)—see under Diptera general on p. 131.

Muscidae Acalypterae et Scatophagidae. *Faune de France* **28**, 832 pp, 897 figs, 27 pls.   (1934) **Séguy, E.**

Additions and corrections to the list of British list of Muscidae Acalyptratae. *Entomologist's mon. Mag.* **46**: 47–8, 124–9, 169–78; **47**: 145–53, 182–7, 229–

34, 253–6.   (1910–11) **Collin, J. E.**
Partly out of date but still useful for families not covered by later works.

Additions and corrections to the list of British acalyptrate Diptera. *Entomologist's mon. Mag.* **110** (1974): 173–81, 8 figs.
                                        (1975) **Cogan, B. H. and Dear, J. P.**
Justifies revised British Check List (Kloet and Hincks (1975) on p. 131) and supplements Collin 1910–11 above.

Danske acalyptrate fluer. 1. Conopidae, Micropezidae, Calobatidae, Megamerinidae og Tanypezidae. *Ent. Meddr.* **31**: 249–64; Danske acalyptrate fluer. 2. Psilidae, Platystomidae og Otitidae (Diptera). *Ent. Meddr.* **32**: 367–88, 22 figs. [Danish].   (1962–64) **Lyneborg, L.**

Catalogue of palaearctic Diptera. Elsevier, Amsterdam, Oxford, New York, Tokyo.   (1984—still appearing) **Soos, A.**, ed.
Catalogues palaearctic species with full synonymies and taxonomic bibliography. To be completed in 14 volumes of which Vols. 9 (Micropezidae–Agromyzidae, 460 pp) and 10 (Clusiidae–Chloropidae, 402 pp) have so far appeared.

Agromyzidae

Agromyzidae. *Handbk Ident. Br. Insects* **10**(5g): 1–136, 360 figs.
                                        (1972) **Spencer, K. A.**

Agromyzidae (Diptera) of economic importance. *Series Ent.* **9**: xi + 1–418, 543 figs, 13 pls.   (1973) **Spencer, K. A.**
Keys and describes pest species.

The Agromyzidae (Diptera) of Fennoscandia and Denmark. *Fauna ent. scand.* **5**: 606 pp, 922 figs.   (1976) **Spencer, K. A.**

The agromyzid fauna of Iceland and the Faroes, with appendices on the *Phytomyza milii* and *robustella* Groups (Diptera, Agromyzidae) *Ent. Meddr* **32**: 393–450, 23 figs (compound).   (1964) **Griffiths, G. C. D.**
Keys. For further notes see *Opusc. Ent.* **33**: 129–38

A revision of the Palaearctic species of the *nigripes* group of the genus *Agromyza* Fallén (Diptera, Agromyzidae). *Tijdschr. Ent.* **106**(2): 113–68.
                                        (1963) **Griffiths, G. C. D.**

Revision of the *Phytomyza syngenesiae* group (Diptera, Agromyzidae), including species hitherto known as *Phytomyza atricornis* Meigen. *Stuttg. Beitr. Naturk.* **177**: 1–28, 23 figs.   (1967) **Griffiths, G. C. D.**

Studies on boreal Agromyzidae (Diptera). V. On the genus *Chromatomyia* Hardy, with revision of Caprifoliaceae-mining species. *Quaest. ent.* **10**: 35–69, illustrated.   (1974) **Griffiths, G. C. D.**

Die Larven der Agromyziden (Diptera) 1–3. *Tijdschr. Ent.* **97**: 115–36; **98**: 257–81; **100**: 73–94, illustr. See also *Tijdschr. Ent.* **98**: 1–27 for register of larvae described earlier by de Meijere.   (1954–57) **Hering, E. M.**

[Series of papers on Agromyzid larvae.] *Proc. R. ent. Soc. Lond.* **A31–3**, illustrated.   (1956–58) **Allen, P.**

Anthomyzidae

The British species of Anthomyzidae (Diptera). *Entomologist's mon. Mag.* **80**: 265–72, 1 fig. (1944) **Collin, J. E.**

Revision of the *Anthomyza* species of Northwest Europe. I. The *gracilis* group. *Entomologica Scand.* **7**: 41–52, 48 figs. II. The *pallida* group. *Entomologica Scand.* **15**: 15–24, 44 figs. (1976, 1984) **Andersson, H.**

Asteiidae

A revision of the British Asteiidae (Diptera) including two additions to the British List. *Proc. Brit. ent. nat. Hist. Soc.* **11**(1–2): 23–34, 9 figs. (1978) **Chandler, P. J.**

Keys; descriptions; British distribution.

Aulacigastridae

Neue Gattungen und Arten der Acalypteratae. *Can. Ent.* **101**: 589–633. (1969) **Hennig, W.**

Keys to world genera.

Braulidae

Die Bienenlaus—Arten. *Angew. Parasit.* **7**: 138–71. (1966) **Örösi Pal, Z.**
Keys to species of the world.

Camillidae

Camillidae. *Fliegen palaearkt. Reg.* **6**(58f): 1–7. (1934) **Duda, O.**
Five new species of Diptera. *Entomologist's mon. Mag.* **69**: 272–5, 10 figs. (1933) **Collin, J. E.**

Includes key to British Camillidae.

Canaceidae

A revision of the dipterous family Canaceidae. *Occ. Pap. Bernice P. Bishop Mus.* **20**: 245–75. (1951) **Wirth, W. W.**
Keys to genera and species of the world.

Carniidae

Some species of the genus *Meonura* (Diptera). *Entomologist's mon. Mag.* **66**: 82–9, 11 figs. (See also *Entomologist's mon. Mag.* **73**: 250–2.) (1930) **Collin, J. E.**

Chamaemyiidae

A new species of *Leucopis* (*Leucopella*) from Chile and a key to the World genera and subgenera of Chamaemyiidae (Diptera). *Can. Ent.* **92**: 51–8. (1960) **McAlpine, J. F.**

A short synopsis of British Chamaemyiidae (Dipt.). *Trans. Soc. Br. Ent.* **15**: 103–15, 12 figs. (1963) **Smith, K. G. V.**

The British species of *Chamaemyia* Mg. (*Ochtiphila* Fln.). *Trans. Soc. Br. Ent.* **17**: 121–8, 9 figs.   (1966) **Collin, J. E.**

A revision of the subgenus *Neoleucopis* (Diptera: Chamaemyiidae). *Can. Ent.* **103**: 1831–1847, 40 figs.   (1971) **McAlpine, J. F.**

Palaearctic species of *Parochthiphila*. *Ent. rev. Wash.* **47**: 388–99, 13 figs.
(1968) **Tanasiychuk, V. N.**

Palaearctic species of the genus *Chamaemyia* Panzer. *Ent. rev. Wash.* **49**: 128–40, 11 figs.   (1970) **Tanasiychuk, V. N.**
Both the above two papers were originally published in Russian in *Ent. Obozr.*

Chyromyidae

Chiromyidae. *Fliegen palaearkt. Reg.* **5**(53c): 51–4.   (1927) **Czerny, L.**
Keys to genera and species.

The palaearctic species of the genus *Aphaniosoma* Beck. (Diptera, Chiromyidae). *Ann. Mag. nat. Hist.* (12)**2**: 127–47, 12 figs.
(1949) **Collin, J. E.**

Chloropidae

Taxonomic and phylogenetic studies on Chloropidae (Diptera), with special reference to Old World genera. *Entomologica scand.* Suppl. **8**: 1–200, 119 figs (compound).   (1977) **Andersson, H.**
Gives keys to most Old World genera.

The British genera and species of Oscinellinae (Diptera, Chloropidae). *Trans. R. ent. Soc. Lond.* **97**: 117–48, 6 figs.   (1946) **Collin, J. E.**

A revision of the British species of *Cetema* Hendel (Diptera, Chloropidae), with two species new to science. *Entomologist* **99**: 116–20.   (1966) **Collin, J. E.**

British *Meromyza* (Dipt., Chloropidae). *Entomologist's mon. Mag.* **117**: 177–97, 89 figs.   (1981) **Ismay, J. W.**
Keys, host plants, distribution.

Platycephala umbraculata (Fabr., 1794) (Dipt., Chloropidae) new Norway. *Fauna norv.* (B) **30**: 111–12.   (1983) **Greve, L.**
Includes key to Norwegian species of genus.

Bemerkungen über einige paläarktische Arten der Gattung *Chlorops* Meigen und Beschreibung einer neuen Art aus der Tschechoslowakei (Diptera, Chloropidae). *Cas. ceske Spol. ent.* **57**: 387–97, 12 figs.   (1960) **Zuska, J.**
Gives key to species.

The immature stages of *Gaurax* (=*Botanobia*) *dubius* Macquart (Dipt., Chloropidae) with notes on the specific status of *G. fascipes* Becker. *Entomologist's mon. Mag.* **100** (1974): 237–9, 8 figs.   (1965) **Smith, K. G. V.**
Gives key to British species.

The genus *Lipara* Meigen (Diptera, Chloropidae), systematics, morphology, behaviour and ecology. *Tijdschr. Ent.* **117**: 1–25, 22 figs.
(1974) **Chvála, M., Doskocil, J., Mook, J. H., and Pokorny, V.**

Clusiidae

An identification guide to British Clusiidae. *Proc. Trans. Br. ent. nat. Hist. Soc.*
15: 89–93, 2 figs. (1982) **Stubbs, A. E.**
Keys to genera and species.

A review of the Australian species of Clusiidae (Diptera, Acalyptrata). *Rec.
Aust. Mus.* **25**: 63–94. (1960) **McAlpine, D. K.**
Keys to world genera.

Coelopidae

Studies on the family Coelopidae (Diptera). *Trans. R. ent. Soc. Lond.* **112**: 109–
40, 24 figs. (1960) **Egglishaw, H. J.**
Systematics and biology, descriptions, and illustrations of larvae.

The European species of the genus *Coelopa* (Dipt., Coelopidae). *Entomologist's
mon. Mag.* **96**: 8–13, 3 figs. (1960) **Burnet, B.**

Diastatidae

A revision of the genus *Campichoeta* Macquart (Diptera: Diastatidae). *Can. Ent.*
94(1): 1–10, 18 figs. (1962) **McAlpine, J. F.**
Keys to world species.

Drosophilidae

Drosophila. *Invertebrate types.* Ginn, London. 144 pp, 53 figs, 8 pls.
(1972) **Shorrocks, B.**
A useful introduction to the principle genus of this important family. Taxonomy,
biology, genetics, and keys to the commonest species (22) of *Drosophila.* (Care should
be taken to avoid confusion with the equally common drosophilid genus *Scaptomyza.*)

The distribution and biology of Drosophilidae (Diptera) in Scotland, including a new species of '*Drosophila*'. *Trans. R. Soc. Edinb.* **62**: 603–54.
(1954) **Basden, E. B.**
Keys to adults of species occurring in Scotland.

On the genera *Scaptomyza* Hardy and *Parascaptomyza* Duda (Dipt.,
Drosophilidae). *Notul. ent.* **35**: 74–91, 32 figs. (1955) **Hackmann, W.**

European species of the subgenus *Amiota* s.str. (Diptera, Drosophilidae). *Acta
ent. bohemoslovaca* **77**: 328–46, 39 figs. (1980) **Máca, J.**
Keys to European species.

European species of the *Drosophila* subgenus *Lordiphosa* (Diptera,
Drosophilidae). *Acta ent. bohemoslovaca* **75**: 404–20, 56 figs.
(1978) **Laštovka, P. and Máca, J.**
Keys to European species, descriptions, distribution.

On the biology and karyology of *Chymomyza costata* Zetterstedt, with reference
to the taxonomy and distribution of various species of *Chymomyza* (Dipt.,
Drosophilidae). *Ann. ent. fenn.* **36**: 1–9, 21 figs, 3 maps.
(1970) **Hackmann, W.** *et al.*

Dryomyzidae

A revision of the family Dryomyzidae (Diptera, Acalyptratae). *Pap. Mich. Acad. Sci.* **42** (1956): 55–68. (1957) **Steyskal, G. C.**

Ephydridae

The European species of *Limnellia* (Dip., Ephydridae). *Entomologica scand.* **2**: 53–9, 16 figs. (1971) **Andersson, H.**

The British species of *Ephydra* (Dip., Ephydridae). *Entomologist's mon. Mag.* **99** (1963): 147–52, 3 figs. (1964) **Collin, J. E.**
Use in conjunction with the next entry, Wirth (1975).

A revision of the brine flies of the genus *Ephydra* of the old world (Diptera, Ephydridae). *Entomologica scand.* **6**: 11–44, 58 figs. (1975) **Wirth, W. W.**

Some new species of the dipterous genus *Scatella* Dsv. and the differentiation of *Stictoscatella* gen. nov. (Ephydridae). *Entomologist's mon. Mag.* **66**: 133–9, 7 figs. (1930) **Collin, J. E.**

A contribution towards the knowledge of the male genitalia of species of *Hydrellia*. *Boll. Mus. civ. Stor. nat. Venezia* **16**: 7–18, 26 pls. (1966) **Collin, J. E.**

The British species of *Psilopa* Fln. and *Discocerina* Mcq. (Dipt., Ephydridae). *Entomologist's mon. Mag.* **79**: 145–51. (1943) **Collin, J. E.**

British *Ochthera* (Diptera, Ephydridae). *Entomologist's mon. Mag.* **121**: 151–4, 15 figs. (1985) **Irwin, A. G.**
Keys, British distribution.

Notes on Scandinavian Ephydridae (Diptera Brachycera). *Ent. Tidschr.* **95**: 186–9. (1974) **Dahl, R. G.**

Heleomyzidae

Monographie der Helomyziden (Dipteren). *Abh. zool.-bot. Ges. Wien* **15**: 1–166. (Supplements 1–10 in *Konowia*, 1926–37). (1924) **Czerny, L.**
Keys to genera and species.

The British species of Helomyzidae (Diptera). *Entomologist's mon. Mag.* **79**: 234–51. (See also *J. Soc. Br. Ent.* **4**: 37–9). (1943) **Collin, J. E.**
Trixoscelidae from southern Spain. *Entomologica scand.* **1**: 127–39, 36 figs. (1970) **Hackmann, W.**
Includes key to European species, nowadays usually regarded as a subfamily of the Heleomyzidae.

Lauxaniidae

A short synopsis of the British Sapromyzidae (Diptera). *Trans. R. ent. Soc. Lond.* **99**: 225–42, 6 figs. See also *Entomologist* **99**: 144. (1948) **Collin, J. E.**

Lonchaeidae and Pallopteridae

A revision of the British (and notes on other) species of Lonchaeidae (Diptera). *Trans. Soc. Br. Ent.* **11**: 181–207, 24 figs. (1953) **Collin, J. E.**
Use in conjunction with Hackmann and Morge (the next three entries).

The Lonchaeidae (Dipt.) of eastern Fennoscandia. *Notul. ent.* **36**: 89–115, illustrated.   (1956) **Hackmann, W.**

Monographie der palaearktischen Lonchaeidae. *Beitr. Ent.* **9**: 1–371, 909–45; **12**: 381–434, illustrated.   (1959–62) **Morge, G.**

Die Lonchaeidae und Pallopteridae Österreichs und der agrenzenden Gebiete. *Naturk. J. Stadt Linz* **1963**: 123–312; **1967**: 141–212, 241 figs.
                                                                         (1963–67) **Morge, G.**

A revision of *Neosilba* McAlpine with a key to the world genera of Lonchaeidae (Diptera). *Can. Ent.* **114**: 105–137, 54 figs.
                                      (1982) **McAlpine, J. F. and Steyskal, G. C.**

Micropezidae

British Micropezidae (Diptera). *Entomologist's Rec. J. Var.* **57**: 115–19.
                                                                      (1945) **Collin, J. E.**

Odiniidae

On the European species of the genus *Odinia* Robineau-Desvoidy (Diptera: Odiniidae). *Proc. R. ent. Soc. Lond.* **B21**: 110–16.   (1952) **Collin, J. E.**

Opomyzidae

The British species of Opomyzidae (Diptera). *Entomologist's Rec. J. Var.* **57**: 13–16. (See also Andrewes, *Entomologist's mon. Mag.* **100**: 167.)
                                                                      (1945) **Collin, J. E.**

The genus *Opomyza* (Fallén) (Dip. Opomyzidae) in Norway. *Fauna norv.* (B) **28**: 96–9, 2 figs (compound).   (1981) **Greve, L.**
Keys to Norwegian genera of family and species of genus *Opomyza*.

The Opomyzidae of eastern Fennoscandia. *Notul. ent.* **38**: 114–26, 13 figs.
                                                                   (1958) **Hackmann, W.**

Otitidae and Platystomatidae

Irish Otitidae and Platystomatidae (Diptera) including a key to the genera known in Ireland and/or Great Britain. *Ir. Nat. J.* **21**(3): 130–6.
                                      (1983) **Speight, M. C. D. and Chandler, P. J.**

Pallopteridae

Die Lonchaeidae und Pallopteridae Österreichs und der agrenzenden Gebiete. *Naturk. J. Stadt Linz* **1963**: 123–312; **1967**: 141–212, 241 figs.
                                                                         (1963–67) **Morge, G.**

Morgea freidbergi new species, a living sister-species of the fossil species *M. mcalpinei*, and a key to World genera of Pallopteridae (Diptera). *Can. Ent.* **113**(2): 81–91, 8 figs.   (1981) **McAlpine, J. F.**

Piophilidae

A revised classification of the Piophilidae, including 'Neottiophilidae' and 'Thyreophoridae' (Diptera: Schizophora). *Mem. ent. Soc. Can.* **103**: 1–66.
                                                                   (1977) **McAlpine, J. F.**
Keys to genera and some species.

A review of the Czechoslovak species of the family Piophilidae with special reference to their importance to food industry (Diptera, Acalyptrata). *Acta ent. bohemoslovaca* **62**(2): 141–57, 11 figs. (1965) **Zuska, J. and Laštovka, P.**

Psilidae

The British species of Psilidae (Diptera). *Entomologist's mon. Mag.* **80**: 214–24, 2 figs. (1944) **Collin, J. E.**

Comparative external morphology of *Psila rosae* (F.) and *P. nigricornis* Mg. (Dipt., Psilidae), third instar larvae and puparia. *Entomologist's mon. Mag.* **97**: 124–7, 4 figs (compound). (1961) **Osborne, P.**
Separates some important pests.

Sciomyzidae

Danish acalyptrate flies. 3. Sciomyzidae (Diptera). *Ent. Meddr* **34**: 61–101, 56 figs. (1965) **Knutson, L. V. and Lyneborg, L.**
Includes all genera and species known from the British Isles. See also *Entomologist's Rec. J. Var.* **78**: 227–30.

Bestimmungstabelle der mitteleuropäischen Sciomyzidae (Diptera) *Ent. Nachr. Dresden* **21**: 33–64, 71 figs. (1977) **Roskošný, R. and Jeremies, M.**
Useful genitalia illustrations include British and N. European species.

[Československé druhy malakofágní čeledi Sciomyzidae (Diptera).] *Folia prirod. fak. Univ. J. E. Purkyne v Brne*, 7, Biol., **15**(4): 1–111. [Czech].
(1966) **Rozkošný, R.**
Keys, illustrations, biology.

The subfamilies of Sciomyzidae of the World (Diptera: Acalyptratae). *Ann. ent. Soc. Am.* **58**: 593–4. (1965) **Steyskal, G. C.**
Keys to subfamilies.

Zur Morphologie und Biologie der Metamorphosestadien mitteleuropäischer Sciomyziden (Diptera). *Acta Sci. nat. Brno* **1**(4): 117–60.
(1967) **Rozkošný, R.**

[Series of well illustrated papers published independently or jointly on larvae of Holarctic Sciomyzidae in various American and British journals: refs in *Zool. Rec.*]
(1959 on) **Foote, B. A., Berg, C. O., Neff, S. E., and Knutson, L.**

Sepsidae

Sepsidae. *Fliegen palaearkt. Reg.* **5**(39a): 1–91. (1949) **Hennig, W.**
The only comprehensive work available on Palaearctic genera and species, but now rather out of date.

Monographie der Sepsiden (Dipt.). I and II. *Annln naturh. Mus. Wien* **39**: 1–153, **40**: 1–110. (1926) **Duda, O.**
Partly out of date, but still the only comprehensive review of world genera.

Notes on the Palaearctic species of the genus *Nemopoda* Robineau-Desvoidy (Diptera, Sepsididae). *Acta ent. bohemoslovaca* **62**: 308–13, 16 figs.

(1965) **Zuska, J.**

Keys to Palaearctic species.

Sepsidae. Diptera Cyclorrhapha, Acalyptrata. *Handbk. Ident. Br. Insects* **10**(5c): 1–35, 123 figs.

(1979) **Pont, A. C.**

Zur Morphologie einiger Larven der Familien Borboridae und Sepsiden (Diptera). *Mitt. zool. Mus. Berl.* **38**: 415–50, 59 figs. (1962) **Schumann, H.**

Sphaeroceridae

Lesser dung-flies (Diptera: Sphaeroceridae). *Handbk Ident. Br. Insects* **10**(5e).

(1987) **Pitkin, B. R.**

Systematics of *Crumomyia* Macquart and *Alloborborus* Duda (Diptera: Sphaeroceridae). *Syst. Ent.* **10**: 167–225, 153 figs.

(1985) **Norrbom, A. L. and Kim, K. C.**

Keys Holarctic genera of Copromyzinae.

Taxonomy and phylogenetic relationships of *Copromyza* Fallén (s.s.) (Diptera: Sphaeroceridae). *Ann. ent. Soc. Am.* **78**(3): 331–47, 57 figs.

(1985) **Norrbom, A. L. and Kim, K. C.**

Keys world species.

A contribution to the knowledge of species of the subfamily Ceropterinae (Diptera: Sphaeroceridae). *Acta zool. hung.* **23**: 371–85, 9 figs.

(1977) **Papp, L.**

Keys to genera of Ceropterinae.

Revision of the subgenus *Leptocera* (s.str.) of Europe (Diptera, Sphaeroceridae). *Ent. Abh. Mus. Tierk. Dresden* **46**(1): 1–44.   (1982) **Roháček, J.**
Keys to European species.

A monograph and reclassification of the previous genus *Limosina* Macquart (Diptera, Sphaeroceridae) of Europe. *Beitr. Ent.* **32**: 195–282; **33**: 3–195.

(1982–83) **Roháček, J.**

Includes keys to subfamilies of Sphaeroceridae and all genera and most species of Limosininae.

Zur Morphologie einiger Larven der Familien Borboridae und Sepsiden (Diptera). *Mitt. zool. Mus. Berl.* **38**: 415–50, 59 figs. (1962) **Schumann, H.**

Description of the puparia of twenty-three British species of Sphaeroceridae (Diptera, Acalyptratae). *Trans. R. ent. Soc. Lond.* **126**(1): 41–56, 61 figs.

(1974) **Okely, E. F.**

Keys to genera and species.

Tephritidae

Trypetidae. *Fliegen palaearkt. Reg.* **5**(49): 1–221.   (1927) **Hendel, F.**
Tephritid flies (Diptera: Tephritidae). *Handbk Ident. Br. Insects* **10**(5a).

(1987) **White, I. M.**

The biology and identification of Trypetid larvae (Diptera: Trypetidae). *Mem. Am. ent. Soc.* **12**: 1–161, 192 figs.   (1946) **Phillips, V. T.**

Tethinidae

British Tethinidae (Diptera). *Entomologist* **93**: 191–3. See also 1966, *Boll. Mus. civ. Stor. nat. Venezia* **16**: 19–32.   (1960) **Collin, J. E.**

CYCLORRHAPHA—CALYPTRATAE

British species nearly 1000.

Oestridae

Hypodermatidae. *Fliegen palaearkt. Reg.* **8**(64b): 1–160. (1965) **Grunin, K. J.**
Oestridae. *Fliegen palaearkt. Reg.* **8**(64a): 1–96.   (1966) **Grunin, K. J.**

Tachinidae, Rhinophoridae, Sarcophagidae, Calliphoridae

Catalogue of Palaearctic Tachinidae (Diptera). *Stuttg. Beitr. Naturk.* (A, Biologie) **369**: 1–228.   (1984) **Herting, B.**
Catalogues Palaearctic species with full synonymy and bibliography of taxonomic literature. Essential for using older works on this family.

Keys to the British Tachinidae. Buncle, Arbroath. 150 pp, 217 figs. Reprinted from *N. West. Nat.* (1946–47).   (1948) **Day, C.**

Tachinidae and Calliphoridae. *Handbk Ident. Br. Insects* **10**(4a): 1–133, 42 figs (compound).   (1954) **Emden, F. I. van**
Nomenclature of the above two works is now out of date and they should be used in conjunction with Mesnil in Lindner's *Die Fliegen* (1925 on) p. 130 and papers by Herting and Mesnil (1961 on) for which see *Zool. Rec.* Also see Herting (1984), this section, for Tachinidae.

The British Tachinidae of Walker and Stephens (Diptera). *Bull. Br. Mus. nat. Hist.* (Ent.) **30**(5): 269–308.   (1974) **Crosskey, R. W.**
Clarifies many of the early names used in this family.

A key to the females of the British species of *Sarcophaga* (Dipt., Calliphoridae). *J. Soc. Br. Ent.* **5**: 119–23.   (1955) **Day, C. D. and Fonseca, E. C. M. d'A.**

Contribution to the knowledge of the genus *Pollenia* R.-D. (Diptera: Calliphoridae). *Acta zool. hung.* **22**: 327–33, 18 figs.   (1976) **Mihalyi, F.**

The British species of the subfamily Sarcophaginae, with illustrations of the male and female terminalia. *Ann. trop. Med. Parasit.* **29**: 73–90, 517–32; **30**: 187–201, 337–50; **31**: 303–31, illustrated.
                    (1935–37) **Patton, W. S. and Wainwright, C. J.**

Sarcophagidae. *Fauna SSSR* (19)**1**: 1–500, 535 figs. (1937) **Rohdendorf, B. B.**

Revision der Gattung *Onesia* Robineau-Desvoidy, 1830 (Dipt., Calliphoridae). *Beitr. Ent.* **14**: 915–38, 28 figs.   (1964) **Schumann, H.**

Merblätter über angewandte Parasitenkunde und Schädlingsbekämpfung.
18. Die Gattung *Lucilia* (Goldfliegen). *Angew. Parasit.* **12**(4): 1–20, 20 figs.
(1971) **Schumann, H.**

Revision der palaearktischen *Melinda*-Arten (Diptera: Calliphoridae). *Dt. ent. Z.* **20**: 293–314, 13 figs. (1973) **Schumann, H.**

Bemerkungen zum Status der Gattungen *Onesia*, *Melinda* und *Bellardia* (Diptera, Calliphoridae). *Mitt. zool. Mus. Berl.* **49**: 333–44, 6 figs.
(1973) **Schumann, H.**

Revision der palaearktischen *Bellardia*-Arten (Diptera, Calliphoridae). *Dt. ent. Z.* **21**: 231–99, 82 figs. (1974) **Schumann, H.**

A taxonomic study of the females of the British *Lucilia* species (Diptera: Calliphoridae). *Proc. R. ent. Soc. Lond.* **B23**: 29–55, 2 figs (compound).
(1954) **Spence, T.**
Provides the first key to females of *Lucilia*.

Rhinophorinae. *Fliegen palaearkt. Reg.* **9**(64a): 1–36.   (1961) **Herting, B.**
Keys to Palaearctic genera and species, but completely unillustrated.

Études sur les mouches parasites. 2, Calliphoridae. Calliphorines (suite), Sarcophaginae et Rhinophorinae de l'Europe occidentale et méridionale. Recherches sur la morphologie et la distribution géographique des Diptères à larves parasites. *Encycl. ent.* (A) **21**: 1–436.   (1941) **Séguy, E.**
Complements Herting (1961) above by useful illustrations. *See also* Herting for later nomenclature.

The immature stages of Rhinophorinae (Diptera: Calliphoridae) that parasitise British woodlice. *Trans. R. ent. Soc. Lond.* **125**(1): 27–44, 63 figs.
(1973) **Bedding, R.**

The blow-fly genus *Lucilia* Robineau-Desvoidy (Diptera, Calliphoridae) in Norway. *Fauna norvegica* (B) **27**: 39–52.   (1980) **Rognes, K.**

Scathophagidae

A short synopsis of the British Scatophagidae (Diptera). *Trans. Soc. Br. Ent.* **13**: 37–56.   (1958) **Collin, J. E.**

The Scatophagidae (Dipt.) of eastern Fennoscandia. *Fauna fenn.* **2**: 1–67.
(1956) **Hackmann, W.**

Revision of the North European species of *Cosmetopus* Becker (Dipt., Scatophagidae). *Entomologica scand.* **5**: 95–102, 23 figs.   (1971) **Andersson, H.**

Les resultats de l'expedition entomologique Tchecoslovaque–Iranienne à l'Iran en 1973. Diptera: Scatophagidae (Avec la description d'une espèce nouvelle du genre *Coniosternum* d'Albanie et avec le tableau analytique des espèces du genre *Coniosternum* Becker, 1894 paléarctiques). *Acta ent. Mus. natn. Prague* **40**: 95–104.   (1981) **Šifner, F.**
Keys Palaearctic *Coniosternum*.

Anthomyiidae

There is no comprehensive work on the British fauna but the following works
will be found useful. Those of Collin should only be used in conjunction with
Hennig because of outdated nomenclature.

Anthomyiidae. *Fliegen palaearkt. Reg.* **7**(63a): 1–974.   (1966–76) **Hennig, W.**
Comprehensive monograph of Palaearctic genera and species.

Two new British species of Anthomyiidae (Dipt.) with taxonomic notes on
related pests of conifers. *Entomologist's mon. Mag.* **100**: 136–44.
(1964) **Ackland, D. M.**

Notes on the palaearctic species of *Egle* R.-D. (Dipt., Anthomyiidae) with
descriptions of two new species. *Entomologist's mon. Mag.* **105**: 185–92.
(1970) **Ackland, D. M.**

The British species of *Prosalpia* Pok. (Dipt., Anthomyiidae). *Entomologist's mon.
Mag.* **79**: 83–6. See also *J. Soc. Br. Ent.* **4**: 40–1, 169–77; **5**: 94–100; *Proc. R.
ent. Soc. Lond.* **B23**: 95–102; *Entomologist* **96**: 277–83.   (1943) **Collin, J. E.**

The genus *Chiastochaeta* Pokorny (Diptera, Anthomyiidae). *Proc. R. ent. Soc.
Lond.* **B23**: 95–102, 3 figs (compound).   (1954) **Collin, J. E.**

Genera and species of Anthomyiidae allied to *Chirosia* (Diptera). *J. Soc. Br. Ent.*
5: 94–100.   (1955) **Collin, J. E.**

Notes on some British species of *Pegohylemyia* (Dipt., Anthomyiidae) with
descriptions of four new species. *Entomologist's mon. Mag.* **102** (1966): 181–
91, 4 figs (compound).   (1967) **Collin, J. E.**

A survey of the Swedish species of *Hydrophoria* and *Acroptena* by O. Ringdahl. *J.
Soc. Br. Ent.* **4**: 75–82.   (1952) **Fonseca, E. C. M. d'A.**
Translated from the Swedish and keys modified to include all British species.

Translations of Ringdahl's Muscid Tables, additions and corrections. *J. Soc.
Br. Ent.* **5**(1): 17–22.   (1954) **Fonseca, E. C. M. d'A.**

A review of the British sub-families and genera of the family Muscidae. *Trans.
Soc. Br. Ent.* **12**: 113–28.   (1956) **Fonseca, E. C. M. d'A.**
Includes key to genera of Anthomyiidae.

Eight undescribed species of Muscidae (Diptera) from Britain. *Entomologist's
mon. Mag.* **101**: 269–78, 7 figs (compound). (1966) **Fonseca, E. C. M. d'A.**

Eustalomyia hilaris Fallén (Diptera: Anthomyiidae) confirmed as British, with
notes on other species in the genus. *Entomologist's Gaz.* **22**: 55–60, 8 figs.
(1971) **Smith, K. G. V.**
Keys, biology, distribution.

The *Anthomyia pluvialis* complex in Europe (Diptera, Anthomyiidae) *Syst. Ent.*
5: 281–90.   (1980) **Michelsen, V.**
Keys to European species.

A revision of the beet leaf-miner complex, *Pegomya hyoscami* s. lat. (Diptera:
Anthomyiidae). *Entomologica scand.* **11**: 297–309.   (1980) **Michelsen, V.**

Identification of the root maggots (Diptera: Anthomyiidae) attacking cruciferous garden crops in Canada, with notes on biology and control. *Can. Ent.* **83**: 109–20, 32 figs. (1951) **Brooks, A. R.**
Useful for some pest species that also occur in Europe.

Muscidae (including Fanniidae)

Muscidae. *Handbk Ident. Br. Insects* **10**(4b): 1–119, 16 figs (compound), 6 pls. (1968) **Fonseca, E. C. M. d'A.**

Myospila hennigi Gregor and Povolny, 1958 (Dipt., Muscidae), new to Britain, and notes on the European species of *Myospila* Rondani, 1856. *Entomologist's mon. Mag.* **106**: 111–13. (1970) **Pont, A. C.**

Fannia lineata (Stein, 1895), new to Britain (Diptera, Fanniidae). *Entomologist's mon. Mag.* **119**: 229–31. (1983) **Pont, A. C.**
Includes revised couplets to Fonseca (1968) above.

Phaonia mediterranea Hennig (Dipt., Muscidae), new to Britain. *Entomologist's mon. Mag.* **108**(1972): 238–9. (1973) **Pont, A. C.**
Includes revised couplets to Fonseca (1968) above.

The biology of the Muscidae of the world. *Séries Ent.* **29**: 1–550, 160 figs (compound). (1985) **Skidmore, P.**
Keys to larvae, descriptions, biology.

Zur Larvalsystematik der Muscinae nebst Beschreibung einiger Musciden- und Anthomyidenlarven. *Dt. ent. Z.* **10**: 134–51, 12 pls. (1963) **Schumann, H.**

Taxonomy of European *Fannia* larvae (Diptera, Fanniidae). *Stuttg. Beitr. Naturk.* **215**: 1–28, 20 figs. (1970) **Lyneborg, L.**

PUPIPARA

Keds, flat-flies and bat-flies (Diptera, Hippoboscidae and Nycteribiidae). *Handbk Ident. Br. Insects* **10**(7): 1–40, 40 figs. (1984) **Hutson, A. M.**

An illustrated catalogue of the Nycteribiidae (Diptera Pupipara) in the Rothschild Collection and the British Museum (Natural History). Cambridge University Press, Cambridge. 506 pp, 898 figs, 5 pls, 6 maps. (1966) **Theodor, O.**

An annotated bibliography of British Nycteribiidae. *Angew. Parasit.* **13**(2): 106–13. (1972) **Thompson, G. B.**

An annotated bibliography of batflies (Diptera: Streblidae; Nycteribiidae). *Pacif. Insects Monogr.* **28**: 119–211. (1971) **Maa, T. C.**

SIPHONAPTERA

British species about 57.

Katalog der palaearktischen Aphanipteren. F. Wagner, Vienna. 55 pp. Supplements
by same author in *Konowia* **10**: 96–100 (1931); **12**: 212–16 (1933); **14**: 217–
24 (1935); **17**: 8–18 (1938). See also Ioff, I. G. and Rostigayev, B. A. (1950)
Mater. Pozn. Fauny Flory SSSR (N.S.) Zool. **15**: 166–87. [Russian].

(1930) **Wagner, J.**

Siphonaptera. *Handbk Ident. Br. Insects* **1**(16): 1–94, 200 figs.

(1957) **Smit, F. G. A. M.**

*An illustrated catalogue of the Rothschild collection of fleas (Siphonaptera) in the British
Museum (Natural History).* 5 vols. British Museum (Natural History),
London. 1, Tungidae, Pulicidae xv + 361 pp (1953); 2, Coptopsyllidae,
Vermipsyllidae, Stephanocircidae (Macropsyllidae), Ischnopsyllidae,
Hypsophthalmidae (=Chimaeropsyllidae), and Xiphiopsyllidae, xi + 445
pp (1956); 3, Hystrichopsyllidae, viii + 560 pp (1962); 4, Hystrichopsyl-
lidae, viii + 549 pp (1966); 5, Leptopsyllidae, Ancistropsyllidae, viii + 530
pp (1971). (1953–71) **Hopkins, G. H. E. and Rothschild, M.**

The distribution and general ecology of the Irish Siphonaptera. *Proc. R. Ir.
Acad.* **64B**(23): 413–63, 2 maps.

(1966) **Claasens, A. J. M. and O'Rourke, F. J.**

The recorded distribution and hosts of Siphonaptera in Britain. *Entomologist's
Gaz.* **8**(1): 45–75. (1957) **Smit, F. G. A. M.**

Catalogue provisoire des Siphonaptères de la faune française. *Ann. Soc. ent. Fr.*
(N.S.) **4**(3): 615–35. (1968) **Beaucournu, J. C.**

Additions au catalogue provisoire des Siphonaptères de la faune française.
Bull. Soc. ent. Fr. **76**(1–2): 46–8. (1971) **Beaucournu, J. C. and Gilot, B.**

Catalogus der Nederlandse Siphonaptera. *Tijdschr. Ent.* **105**(3): 45–96, 8 figs.

(1962) **Smit, F. G. A. M.**

De vlooien (Siphonaptera) van de Benelux-landen. *Wet. Meded. K. ned. natuurh.
Veren.* **72**: 1–48, 170 figs. (1968) **Smit, F. G. A. M.**

Lopper. *Danm. Fauna* **60**: 1–125, 182 figs. (1954) **Smit, F. G. A. M.**

The Kemner collection of Siphonaptera in the Entomological Museum, Lund,
with a check-list of the fleas of Sweden. *Entomologica scand.* **2**(4): 269–86, 13
figs. (1971) **Brinck-Lindroth, G. and Smit, F. G. A. M.**

A catalogue of the Siphonaptera of Finland with distribution maps of all
Fennoscandian species. *Ann. Zool. fenn.* **6**(1): 47–86, 60 figs, 1 pl.

(1969) **Smit, F. G. A. M.**

Zur Kenntnis der Flöhe Deutschlands. I. Zur Taxonomie der Vogelflöhe
(Insecta, Siphonaptera). *Dt. Ent. Z.* (N.F.) **14**(I–II): 81–108, 25 figs.

(1967) **Peus, F.**

Zur Kenntnis der Flöhe Deutschlands. II. Faunistik und Ökologie der
Vogelflöhe (Insecta, Siphonaptera). *Zool. Jb. Syst.* **95**(4): 571–633.

(1968) **Peus, F.**

Zur Kenntnis der Flöhe Deutschlands (Insecta, Siphonaptera). III. Faunistik und Ökologie der Säugetierflöhe. Insectivora, Lagomorpha, Rodentia. *Zool. Jb. Syst.* **97**: 1–54. (1970) **Peus, F.**

Zur Kenntnis der Flöhe Deutschlands (Schluss) (Insecta, Siphonaptera). IV. Faunistik und Ökologie der Säugertierflöhe. *Zool. Jb. Syst.* **99**(3): 408–504, 2 maps. (1972) **Peus, F.**

[Order Siphonaptera (Aphaniptera, Suctoria)—Fleas.] In [Keys to the insects of the European part of the USSR (ed. G. Ya Bei-Bienko).] *Opred. Faune SSSR* **103**: 799–844. [Russian]. (1970) **Scanlon, O. I.**

Siphonaptera (Fleas). In *Insects and other Arthropods of medical importance* (ed. K. G. V. Smith), pp. 325–71. British Museum (Natural History), London. (1973) **Smit, F. G. A. M.**

LEPIDOPTERA

British species about 2500. Traditionally these have been divided into Rhopalocera (butterflies) and Heterocera (moths), or into Macrolepidoptera for the larger species and Microlepidoptera for the smaller species. A modern classification into suborders and superfamilies blurs these divisions, but so many works have been written using the older groupings that for the purposes of this bibliography the old names Rhopalocera, Macro-heterocera (larger moths), and Microlepidoptera have been retained.

General

Includes economic, immature stages and structure, where appropriate, also those works covering more than one major subgroup.

A revised handbook of the British Lepidoptera, 2nd edn. Watkins and Doncaster, London. 914 pp. (1928) **Meyrick, E.**

The Lepidoptera of the British Isles. Lovell Reeve, London. 11 vols, about 300 pp per vol., 469 pls. (1892–1907) **Barrett, C. G.**

A natural history of the British Lepidoptera. Swann, Sonnenschein, London. Vols 1–5 and 8–10, 3648 pp. (1899–1909) **Tutt, J. W.**

A check list of British insects, part 2: Lepidoptera, 2nd edn. (Revised by Bradley, J. D., Fletcher, D. S., and Whalley, P. E. S.) *Handbk Ident. Br. Insects* **11**(2): 1–153. (1972) **Kloet, G. S. and Hincks, W. D.**

Addenda and corrigenda to the Lepidoptera part of Kloet and Hincks Check List of British Insects (2nd edn). *Entomologist's Gaz.* **25**: 219–22. (1972) **Bradley, J. D., Fletcher, D. S., and Whalley, P. E. S.**

A recorder's log book or label list of British butterflies and moths. Curwen Books, London. 136 pp. (1979) **Bradley, J. D. and Fletcher, D. S.**

A recorder's log book or label list of British butterflies and moths. Index. Leicestershire Museums Service, Leicester. 59 pp. (1983) **Hall-Smith, D. H.**

Macrolepidoptera of the world. A. Kernen, Stuttgart. Vols 1–4 and suppls Rhopalocera and Macro-Heterocera—Fauna Palaearctica, numerous pls. (1906–54) **Seitz, A.**

Liste Systématique et Synonymique des Lépidoptères de France, Belgique et Corse. *Alexanor (Suppl.):* 1–334. (1980) **Leraut, P.**

Catalogue des Lépidoptères de France et de Belgique. Le Carriol, Lot. 2 vols, 800 and 696 pp. (1923–49) **Lhomme, L.**

Catalogus der Nederlandske Macrolepidoptera. *Tijdschr. Ent.* **79–112**, suppls 1–16, illustr. (1936–70) **Lempke, B. J.**

Naamlijst van de Nederlandse Lepidoptera. K. Nederl Natuurh. Ver., Amsterdam. 99 pp. (1976) **Lempke, B. J.**

Katalog over de danske Sommerfugle. *Ent. Meddr* **52** (2–3): 1–163. (1985) **Karsholdt, O.** *et al.*

Wir bestimmen Schmetterlinge. Neumann–Neudamm, Leipzig. 792 pp, 60 pls, numerous text figs. (1984) **Koch, M.**
Useful general account with tabulated information and colour illustrations of many species—nomenclature rather out of date.

Die Schmetterlinge Mitteleuropas. Franckh'sche, Stuttgart. 5 vols, 1161 pp. (1954–81) **Forster, W. and Wohlfahrt, T. A.**

Die Grosschmetterlinge Mittel-deutschlands. Urania, Jena. 5 vols, about 4000 pp. (1951–55) **Bergmann, A.**
Illustrations of adults and habitats.

Svenska Fjärilar. Systematisk bearbetning av Sveriges Storfjärilar, Macrolepidoptera. Nordisk Familjeboks, Stockholm. 440 pp, 1800 illustr, 12 maps. [Swedish]. (1935–41) **Nordström, F., Wahlgren, E., and Tullgren, A.**

Catalogue of the Lepidoptera of Norway. Norsk Lepidopterologisk Selskap, Oslo. 3 vols, 89 pp, figs, maps. (1958–72) **Opheim, M.**

A field guide to caterpillars of butterflies and moths in Britain and Europe. Collins, London. 296 pp, 38 pls. (1986) **Carter, D. J. and Hargreaves, B.**
General account with illustrations of larvae on their foodplants.

The genitalia of the British Rhopalocera and the larger moths. Pierce, Oundle. 66 pp, 21 pls. (1941) **Pierce, F. N. and Beirne, B. P.**
Reprinted (1968) by Classey, Feltham.

The larvae of the British butterflies and moths (ed. H. T. Stainton and G. T. Porritt). Ray Society, London. 9 vols, illustrated in col. (1886–1901) **Buckler, W.**

Larval food plants. Watkins and Doncaster, London. 126 pp. (1949) **Allen, P. B. M.**

Die Raupen der Schmetterlinge Europas. Schweizerbart'sche, Stuttgart. Vol. 4, 60 col. pls. (1910) **Spuler, A.**
Over 2200 excellent coloured pictures of larvae, pupae, and some ova.

Pest Lepidoptera of Europe. *Series Ent.* **31**: 1–431, 79 figs, 40 maps, 41 pls. (1984) **Carter, D. J.**
General account with emphasis on immature stages.

Lépidoptères (Macrolepidoptera and Pyraloidea). *Entomologie appliqué à l'agriculture.* Masson et Cie, Paris. Vol. 2, 575 pp, 480 figs. (1972) **Balachowsky, A. S.**

Systematic

RHOPALOCERA (Butterflies: superfamilies Papilionoidea and Hesperioidea)

The complete book of British butterflies. Ward Lock, London. 384 pp, 160 + figs, 32 col. pls. (1934) **Frohawk, F. W.**

South's British butterflies. Warne, London. 210 pp, 48 col. pls. (1973) **Howarth, T. G.**

Colour identification guide to butterflies of the British Isles, revised edn. Viking, Harmondsworth. 46 pp, 48 col. pls. (1984) **Howarth, T. G.**

A complete guide to British butterflies. Cape, London. 159 pp, many col. figs. (1982) **Brooks, M. and Knight, C.**
Basic text but with good colour photographs of life histories.

A field guide to the butterflies of Britain and Europe. Collins, London. 380 pp, 60 col. pls, maps. (1970) **Higgins, L. G. and Riley, N. D.**
Republished in several other European languages.

Classification of European butterflies. Collins, London. 320 pp, 402 figs. (1976) **Higgins, L. G.**

Concise bibliography of European butterflies. *Butterflies of Europe.* Aula-Verlag, Wiesbaden. Vol. 1, 447 pp. (1985) **Kudrna, O.**

The butterflies of Scandinavia in nature. Skandinavisk Bogforlag, Odense. 215 pp, 102 pls. (1982) **Henriksen, H. J. and Kreuzer, I. B.**
Good general account well illustrated with colour photographs of habitats and life histories.

De fennoskandiska dagfjärilarnas utbredning. Lepidoptera diurna (Rhopalocera och Hesperoidea). *Acta Univ. Lund.* (NF.) **51**(1): 176 pp, 119 maps. (1955) **Nordström, F.**

A guide to the genera and species of Parnassiinae (Lepidoptera, Papilionidae). *Bull. Br. Mus. nat. Hist.* (Ent.) **31**: 71–105, 32 figs, 16 pls. (1975) **Ackery, P. R.**

The caterpillars of the British butterflies. Warne, London. 288 pp, 32 pls. (1944) **Stokoe, W. J. and Stovin, G. H. T.**

Die Raupen der europäischen Tagfalter. Sciences Nat., Venette. 47 pp, 19 pls.
(1985) **Bodi, E.**
Excellent colour photographs of larvae.

MACRO-HETEROCERA

Larger moths; main superfamilies are Sphingoidea, Bombycoidea, Noctuoidea and Geometroidea, but Hepialoidea, Cossoidea, Zygaenoidea, and the family Sesiidae, although related to Microlepidoptera, have traditionally been placed here.

The moths of the British Isles (revised edn. by H. M. Edelsten, D. S. Fletcher, and R. J. Collins). Warne, London. Vol. 1, 427 pp, 148 pls; Vol. 2, 379 pp, 141 pls. (1961) **South, R.**

Colour identification guide to moths of the British Isles. Viking, Harmondsworth. 267 pp, 57 figs, 42 pls. (1984) **Skinner, B.**
Excellent colour plates of British Macrolepidoptera.

Sphingidae-Noctuidae (Part 1). *The moths and butterflies of Great Britain and Ireland*. Curwen, London. Vol. 9, 288 pp, 19 figs, 13 pls, 203 maps.
(1979) **Heath, J.** *et al.*

Noctuidae (Part 2) and Agaristidae. *The moths and butterflies of Great Britain and Ireland*. Harley Books, Colchester. Vol. 10, 459 pp, 23 figs, 13 pls, 225 maps.
(1983) **Heath, J.** *et al.*

An identification guide to the British pugs Lepidoptera: Geometridae. British Entomological and Natural History Society, London. 42 pp, 16 pls.
(1981) **Agassiz, D.** *et al.*

Noctuelles et géomètres d'Europe. Meyrin, Geneva. 4 vols, about 1000 pp, 81 + 70 pls. (1910–20) **Culot, J.**

Nordens Eupithecier. Gravers Andersen, Aarhus. 147 pp, figs, maps, 13 pls. [Danish: English diagnoses]. (1948) **Juul, K.**

De Danske Spindere. Notodontidae, etc. Universitets forlaget, Aarhus. 270 pp, 24 col. pls. (1960) **Hoffmeyer, S.**

Nordens Malère. *Danmarks Dyreliv*. Faunabøger, Copenhagen. Vol. 2, 332 pp, 358 figs, 24 pls. (1984) **Skou, P.**
Good, well illustrated account of Scandinavian Drepanidae and Geometridae.

Danske Natsommerfugle. *Dansk. faun. Bibliot.* **1**: 272 pp, 6 pls, numerous maps and text figs. (1981) **Fibiger, M. and Svendson, P.**
General account of Danish Macrolepidoptera with distribution maps.

De Danske Ugler. Noctuidae. Universitets forlaget, Aarhus. 387 pp, 33 col. pls.
(1962) **Hoffmeyer, S.**

De Danske Malere. Geometridae, 2nd edn. Universitets forlaget, Aarhus. 361 pp, 25 col. pls. (1966) **Hoffmeyer, S.**

De fennoskandiska svärmarnas och spinnarnas utbredning. (Sphinges, Bombycimorpha etc). *Acta Univ. Lund.* (N.F.) **57**(4): 87 pp, 181 maps.
(1961) **Nordström, F.** *et al.*

Suomen Perhoset. Yökköset 1 and 2. Otava, Helsinki. Vol. 1, 256 pp, 18 pls; Vol. 2, 304 pp, 17 pls; both with numerous maps and text figs.
(1977–79) **Mikkola, K. and Jalas, I.**
These two volumes cover Finnish Noctuidae but are very useful for Scandinavian species in general with useful sketches of genitalia.

Guides des papillons nocturnes d'Europe et d'Afrique du nord. Héterocères (Partim). Delachaux et Niestlé, Paris. 228 pp, 40 pls.
(1978) **Rougeot, P.-C. and Viette, P.**
Useful account of Sphingidae, Lasiocampidae, Saturniidae, and Notodontidae. German language edition (1983) (Bauer).

The Sesiidae (Lepidoptera) of Fennoscandia and Denmark. *Fauna ent. Scand.* **2**: 91 pp, 144 figs, 1 map, col. pls. (1974) **Fibiger, M. and Kristensen, N. P.**

The genitalia of the group Noctuidae of the Lepidoptera of the British Isles. Duncan, Liverpool. 88 pp, 32 pls. Reprinted (1967) by Classey, Feltham.
(1909) **Pierce, F. N.**

The genitalia of the group Geometridae of the Lepidoptera of the British Isles. Pierce, Liverpool. 88 pp, 48 pls. Reprinted 1968 by Classey, Feltham.
(1914) **Pierce, F. N.**

The genitalia of the group Noctuidae of the Lepidoptera of the British Islands. An account of the morphology of the female reproductive organs, 2nd edn. Classey, Feltham. 64 pp, 15 pls.
(1952) **Pierce, F. N.**

The entomologist's logbook, and dictionary of the life histories and food plants of the British Macrolepidoptera. Routledge, London. 374 pp.
(1913) **Scorer, A. G.**

The caterpillars of the British moths. Warne, London. Vol. 1, 408 pp, 90 pls; Vol. 2, 380 pp, 51 pls.
(1949) **Stokoe, W. J. and Stovin, G. H. T.**

Die Larvalsystematik der Eulen. Noctuidae. *Abh. Larvalsyst. Insekten.* **4**: 406 pp, 488 figs.
(1960) **Beck, H.**

MICROLEPIDOPTERA

Smaller moths; main superfamilies, some of which are often divided in recent classifications: Micropterigoidea, Eriocranioidea, Nepticuloidea, Incurvarioidea, Copromorphoidea, Tineoidea, Gelechioidea, Tortricoidea, Pyraloidea—see also note at head of the preceding section, Macro-Heterocera.

Micropterigidae-Heliozidae. *The moths and butterflies of Great Britain and Ireland.* Curwen and Blackwell, London. Vol. 1, 343 pp, 85 figs, 152 maps, 13 pls. Vol. 2, Cossidae-Heliodinidae. Harley Books, Colchester. 460 pp, 123 figs, 14 pls, 223 maps.
(1976 and 1985) **Heath, J.** *et al.*

Illustrated papers on British Microlepidoptera. British Entomological and Natural History Society, London. 170 pp, 12 pls.   (1978) **Jacobs, S. N. A.** *et al.*

Klein-Schmetterlinge oder Micro-lepidoptera. I. Die Wickler (s. str.) (Tortricidae). *Tierwelt Dtl.* **48**: 233 pp, many figs, 22 pls. (1961) **Hannemann, H. J.**

Klein-Schmetterlinge oder Micro-lepidoptera. II. Cochylidae, Carposinidae, Pyraloidea. *Tierwelt Dtl.* **50**: 401 pp, many figs, 22 pls.

(1964) **Hannemann, H. J.**

Kleinschmetterlinge oder Microlepidoptera III. Federmotte (Pterophoridae), Gespinstmotten (Yponomeutidae), Echte Motten (Tineidae). *Tierwelt Dtl.* **63**: 273 pp, 17 pls.   (1977) **Hannemann, H. J.**

A field guide to the smaller British Lepidoptera. British Entomological and Natural History Society, London. 271 pp.   (1979) **Emmet, A. M.**
Useful tabulations of biological data but no illustrations.

Part 1 (Microlepidoptera). *Entomologie appliqué à l'agriculture. Vol. 2. Lépidoptères.* Masson et Cie, Paris. 1057 pp, 373 figs, 4 pls. (1966) **Balachowsky, A. S.**

Die Larvalsystematik einiger Kleinschmetterlingefamilien. *Abh. Larvalsyst. Insekten* **2**: 145 pp, 212 figs.   (1958) **Werner, K.**

Keys for the identification of the Lepidoptera infesting stored food products. *Proc. zool. Soc. Lond.* **113**: 55–148, 287 figs, 5 pls.

(1943) **Corbet, A. S. and Tams, W. H. T.**
Descriptions and illustrations of superficial appearance, wing venation, and genitalia structure. Distribution, larval foods.

The larvae of the Lepidoptera associated with stored products. *Bull. ent. Res.* **34**: 163–212, 128 figs.   (1943) **Hinton, H. E.**
Morphology and chaetotaxy.

Clothes moths and carpet beetles, revised edn. Economic Ser. no. 14. British Museum (Natural History), London. 15 pp, 9 figs.

(1967) **Whalley, P. E. S.** *et al.*

Common insect pests of stored food products, 6th edn. Economic Ser. no. 15. British Museum (Natural History), London. 69 pp, 130 figs.

(1980) **Freeman, P.**, ed.
Keys to identification of adults and larvae.

NEPTICULOIDEA, TINEOIDEA, GELECHIOIDEA

The male genitalia of the British Stigmellidae. *Proc. R. Ir. Acad.* **50**: 191–218, 81 figs.   (1945) **Beirne, B. P.**

The natural history of the Tineina. J. van Voorst, London. 13 vols, col. pls.

(1855–73) **Stainton, H. T.**

Beiträge zur Insekten Fauna der DDR: Lepidoptera—Acrolepiidae. *Beitr. Ent.* **20**: 209–22, 24 figs, 2 col. pls.   (1970) **Gaedike, R.**

Beiträge zur Insekten Fauna der DDR: Lepidoptera—Argyresthiidae. *Beitr. Ent.* **19**: 693–752, 34 figs, 2 col. pls.   (1969) **Friese, G.**

Glyphipterigidae. *Microlepidoptera Palaearctica.* Braun, Karlsruhe. Vol. 7 (2 parts), 436 pp, 26 figs, 175 pls.   (1986) **Diakonoff, A.**

Key to the British and French species of *Phyllonorycter* Hübner (Lep., Gracillariidae). *Entomologist's Gaz.* **29**: 3–33, figs.   (1969) **Bradley, J. D.** *et al.*

Beiträge zur Insekten Fauna der DDR: Lepidoptera—Tineidae. *Beitr. Ent.* **19**: 311–88, 205 figs, 2 col. pls.   (1969) **Petersen, G.**

Die Genitalien der paläarktischen Tineiden (Lepidoptera, Tineidae). *Beitr. Ent.* **7**: 55–176, 338–79, 557–95, 247 figs, 6 pls; **8**: 111–18, 398–430, 20 figs, 2 pls.   (1957–58) **Petersen, G.**

Lepidoptera. Tineidae Part 2 Nemapogoninae. *Fauna SSSR* **4**(2). Translated by Israel program for scientific translations, Jerusalem. 436 pp, 385 figs.
                    (1968) **Zagulyaev, A. K.**

True moths (Tineidae), part 3 subfamily Tineinae. *Fauna of the USSR Lepidoptera.* Indian National Scientific Documentation Centre, New Delhi. Vol. 4, No. 3, 443 pp, 231 figs.   (1975) **Zagulyaev, A. K.**
English translation of *Fauna SSSR* **78**: 1–267 (1960).

The genitalia of the tineid families of the Lepidoptera of the British Isles. Pierce, Oundle. 116 pp, 58 pls.   (1935) **Pierce, F. N. and Metcalfe, J. W.**

The larvae of the species of Tineidae of economic importance. *Bull. ent. Res.* **47**: 251–346, 216 figs.   (1955) **Hinton, H. E.**

Beiträge zur Insekten Fauna der DDR: Lepidoptera—Coleophoridae. *Beitr. Ent.* **24**: 153–278, 363 figs, 2 col. pls.   (1974) **Patzak, H.**

The Elachistidae (Lepidoptera) of Fennoscandia and Denmark. *Fauna Ent. Scand.* **6**: 299 pp, genitalia figs, col. pls.
                    (1977) **Traugott-Olsen, E. and Schmidt Nielsen, E.**

Ethmiidae. *Microlepidoptera Palaearctica.* Fromme, Vienna. Vol. 2 (2 parts), 185 pp, 106 pls.   (1967) **Sattler, K.**

Beiträge zur Insekten Fauna der DDR: Lepidoptera—Ethmiidae. *Beitr. Ent.* **23**: 291–312, 27 figs, 2 col. pls.   (1973) **Friese, G.**

Lecithoceridae. *Microlepidoptera Palaearctica.* Fromme, Vienna. Vol. 5 (2 parts), 306 pp, 168 figs, 93 pls.   (1978) **Gozmány, L.**

Natürliche Gruppierung der Europäischen Arten der Gattung *Depressaria* s.l. (Lep., Oecophoridae). *Mitt. zool. Mus. Berl.* **29**: 269–373, 130 figs.
                    (1953) **Hannemann, H. J.**

The Scythrididae (Lepidoptera) of northern Europe. *Fauna ent. Scand.* **13**: 137 pp, 136 figs.   (1984) **Bengtsson, B. A.**

TORTRICHOIDEA

British tortrichoid moths. Cochylidae and Tortricidae: Tortricinae. Ray Society, London. 251 pp, 47 pls, some coloured.
                    (1973) **Bradley, J. D., Tremewan, W. G., and Smith, A.**
Includes half-tone plates of larval habitats..

British tortricoid moths. Torticidae: Olethreutinae. Ray Society, London. 336 pp, 54

figs, 43 pls.   (1979) **Bradley, J. D., Tremewan, W. G., and Smith, A.**
Excellent, well illustrated general account.

Die Gattungen der palaearktischen Tortricidae. 8 parts. *Tijdschr. Ent.* **97–111**,
numerous figs and pls.   (1954–68) **Obraztsov, N. S.**

Cochylidae. *Microlepidoptera Palaearctica.* Fromme, Vienna. Vol. 3 (2 parts),
528 pp, 161 pls.   (1970) **Razowski, J.**

Tortricini. *Microlepidoptera Palaearctica.* Braun, Karlsruhe. Vol. 6 (2 parts), 376
pp, 138 figs, 101 pls.   (1984) **Razowski, J.**

De Nederlandse Bladrollers (Tortricidae). *Monografieën ned. ent. Vereen.* **3**: 200
pp, 99 pls.   (1968) **Bentinck, G. A. and Diakonoff, A.**

The genitalia of the group Tortricidae of the Lepidoptera of the British Isles. Pierce,
Oundle. 101 pp, 34 pls. Reprinted (1968) by Classey, Feltham.
                                (1922) **Pierce, F. N. and Metcalfe, J. W.**

Die Larvalsystematik der Wickler. Tortricidae. *Abh. Larvalsyst. Insekten* **3**: 269
pp, 276 figs.   (1958) **Swatschek, B.**

PYRALOIDEA and PTEROPHOROIDEA

British pyralid and plume moths. Warne, London. 208 pp, 189 figs, 16 col. pls.
                                                     (1954) **Beirne, B. P.**

British pyralid moths. Harley Books, Colchester. 175 pp, 12 figs, 8 pls.
                                                       (1986) **Goater, B.**
Good general account with excellent illustrations.

Crambinae. *Microlepidoptera Palaearctica.* Fromme, Vienna. Vol. 1 (2 parts),
553 pp, 133 pls.   (1965) **Bleszynski, S.**

Nordeuropas Pyraliden. *Danmarks Dyreliv.* Faunabøger, Copenhagen. Vol. 3,
287 pp, 264 figs, 8 pls, 211 maps.   (1986) **Palm, E.**

Beiträge zur Insekten Fauna der DDR: Lepidoptera—Galleriidae. *Beitr. Ent.*
23: 313–24, 14 figs, 2 col. pls.   (1973) **Petersen, G.**

Phycitinae. *Microlepidoptera Palaearctica.* Fromme, Vienna. Vol. 4 (2 parts), 752
pp, 170 pls.   (1973) **Roesler, R. U.**

The genitalia of the British pyrales with the deltoids and plumes. Pierce, Oundle. 69 pp,
30 pls. Reprinted (1968) by Classey, Feltham.
                                (1938) **Pierce, F. N. and Metcalfe, J. W.**

Die Raupen mitteleuropäischen Pyraustinae. *Beitr. Ent.* **5**: 521–639, 279 figs.
                                                  (1955) **Bollmann, H.-G.**

Die Larvalsystematik der Zünsler. Pyralidae. *Abh. Larvalsyst. Insekten* **5**: 163
pp, 219 figs.   (1960) **Hasenfuss, I.**

TRICHOPTERA

British species 199.

Caddis larvae: larvae of the British Trichoptera. Hutchinson, London. 476 pp, 980 figs. (1967) **Hickin, N. E.**

A key to the adults of the British Trichoptera. *Scient. Publs Freshwat. biol. Ass.* **28**: 1–151, figs, 5 pls. (1973) **Macan, T. T.**

Atlas of European Trichoptera. *Series entomologica.* Junk, The Hague. Vol. 24, 298 pp, many figs. (1983) **Malicky, H.**
Genitalia figures of all European species grouped in 288 pls, no text.

The identification of British limnephilid larvae (Trichoptera). *Systematic Entomology* **1**: 147–67, 116 figs. (1976) **Hiley, P. D.**

A key to larvae of the family Leptoceridae (Trichoptera) in Great Britain and Ireland. *Freshwat. Biol.* **11**: 273–97. (1981) **Wallace, I. D.**

[Fauna of the USSR. Trichoptera 1–2. *Zool. Inst. Acad. Sci. USSR* (N.S.), nos 88 and 95.] Vol. 1, 638 pp, 773 figs; Vol. 2, 700 pp, 877 figs. [Russian]. English translation (1970–71) by the Israel Program for scientific translations, Jerusalem. (1964 and 1966) **Lepneva, S. G.**
A very good monograph of the Russian species; much overlap with western Europe; contains descriptions of many Western European larvae and pupae.

Trichoptera. In *Limnofauna Europaea* (ed. J. Illies), pp. 285–309. G. Fischer, Stuttgart. (1967) **Botosaneanu, L.**
Distribution of all European species of Trichoptera.

Caseless caddis larvae of the British Isles. *Scient. Publs Freshwat. biol. Ass.* **43**: 1–92, 137 figs, 4 pls. (1981) **Edington, J. M. and Hildrew, A. G.**
Volume on the case-building species *in preparation.*

HYMENOPTERA

British species about 6000.

Hymenoptera. Introduction and keys to families, 2nd edn. *Handbk Ident. Br. Insects* **6**(1): iv + 100, 197 + xxii figs. (1977) **Richards, O. W.**

A check list of British insects. Hymenoptera, 2nd edn (Revised M. G. Fitton *et al.*) *Handbk Ident. Br. Insects* **11**(4): ix + 1–159. (1977) **Kloet, G. S. and Hincks, W. D.**

Die Hymenopteren Nord- und Mitteleuropas, 2nd edn. G. Fischer, Jena. 1062 pp, 127 figs. (1930) **Schmiedeknecht, O.**
Keys to genera of adults and species in some groups. Rather out of date.

The Hymenoptera associated with spiders in Europe. *Zool. J. Linn. Soc.* (in press) (1987) **Fitton, M. G., Shaw, M. R., and Austin, A. D.**
Includes keys to genera for all groups except Pompilidae.

The nocturnal Ichneumonoidea of the British Isles, including a key to genera. *Entomologist's Gaz.* **27**: 35–49, 20 figs.
(1976) Gauld, I. D. and Huddleston, T.

SYMPHYTA

British species about 480.

Hymenoptera Symphyta. *Handbk Ident. Br. Insects* **6**(2b): 51–137, 212 figs; **6**(2c): 139–252, 474 figs.   (1952 and 1958) **Benson, R. B.**
Basic identification guides that need taxonomy modifying to take account of more recent work (Benson 1960 and Smith 1969–79).

Symphyta (except Tenthredinidae). *Handbk Ident. Br. Insects* **6**(2a): 1–67, 148 figs.   (1981) **Quinlan, J. and Gauld, I. D.**

Studies in *Pontania* (Hym. Tenthredinidae). *Bull. Br. Mus. nat. Hist.* (Entom.) **8**: 367–87, 8 figs.   (1960) **Benson, R. B.**

The classification of *Rhogogaster* Konow. *Proc. R. ent. Soc. Lond.* (B) **34**: 105–12, 21 figs.   (1965) **Benson, R. B.**

Die Larvalsystematik der Blattwespen. *Abh. Larvalsyst. Insekten* **1**: 1–339, 435 figs.   (1957) **Lorenz, H. and Kraus, M.**

Die Tenthredinoidea Mitteleuropas. Friedländer, Berlin. 790 pp, 154 figs. (Originally published in Beihefte, *Dt. ent. Z.*, 1912–17).   (1918) **Enslin, E.**

Nearctic Sawflies pts I–IV. *US Dept. Agric. techn. Bull.* **1379**, **1398**, **1420**, **1595**: 483 pp, illustrated.   (1969–79) **Smith, D. R.**
Needed for modern taxonomic concepts.

APOCRITA—PARASITICA

ICHNEUMOIDEA

Owing to complexities of classification and the large numbers of species involved, it has not been possible to group papers satisfactorily and they are therefore presented in alphabetical order by author and chronologically within author.

Ichneumonidae

British species over 2000.

Révision des espèces ouest-paléarctiques du genre *Trieces* (Hym., Ichneumonidae). *Annls Soc. ent. Fr.* (N.S.) **9**: 975–88, 28 figs.
(1973) Aeschlimann, J. P.

Révision des espèces ouest-paléarctiques du genre *Triclistus* Foerster (Hymenoptera: Ichneumonidae). *Mitt. Schweiz. ent. Ges.* **46**: 219–52, 91 figs.
(1973) Aeschlimann, J. P.

Révision des espèces ouest-paléarctiques du genre *Chorinaeus* (Hym., Ichneumonidae). *Annls Soc. ent. Fr.* (N.S.) **11**: 723–44, 38 figs.
(1975) **Aeschlimann, J. P.**

Banchinae et suppl aux Pimplinae. *Les Ichneumonides ouest-paléarctiques et leurs hôtes.* Quatre Feuilles Editeur, Paris. Vol. 2, 318 pp. (1978) **Aubert, J. E.**
Includes keys to species of *Glypta, Lissonata,* and *Apophua.*

Révision des Ichneumonides *Stenomacrus* sensu lato. *Mitt. Münch. ent. Ges.* **71**: 139–59. (1981) **Aubert, J. F.**

Révision préliminaire des Ichneumonides Orthocentrinae européenes (Hym., Ichneumonidae). *Eos* **52**: 7–28. (1978) **Aubert, J. F.**
Includes key to species of *Orthocentrus.*

British species of Diplazonini (Bassini auct.) with a study of the genital and post-genital abdominal sclerites in the male (Hym. Ichneum.). *Trans. R. ent. Soc. Lond.* **91**: 661–712, 11 figs. (1941) **Beirne, B. P.**
Needs to be used with caution. See Fitton and Rotheray (1982), this section.

A systematic study of the genus *Ophion* in Britain. *Tijdschr. Ent.* **125**: 57–97, 45 figs. (1982) **Brock, J. P.**
Includes key.

Révision des espèces ouest-paléarctiques du genre *Netelia* Gray (Hym., Ichneumonidae). *Studi Sassaresi. Annl. Fac. Agr. Univ. Sassari* **23**: 1–126, 21 pls.
(1975) **Delrio, G.**

A review of the British species of *Tryphon* Fallén (Hym., Ichneumonidae). *Entomologist's mon. Mag.* **110**: 153–71, 41 figs. (1975) **Fitton, M. G.**

The British Acaenitinae (Hymenoptera: Ichneumonidae). *Entomologist's Gaz.* **32**: 185–92, 5 figs. (1981) **Fitton, M. G.**

A review of the British Collyriinae, Eucerotinae, Stilbopinae and Neorhacodinae (Hymenoptera: Ichneumonidae). *Entomologist's Gaz.* **35**: 185–95, 21 figs. (1984) **Fitton, M. G.**

The ichneumon fly genus *Banchus* (Hymenoptera) in the Old World. *Bull. Brit. Mus. nat. Hist.* (Entom.) **51**: 1–60, 129 figs. (1985) **Fitton, M. G.**

The British species of *Cidaphus* (Hymenoptera: Ichneumonidae). *Entomologist's Gaz.* **36**: 293–7, 6 figs. (1985) **Fitton, M. G.**

A review of the British Cremastinae (Hymenoptera: Ichneumonidae), with keys to the species. *Entomologist's Gaz.* **31**: 63–71, 22 figs.
(1980) **Fitton, M. G. and Gauld, I. D.**

The taxonomy and biology of the British Adelognathinae (Hymenoptera: Ichneumonidae). *J. nat. Hist.* **16**: 275–83, 18 figs.
(1982) **Fitton, M. G., Gauld, I. D., and Shaw, M. R.**

A key to the European genera of diplazontine ichneumonflies, with notes on the British fauna. *Syst. Ent.* **7**: 311–20, 27 figs.
(1982) **Fitton, M. G. and Rotheray, G. E.**

Notes on the British Ophionini (Hym., Ichneumonidae) including a provisional key to species. *Entomologist's Gaz.* **24**: 55–65, 6 figs.
(1973) **Gauld, I. D.**

The British species of Phrudinae (Hym., Ichneumonidae). *Entomologist's mon. Mag.* **115**: 197–9, 10 figs. (1979) **Gauld, I. D. and Fitton, M. G.**

Keys to the British xoridine parasitoids of wood-boring beetles (Hymenoptera: Ichneumonidae). *Entomologist's Gaz.* **32**: 259–67, 29 figs.
(1981) **Gauld, I. D. and Fitton, M. G.**

Ichneumonidae, subfamilies Orthopelmatinae and Anomaloninae. *Handbk Ident. Br. Insects* **7**(2b): 1–32, 82 figs.
(1977) **Gauld, I. D. and Mitchell, P. A.**

Die Arten der Gattung *Glyptorhaestus* Thomson (Hymenoptera, Ichneumonidae). *Zeit. Arbeit. Österreich. Ent.* **27**: 39–46, 4 figs. (1975) **Hinz, R.**

Die europäischen Arten der Gattung *Trematopygodes* Aubert (Hymenoptera, Ichneumonidae). *Nachr. Bayerisch. Ent.* **29**: 89–93, 3 figs. (1980) **Hinz, R.**

Die paläarktischen Arten der Gattung *Trematopygus* Holmgren (Hymenoptera, Ichneumonidae). *Spixiana* **8**: 265–76, 29 figs. (1985) **Hinz, R.**

Untersuchungen zur Systematik einiger *Phygadeuon*-Arten aus der Verwandtschaft des *P. vexator* Thunberg und des *P. fumator* Gravenhorst (Hymenoptera, Ichneumonidae). *Opusc. zool. Nr.* **98**: 1–22, 29 figs.
(1967) **Horstmann, K.**

Typenrevision der europäischen Arten der Gattung *Diadegma* Foerster (syn. *Angitia* Holmgren). *Beitr. Ent.* **19**: 413–72, 122 figs. (1969) **Horstmann, K.** Includes a key to species.

Revision der europäischen Tersilochinen 1. (Hymenoptera: Ichneumonidae). *Veroff. zool. Staatssamml. Münch.* **15**: 45–138, 187 figs. (1971) **Horstmann, K.**

Revision der europäischen Arten der Gattung *Lathrostizus* Foerster (Hymenoptera, Ichneumonidae). *Mitt. Dt. ent. Ges.* **30**: 8–12 and 16–18, 10 figs. (1971) **Horstmann, K.**

Revision der Gattung *Nepiesta* Foerster (mit einer Übersicht über die Arten der Gattung *Leptoperilissus* Schmiedeknecht) (Hymenoptera, Ichneumonidae). *Polskie Pismo ent.* **43**: 729–41, 4 figs. (1973) **Horstmann, K.**

Revision der westpaläarktischen Arten der Schlupfwespen-Gattung *Bathyplectes* und *Biolysia* (Hymenoptera: Ichneumonidae). *Ent. Germanica* **1**: 58–81, 34 figs. (1974) **Horstmann, K.**

Typenrevision der von E. Zilahi-Kiss beschriebenen Hemitelinen mit Bemerkungen zu den Gattungen *Hemiteles* Grav. (s.str.), *Gnotus* Foerster und *Xiphulcus* Townes (Hymenoptera, Ichneumonidae). *Annls hist.-nat. Mus. nat. Hung.* **66**: 339–46. (1974) **Horstmann, K.** Includes keys to species.

Neubearbeitung der Gattung *Nemeritis* Holmgren (Hymenoptera, Ichneumonidae). *Polskie Pismo ent.* **45**: 251–65, 19 figs. (1975) **Horstmann, K.** Includes key to European species.

Nachträg zur Revision der europäischen *Dichrogaster*-Arten (Hymenoptera, Ichneumonidae). *Zeitschr. Arbeitsg. Österreich. Ent.* **28**: 55–61, 14 figs. (1976) **Horstmann, K.** Includes a key.

Revision der europäischen Arten der Gattung *Ceratophygadeuon* Viereck (Hymenoptera, Ichneumonidae). *Zeitschr. Arbeitsg. Österreich. Ent.* **31**: 41–8, 13 figs. (1979) **Horstmann, K.**

Revision der europäischen Arten der Gattung *Aclastus* Foerster (Hymenoptera, Ichneumonidae). *Polskie Pismo ent.* **50**: 133–58, 50 figs. (1980) **Horstmann, K.**

Revision der europäischen Tersilochinae 2. (Hymenoptera: Ichneumonidae). *Spixiana* Suppl. **4**: 1–76, 150 figs. (1981) **Horstmann, K.**

Die westpaläarktischen Arten der Gattung *Chirotica* Foerster, 1869 (Hymenoptera, Ichneumonidae). *Entomofauna* **4**: 1–33. (1983) **Horstmann, K.**

Revision der paläarktischen Arten der Gattung *Hidryta* Foerster (Hymenoptera, Ichneumonidae). *Zeitschr. Arbeitsg. Österreich. Ent.* **35**: 113–17, 4 figs. (1984) **Horstmann, K.**

Revision der mit *difformis* (Gmelin, 1790) verwandten westpaläarktischen Arten der Gattung *Campoplex* Gravenhorst, 1829 (Hymenoptera, Ichneumonidae). *Entomofauna* **6**: 129–63, 18 figs. (1985) **Horstmann, K.**

Revision of the European species of the genus *Hadrodactylus* Foerster (Hymenoptera: Ichneumonidae). Part 1. *Entomologica scand.* **10**: 303–13; **12**: 231–9. (1979 and 1981) **Idar, M.**

A revision of the genus *Atractodes* (Hymenoptera, Ichneumonidae) of the western palaearctic region. *Acta ent. Fenn.* **34**: 1–44, 65 figs. (1979) **Jussila, R.**

[Ichneumonidae (subfamily Tryphoninae) tribe Tryphonini.] *Fauna SSSR* (N.S.) **106** (Hymenoptera) **3**(1): 1–320, 417 figs. [Russian]. English translation (1981) by Amerind, New Delhi. (1973) **Kasparyan, D. R.**

[Hymenoptera, Ichneumonidae. *Keys to insects of the European parts of the USSR.* Vol. 3, pp. 1–688, 311 pls.] [Russian]. (1981) **Kasparyan, D. R.** A valuable compendium with some original work; derived mainly from studies published elsewhere.

A review and a revision in greater part, of the Cteniscini of the Old World (Hym., Ichneumonidae). *Bull. Brit. Mus. nat. Hist.* (Ent.) **2**: 305–460, 86 figs, 7 pls. (1952) **Kerrich, G. J.**

An introduction to the Mesochorinae (Hymenoptera, Ichneumonidae). *Proc. Trans. Brit. ent. nat. Hist. Soc.* **14**: 93–7, 10 figs. (1981) **Lawton, F. D.**

Notes on British Metopiini (Hym., Ichneumonidae). *Entomologist's mon. Mag.* **72**: 83–6, 3 figs. (1936) **Perkins, J. F.**

A synopsis of the British Pimplini, with notes on the synonymy of the European species. *Trans. R. ent. Soc. Lond.* **91**: 637–59, 52 figs.
(1941) **Perkins, J. F.**

Hymenoptera Ichneumonoidea. Ichneumonidae, key to subfamilies, Ichneumoninae 1–2, Alomyinae, Agriotypinae and Lycorininae. *Handbk Ident. Br. Insects* **7**(2a i–ii): 1–213, 758 figs.
(1959–60) **Perkins, J. F.**

A study of the genus *Trychosis* Foerster in Europe (Hymenoptera, Ichneumonidae, Cryptinae). *Zoolog. Verhandel.* **79**: 1–40, 35 figs.
(1966) **Rossem, G. van**

A revision of the genus *Cryptus* Fabricius s. str. in the western palearctic region, with keys to genera of Cryptina and species of *Cryptus* (Hymenoptera, Ichneumonidae). *Tijdschr. Ent.* **112**: 299–374, 48 figs, 1 pl.
(1969) **Rossem, G. van**

A study of the genus *Meringopus* in Europe and of some related species from Asia (Hymenoptera, Ichneumonidae, Cryptinae). *Tijdschr. Ent.* **112**: 165–96, 9 figs., 3 pls.
(1969) **Rossem, G. van**

The genus *Buathra* Cameron in Europe (Hymenoptera, Ichneumonidae). *Tijdschr. Ent.* **114**: 201–7, 1 fig., 1 pl.
(1971) **Rossem, G. van**

A revision of some western palearctic Oxytorine genera (Hymenoptera, Ichneumonidae). *Spixiana* Suppl. **4**: 79–135, 3 figs, 2 pls.
(1981) **Rossem, G. van**

A revision of the western palearctic oxytorine genera. Part II. Genus *Eusterinx* (Hymenoptera, Ichneumonidae). *Spixiana* **5**: 149–70.
(1982) **Rossem, G. van**

A revision of the western palearctic oxytorine genera. Part III. Genus *Proclitus* (Hymenoptera, Ichneumonidae). *Contr. Am. ent. Inst.* **20**: 153–65.
(1983) **Rossem, G. van**

A revision of the western palearctic oxytorine genera. Part IV. Genus *Megastylus* (Hymenoptera, Ichneumonidae). *Entomofauna* **4**: 121–32, 6 figs.
(1983) **Rossem, G. van**

A revision of the world species of *Sinophorus* (Ichneumonidae). *Mem. Am. ent. Inst.* **38**: 1–403, 170 figs.
(1984) **Sanborne, M.**

Revision of European species of the genus *Bathythrix* Foerster (Hymenoptera, Ichneumonidae). *Annls zool. Inst. Zool. Polska Akad. Nauk. Warszawa* **35**(23): 319–65, 66 figs. [Polish and Russian summaries.] (1980) **Sawoniewicz, J.**

A catalogue and reclassification of the eastern palaearctic Ichneumonidae. *Mem. Am. ent. Inst.* **5**: v + 661.
(1965) **Townes, H. K., Momoi, S., and Townes, M.**
Keys to genera. Nomenclature does not always follow the *International Code of Zoological Nomenclature*.

The genera of the Ichneumonidae 1–4. *Mem. Am. ent. Inst.* **11–13, 17**: 1516 pp, 847 figs.
(1969–71) **Townes, H. K.**

Comment above applies. The only modern work on world genera (excluding Ichneumoninae).

Revisions of twenty genera of Gelini (Ichneumonidae). *Mem. Am. ent. Inst.* **35**: 1–281, 172 figs. (1983) **Townes, H. K.**

Braconidae

British species over 600.

A preliminary key to the subfamilies of the Braconidae (Hymenoptera). *Tijdschr. Ent.* **119**: 33–78, 123 figs. (1976) **Achterberg, C. van**
Good comprehensive key but more difficult to use than that of Marsh (1963), this section.

A revision of the tribus Blacini (Hymenoptera, Braconidae, Helconinae). *Tijdschr. Ent.* **118**: 159–322, 476 figs. (1976) **Achterberg, C. van**

A revision of the subfamily Zelinae auct. (Hymenoptera, Braconidae). *Tijdschr. Ent.* **122**: 241–479, 900 figs. (1979) **Achterberg, C. van**

Revisionary notes on the palaearctic genera and species of the tribe Exothecini Foerster (Hymenoptera, Braconidae). *Zool. Mededel.* **57**: 339–55, 32 figs.
(1983) **Achterberg, C. van**

Essay on the phylogeny of Braconidae (Hymenoptera: Ichneumonoidea). *Ent. Tidskr.* **105**: 41–58, 17 figs. (1984) **Achterberg, C. van**
No keys. Diagnoses of subfamilies and discussion of their relationships.

The present state of nomenclature of wing venation in the Braconidae (Hymenoptera); its origins and comparison with related groups. *J. Ent.* (B) **43**: 63–72, 10 figs. (1974) **Eady, R. D.**

A revision of the genus *Macrocentrus* Curtis (Hym., Braconidae) in Europe with descriptions of four new species. *Entomologist's Gaz.* **15**: 97–127, 92 figs.
(1964) **Eady, R. D. and Clarke, J. A. J.**

Hymenoptera, Braconidae (Opiinae I). *Das Tierreich* **91**: xii + 620 pp, 63 figs.
(1973) **Fischer, M.**
Brings together all Fischer's previous work on palaearctic Opiinae.

The Alysiinae (Hym. Braconidae) parasites of the Agromyzidae (Diptera). Parts I–V. *Beitr. Ent.* **14**, **16–18**: 516 pp, 185 figs.
(1964–68) **Griffiths, G. C. D.**
A comprehensive review of the taxonomy of the tribe Dacnusini but unsuitable as an identification manual.

A revision of the western palaearctic species of the genus *Meteorus* (Hymenoptera: Braconidae). *Bull. Br. Mus. nat. Hist.* (Ent.) **41**: 1–58, 58 figs.
(1980) **Huddleston, T.**

The palaearctic species of *Ascogaster* (Hymenoptera: Braconidae). *Bull. Br. Mus. nat. Hist.* (Ent.) **49**: 341–92, 79 figs. (1984) **Huddleston, T.**

The European species of *Leiophron* Nees and *Peristenus* Foerster (Hymenoptera:

Braconidae, Euphorinae). *Trans. R. ent. Soc. Lond.* **126**: 207–38, 50 figs.
                                                    (1974) **Loan, C. C.**

Contributions to our knowledge of the British Braconidae. (In parts). *Ento-mologist* **47–57**, 139 pp, 2 pls.   (1914–24) **Lyle, G. T.**

Hymenopterorum Catalogus. Part 3 (nov. edn) Aphidiidae. Junk, Gravenhage. 103 pp.   (1968) **Mackauer, M.**
No keys; a collation of the literature and published host records.

A key to the nearctic subfamilies of the family Braconidae (Hymenoptera). *Annls ent. Soc. Am.* **56**: 522–7, 29 figs.   (1963) **Marsh, P. M.**
Designed for the North American fauna but usable for European. Simpler to use than van Achterberg (1976), this section.

Monograph of British Braconidae. 8 parts. *Trans. ent. Soc. Lond.*, 424 pp, 9 pls.
                                                (1885–99) **Marshall, T. A.**
The only comprehensive manual specifically dealing with the British Braconid fauna; some parts still useful but mostly superseded by more up-to-date works.

A revision of the European Dacnusini (Hym., Braconidae, Dacnusinae). 8 parts. *Entomologist's mon. Mag.* **79–90**: 145 pp, 348 figs.
                                                (1943–54) **Nixon, G. E. J.**
Taxonomy and nomenclature in part superseded by Griffiths (1964–68), this section, but keys much more practical.

A reclassification of the tribe Microgasterini (Hymenoptera: Braconidae). *Bull. Br. Mus. nat. Hist.* (Ent.) Suppl. **2**: 1–284, 346 figs.
                                                (1965) **Nixon, G. E. J.**
Includes keys to genera.

A revision of the genus *Microgaster* Latreille (Hymenoptera: Braconidae). *Bull. Br. Mus. nat. Hist.* (Ent.) **22**: 31–72, 33 figs.   (1968) **Nixon, G. E. J.**

A revision of the NW European species of *Microplitis* Förster (Hymenoptera: Braconidae). *Bull. Br. Mus. nat. Hist.* (Ent.) **25**(1): 1–30, 29 figs.
                                                (1970) **Nixon, G. E. J.**

[Four papers revising the north-western European species of groups of *Apanteles* Förster]. *Bull. ent. Res.* **61**, 63–5, 226 pp, 260 figs.
                                                (1972–76) **Nixon, G. E. J.**

A revision of the European Agathidinae (Hymenoptera: Braconidae). *Bull. Br. Mus. nat. Hist.* (Ent.) **52**: 183–242, 68 figs.   (1986) **Nixon, G. E. J.**

A contribution towards knowledge of the world literature regarding Braconidae (Hymenoptera: Braconidae). *Beitr. Ent.* **15**: 243–500.
                                                (1965) **Shenefelt, R. D.**
No keys; papers on Braconidae listed under authors.

Hymenoptera Catalogus (nov. edn). Parts 4–7, 9–13, and 15–16 (Braconidae parts 1–11). Junk, Gravenhage. Over 2000 pp. (1969–80) **Shenefelt, R. D.**
No keys; a collation of the literature and published host records.

The morphology of the head of larval Hymenoptera with special reference to the head of Ichneumonoidea, including a classification of the final instar of

the Braconidae. *Trans. R. ent. Soc. Lond.* **103**: 27–84, 34 figs.
(1952) Short, J. R. T.

Aphid parasites of Czechoslovakia. A review of the Czechoslovak Aphidiidae (Hymenoptera). Junk, The Hague. 247 pp, 21 pls.   (1966) **Stary, P.**
Keys to genera and species of Aphidiinae (as Aphidiidae).

A review of the *Aphidius*-species (Hymenoptera, Aphidiidae) of Europe. *Ann. zool. bot. Bratislava* **84**: 1–85, 25 figs (compound), 3 pls.   (1973) **Stary, P.**

EVANOIDEA, STEPHANOIDEA, and TRIGONALYOIDEA

British species 9.

The morphology, taxonomy and biology of the British Evanoidea (Hymenoptera). *Trans. R. ent. Soc. Lond.* **102**: 247–301, 72 figs.
(1951) Crosskey, R. W.
Includes keys.

Beiträge zur Insektenfauna der DDR: Hymenoptera—Evanoidea, Stephanoidea, Trigonalyoidea. *Faun. Abh. staatl. Mus. Tierk. Dresden* **11**: 161–90, 27 figs.   (1984) **Oehlke, J.**
Includes keys and covers virtually all of NW European fauna.

CYNIPOIDEA

British species about 200.

Cynipidae. *Tierreich* **24**: xxxv + 891 pp, 422 figs.
(1910) Dalla-Torre, K. W. and Kieffer, J. J.
Badly out of date but the only comprehensive account of world species.

Cynipoidea (Hym.) 1905–1950. Ann Arbor, Michigan. 351 pp, 224 figs.
(1952) Weld, L. H.
Privately published and thus very scarce. Keys to world genera: gives reference lists of species described since Dalla-Torre and Kieffer (1910). Becoming dated.

Hymenoptera, Cynipoidea. Key to families and subfamilies, and Cynipinae (including galls). *Handbk Ident. Br. Insects* **8**(1A): 1–81, 371 + vii figs.
(1963) Eady, R. D. and Quinlan, J.
Keys to genera and British species (including galls) of the Cynipinae. Key to families superseded by the next entry, Fergusson (1986).

Charipidae, Ibaliidae and Figitidae. Hymenoptera, Cynipoidea. *Handbk Ident. Br. Insects* **8**(1c): 1–55, 97 figs.   (1986) **Fergusson, N. D. M.**
Keys to families, genera, and British species. Includes notes on collecting, biology, etc. and a comprehensive bibliography.

Hymenoptera, Cynipoidea, Eucoilidae. *Handbk Ident. Br. Insects* **8**(1b): 1–58, 205 figs.
(1978) Quinlan, J.
Keys to British genera and species.

CHALCIDOIDEA

British species over 1500.

Keys to the Chalcidoidea of Czechoslovakia (Insecta: Hymenoptera). *Mem. ent. Soc. Can.* **34**: 1–121, 288 figs.
(1964) **Peck, O., Bouček, Z., and Hoffer, A.**
Keys genera, including almost all British as well as families.

[Hymenoptera II. Chalcidoidea.] *Opred. Nas. Europ. Chasti SSSR* **3**: 28–538, pls 10–192. [Russian].   (1978) **Trjapitzin, V. A.**, ed.
Includes keys to most European Chalcidoidea.

[Chalcidoidea of the USSR.] *Opred. Faune SSSR* **44**: 1–575, 592 figs. [Russian].
(1952) **Nikol'skaya, M. N.**
Keys families, genera, and species in some groups.

Hymenoptera Chalcidoidea. Agaontidae, Leucospidae, Chalcididae, Eucharitidae, Perilampidae, Cleonymidae and Thysanidae. *Handbk Ident. Br. Insects* **8**(2a): 1–40, 79 + v figs. (1958) **Ferrière, C. and Kerrich, G. J.**
Outdated key to families; keys to species.

On the biology of the inhabitants of oak galls of Cynipidae (Hymenoptera) in Britain. *Trans. Soc. Br. Ent.* **14**: 237–68, 26 figs.   (1961) **Askew, R. R.**
Includes key to species of *Torymus*.

The first revision of the European species of the family Chalcididae (Hymenoptera). *Acta ent. Mus. nat. Pragae Suppl.* **1**: 1–108, 150 figs.
(1952) **Bouček, Z.**
Includes key to genera and species.

A contribution to the biology and taxonomy of some palaearctic species of *Tetramesa* Walker (=*Isosoma* Walker; =*Harmolita* Motsch.) (Hymenoptera: Eurytomidae), with particular reference to the British fauna. *Trans. R. ent. Soc. Lond.* **113**(9): 175–216, 105 figs.   (1961) **Claridge, M. F.**
Includes a key to British species of *Tetramesa*.

Notes sur les *Eurytoma* (Hym., Chalcidoidea). I. Les types de Thomson et de Mayr. *Mitt. Schweiz. ent. Ges.* **23**: 377–410, 6 figs.   (1950) **Ferriere, C.**

Mongolian eurytomids (Hymenoptera: Chalcidoidea) II. *Acta zool. Acad. Sci. Hung.* **22**(1–2): 173–87.   (1976) **Szelenyi, O.**
Includes a key to the palaearctic species of *Eurytoma*.

Mongolian eurytomids (Hymenoptera: Chalcidoidea) III. *Acta zool. Acad. Sci. Hung.* **22**(3–4): 397–405, 3 figs.   (1976) **Szelenyi, O.**
Includes key to palaearctic species of *Bruchophagus*.

Les espèces françaises du genre *Perilampus* Latr. (Hym. Perilampidae). *Bull. Soc. ent. Fr.* **52**: 68–74, 12 figs.   (1952) **Steffan, J. R.**

Beiträge zur Kenntnis der dänischen Callimomiden, mit Bestimmungstabellen der europäischen Arten (Hym. Chalc.). *Ent. Meddr* **17**: 232–85, 19 figs.
(1930–31) **Hoffmeyer, E. B.**
Keys to Danish species of Ormyridae and Torymidae.

Poznamky o Československych Perilampidae. Notes on Czechoslovak Perilampidae (Hymenoptera, Chalcidoidea). *Acta faun. ent. Mus. nat. Pragae* **1**: 83–98, 9 figs.   (1956) **Bouček, Z.**
Includes key to central European genera and to Czech species of *Perilampus* and *Chrysolampus*.

The Pteromalidae of north-western Europe (Hymenoptera: Chalcidoidea). *Bull. Br. Mus. nat Hist.* (Ent.) Suppl. **16**: 1–908, 686 figs.
(1969) **Graham, M. W. R. de V.**
A comprehensive work with keys to species.

On European Pteromalidae (Hymenoptera): a revision of *Cleonymus*, *Eunotus* and *Spaniopus* with descriptions of new genera and species. *Bull. Br. Mus. nat. Hist.* (Ent.) **27**: 267–315, illustrated.   (1972) **Bouček, Z.**

[The classification of parasitic Hymenoptera of the family Encyrtidae (Hymenoptera, Chalcidoidea). Part I. Survey of the systems of classification. The subfamily Tetracneminae Howard 1892.] *Ent. Obozr.* **52**(1): 163–75. [Russian]. Translated in *Ent. Rev. Washington* **52**: 118–25.
(1973) **Trjapitzin, V. A.**

Review of the genera of palaearctic encyrtids (Hymenoptera, Encyrtidae). *Trudy. Vses. Ent. Obschck.* **54**: 68–155.   (1971) **Trjapitzin, V. A.**
Key to genera.

Notes on some genera and species of Encyrtidae (Hym., Chalcidoidea), with special reference to Dalman's types. *Ent. Tidskr.* **79**(3–4): 147–75.
(1958) **Graham, M. W. R. de V.**
Key to species of *Bothriothorax* and notes on other genera.

Synonymic and descriptive notes on European Encyrtidae (Hym., Chalcidoidea). *Polskie Pismo ent.* **39**(2): 211–319. (1969) **Graham, M. W. R. de V.**
Key to species of *Encyrtus* and notes on other genera.

On the classification of the Anagyrine Encyrtidae, with a revision of some of the genera (Hymenoptera: Chalcidoidea). *Bull. Br. Mus. nat. Hist.* (Ent.) **20**(5): 141–250, 114 figs, 4 pls.   (1967) **Kerrich, G. J.**
Keys to genera *Aglyptus* and *Ericydnus*, the latter unworkable.

The British species of *Aphelinus* with notes and descriptions of other European Aphelinidae (Hymenoptera). *Syst. Ent.* **1**: 123–46, 15 figs.
(1976) **Graham, M. W. R. de V.**
Includes key to species.

Hymenoptera Aphelinidae d'Europe et du bassin méditerranéen. Masson, Paris. 206 pp, 80 figs.   (1965) **Ferrière, C.**
A practical account of most aphelinids in the area, but a few errors at species level.

Les espèces européennes du genre *Elasmus* Westw. (Hym., Chalc.). *Mitt.*

Schweiz. ent. Ges. **20**: 565–80, 4 figs. (1947) **Ferrière, C.**
Key to European species.

Revision der europäischen Tetracampidae (Hym., Chalcidoidea) mit einem Katalog der Arten der Welt. *Sborn. ent. Odd. nár. Mus. Praze* **32**: 41–90, 35 figs. (1958) **Bouček, Z.**
Keys to European genera and species.

Hymenoptera 2. Chalcidoidea Section (b). *Handbk Ident. Br. Insects* **8**(82b): 1–39, 83 + iii figs. (1968) **Askew, R. R.**
Keys to species of Elasmidae and Eulophidae (Eulophinae, Euderinae only) for the British fauna.

Taxonomy of some European *Chrysonotomyia* Ashmead (Hymenoptera, Eulophidae) with description of *C. longiventris* n. sp. and notes on distribution. *Entomologica scand.* **10**: 27–31, 11 figs. (1979) **Askew, R. R.**
Includes key to NW European species.

A study of central European Eulophidae, I: Eulophinae (Hymenoptera). II: *Diaulinopsis* and *Cirrospilus* (Hymenoptera). *Sborn. ent. Odd. nár. Mus. Praze* **33**: 117–94, 43 figs. (1959) **Bouček, Z.**
Keys to species.

Studien über europäischen Eulophidae, III: Euderinae (Hymenoptera: Chalcidoidea). *Beitr. Ent.* **13**: 257–81, 22 figs. (1963) **Bouček, Z.**
Keys to species.

Studies of European Eulophidae, IV: *Pediobus* Walk. and two allied genera. *Sborn. ent. Odd. nár. Mus. Praze* **36**: 5–90, 78 figs. (1965) **Bouček, Z.**
Key to European species of *Pediobius*.

The British species of *Achrysocharoides* (Hymenoptera, Eulophidae). *Syst. Ent.* **5**: 245–62, 13 figs. (1980) **Bryan, G.**
Includes key to species.

Taxonomy and biology of the palaearctic species of *Chrysocharis* Förster, 1856 (Hymenoptera: Eulophidae). *Entomologica scand.* Suppl. No. **26**: 1–130, 225 figs. (1985) **Hansson, C.**
Includes key to species.

Keys to the British genera and species of Elachertinae, Eulophinae, Entedontinae and Euderinae (Hym., Chalcidoidea). *Trans. Soc. Br. Ent.* **13**(10): 169–204, 26 figs. (1959) **Graham, M. W. R. de V.**

The European Tetrastichinae (Hymenoptera: Eulophidae). *Bull. Br. Mus. nat. Hist.* (Ent.) **54**: (In press). (1987) **Graham, M. W. R. de V.**

Additions and corrections to the British list of Eulophidae (Hym., Chalcidoidea). *Trans. Soc. Br. ent.* **15**(9): 167–275, 59 figs. (1963) **Graham, M. W. R. de V.**
Updates keys to species of some genera in Graham (1959).

Revision of British *Entedon* (Hymenoptera: Chalcidoidea), with descriptions of four new species. *Trans. R. ent. Soc. Lond.* **123**(3): 313–58, 27 figs. (1971) **Graham, M. W. R. de V.**

182   ANIMALIA

The European Trichogramminae. *Ent. Meddr* **12**: 257–354, 21 figs.

(1919) **Kryger, J. P.**

The classification of the Trichogrammatidae (Hymenoptera: Chalcidoidea). *Proc. Calif. Acad. nat. Sci.* **35**: 477–586, 70 figs.

(1968) **Doutt, R. L. and Viggiani, G.**

Includes key to world genera.

Caractérisation morphologique des groupes et espèces du genre *Trichogramma* Westwood. *Les Trichogrammes 1st Symposium Internat. Antibes*, pp. 45–75, 7 figs.

(1982) **Voegele, J. and Pintureau, B.**

Includes keys to species-groups and to species.

The holarctic genera of Mymaridae (Hymenoptera: Chalcidoidea). *Mem. ent. Soc. Wash.* **12**: 1–67, 60 figs. (1984) **Schauff, M. E.**

Includes an excellent key to genera.

The European Mymaridae comprising the genera known up to *c.* 1930. *Ent. Meddr* **26**: 1–97, 45 figs. (1950–51) **Kryger, J. P.**

The genera of the Mymaridae (Hymenoptera: Chalcidoidea). *Ent. Mem. Dept. agr. tech. Services Republ. S. Afr.* **5**: 1–71, 47 figs.

(1961) **Anneke, D. P. and Doutt, R. L.**

Key to most of world genera.

The British species of the genus *Alaptus* Haliday *in* Walker (Hym., Chalc., Mymaridae). *Trans. Soc. Br. Ent.* **13**: 137–48, 8 figs. (1959) **Hincks, W. D.**
Key to British species.

The British species of *Gonatocerus* Nees (Hymenoptera: Mymaridae), egg parasite of Homoptera. *Syst. Ent.* **11**: 213–29, 53 figs.

(1986) **Matthews, M. J.**

Includes key to most European species.

The British species of the genus *Ooctonus* Haliday, with a note on some recent work on fairy flies (Hym., Mymaridae). *Trans. Soc. Br. Ent.* **11**(7): 153–63, 8 figs. (1952) **Hincks, W. D.**
Key to British species.

Notes on some British Mymaridae (Hym.). *Trans. Soc. Br. Ent.* **10**: 167–207, 4 figs, 1 pl. (1950) **Hincks, W. D.**
Key to British species of *Polynema*.

Étude sur les Mymarommidae et les Mymaridae de la Belgique (Hym., Chalcidoidea). *Mém. Mus. r. Hist. nat. Belg.* **108**: 1–248, 24 pls.

(1948) **Debauche, H. R.**

Includes keys to Belgian genera and species of Mymaridae.

PROCTOTRUPOIDEA and CERAPHRONOIDEA

Serphidae, Calliceratidae, Diapriidae, Scelionidae. *Tierreich* **42, 44, 48**: 1813 pp, 608 figs. (1914–26) **Kieffer, J. J.**

[Proctotrupoids (Hymenoptera, Proctotrupoidea) of the USSR.] *Trudy Vses.*

ent. Obshch. **54**: 3–67, 131 figs. [Russian].   (1971) **Kozlov, M. A.**

A revision of the Heloridae (Hymenoptera). *Contr. Am. ent. Inst.* **15**(2): 12 pp, 7 figs.   (1977) **Townes, H.**

Revision der Heloridae (Hymenopt., Proctotrupoidea). *Mitt. Schweiz. ent. Ges.* **28**: 233–50, 3 pls.   (1955) **Pschorn-Walcher, H.**

Hymenoptera. Heloridae et Proctotrupidae. *Insecta Helvetica Fauna* **4**: 1–63, 103 figs.   (1971) **Pschorn-Walcher, H.**

A revision of the Serphidae (Hymenoptera). *Mem. Am. ent. Inst.* **32**: 1–541, 575 figs.   (1981) **Townes, H. and Townes, M.**

A preliminary revision of the British Proctotrupinae (Hym., Proctotrupoidea). *Trans. R. ent. Soc. Lond.* **87**: 431–65, 71 figs.

(1938) **Nixon, G. E. J.**

Hymenoptera, Proctotrupoidea, Diapriidae subfamily Belytinae. *Handbk Ident. Br. Insects* **8**(3d ii): 1–107, 314 figs.   (1957) **Nixon, G. E. J.**
Includes many NW European species.

Diapriidae (Diapriinae) Hymenoptera, Proctotrupoidea. *Handbk Ident. Br. Insects* **8**(3d i): 1–55, 125 figs.   (1980) **Nixon, G. E. J.**

Key to genera of Scelionidae of the holarctic region, with descriptions of new genera and species (Hymenoptera: Proctotrupoidea). *Mem. ent. Soc. Can.* **113**: 1–54, 207 figs.   (1980) **Masner, L.**

Contribution à l'étude des Hyménoptères Proctotrupoidea (Ceraphronidae). Parts II, III, IV, VI. *Bull. Annls Soc. R. ent. Belg.* **99–101**: 155 pp, 156 figs.

(1963–65) **Dessart, P.**

Contribution à l'étude des Hyménoptères Proctotrupoidea IX. Révision du genre *Macrostigma* Rondani, 1877. (Ceraphronidae, Megaspilinae). *Redia* **49**: 157–63, 5 figs.   (1965) **Dessart, P.**

Contribution à l'étude des Hyménoptères Proctotrupoidea (Ceraphronidae). Parts 10 and 12. *Bull. Inst. R. Sci. nat. Belg.* **42**(18): 1–85, 75 figs; **42**(32): 1–16, 2 figs.   (1966) **Dessart, P.**

A revision of the British species of *Dendrocerus* Ratzeburg (Hymenoptera: Ceraphronoidea) with a review of their biology as aphid hyperparasites. *Bull. Br. Mus. nat. Hist.* (Ent.) **41**: 255–314, 63 figs.

(1980) **Fergusson, N. D. M.**

Three new species of *Aphanogmus* (Hymenoptera: Ceraphronidae) from Britain, with a re-description of *A. fumipennis* Thoms., 1858, a species new to Britain. *Trans. Soc. Br. Ent.* **14**: 115–30, 17 figs.   (1960) **Parr, M. J.**

APOCRITA—ACULEATA

General

The Hymenoptera Aculeata of the British Isles. Reeve, London. xii + 391 pp, 52 pls.

(1896) **Saunders, E.**

Scolioidea, Vespoidea and Sphecoidea. *Handbk Ident. Br. Insects* **6**(3b): 118 pp,
    242 figs.   (1980) **Richards, O. W.**
    Does not include Formicidae.

Beiträg zur Insekten-Fauna der DDR: Hymenoptera-Bestimmungstabellen
    bis zu den Unterfamilien. *Beitr. Ent.* **19**: 753–801, 146 figs. (1969) **Oehlke, J.**

A list of the Hymenoptera Aculeata (*sensu lato*) of Ireland. *Proc. R. Ir. Acad.*
    37(B) no. 22: 201–55.   (1927) **Stelfox, A. W.**

Some recent records for Irish Aculeate Hymenoptera. *Entomologist's mon. Mag.*
    69: 47–53.   (1933) **Stelfox, A. W.**

Bees, wasps, ants and allied insects of the British Isles. Warne, London. xxv + 238
    pp, 111 pls.   (1932) **Step, E.**

Hymenopterist's Handbook. *Amateur Ent.* **7** (1943): 13–17, 19–25, 33–5, 55–81.
                                                                   (1945) **Yarrow, I. H. H.**
    Key to genera, biology. References to additions to British List since Saunders
    (1896), this section, excluding Formicoidea.

Larvae and nests

Biology of some stem-nesting aculeate Hymenoptera. *Trans. R. ent. Soc. Lond.*
    122(11): 323–99, 17 figs.   (1971) **Danks, H. V.**
    Includes a key to the nests of British aculeate Hymenoptera in *Rubus* stems and their
    commoner parasites.

The Sphecidae (Hymenoptera) of Fennoscandia and Denmark. *Fauna ent.*
    Scand. **4**(1): 41–51, 25 figs.   (1975) **Lomholdt, O.**
    Key to genera of sphecid larvae.

The nest habits of solitary bees. *Eos Madr.* **11**: 201–309, 13 pls.
                                                                   (1936) **Malyshev, S. I.**

Comparative morphological and systematic studies of bee larvae with a key to
    families of hymenopterous larvae. *Kans. Univ. Sci. Bull.* **35** (8, pt 2): 987–
    1102, 287 figs.   (1953) **Michener, C. D.**

Systematic

CHRYSIDOIDEA

Revision der Familie Chrysididae (Hymenoptera). *Mitt. schweiz. ent. Ges.*
    32(1–2): 1–240, 716 figs.   (1959) **Linsenmaier, W.**

Revision der Familie Chrysididae (Hymenoptera). Zweiter Nachtrag. *Mitt*
    schweiz. ent. Ges. **41**: 1–144, 13 figs.   (1968) **Linsenmaier, W.**

Cuckoo-wasps. Hymenoptera, Chrysididae. *Handbk Ident. Br. Insects* **6**(5): 37
    pp, 95 figs.   (1984) **Morgan, D.**

Hymenoptera. Bethyloidea (excluding Chrysididae). *Handbk Ident. Br. Insects* **6**(3a): 38 pp, 84 figs.   (1976) **Perkins, J. F.**

SCOLIOIDEA (including FORMICIDAE)

Hymenoptera. Formicidae. *Handbk Ident. Br. Insects* **6**(3c): 33 pp, 65 figs.
(1975) **Bolton, B. and Collingwood, C. A.**

Les fourmis (Hymenoptera Formicidae) d'Europe Occidentale et Septentrionale. *Faune de l'Europe et du Bassin Méditerranéen* **3**: 411 pp, 379 figs, 46 photos.
(1968) **Bernard, F.**

British ants. Routledge, London. xv + 436 pp, 93 figs., 18 pls.
(1927) **Donisthorpe, H. St J. K.**

The guests of British ants, their habits and life histories. Routledge, London. xxiii + 244 pp, 55 figs, 16 pls.   (1927) **Donisthorpe, H. St J. K.**

The Formicidae of Fennoscandia and Denmark. *Fauna ent. scand.* **8**: 1–174, 268 figs.   (1979) **Collingwood, C. A.**

Hymenoptera: Formicidae. *Insecta Helvetica* **6**: 298 pp, 627 figs.
(1977) **Kutter, H.**

POMPILOIDEA

The biology of the British Pompilidae (Hymenoptera). *Trans. Soc. Br. Ent.* **6**: 51–114.   (1939) **Hamm, A. H. and Richards, O. W.**

Hymenoptera: Pompilidae. *Insecta Helvetica* **5**: 176 pp, 489 figs.
(1972) **Wolf, H.**

Hymenoptera, Pompiloidea. *Handbk Ident. Br. Insects* **6**: in press.
(1987) **Day, M. C.**

VESPOIDEA

Les guêpes (*Vespa* L. s.l.) de la Suisse. *Bull. Soc. vaud. Sci. nat.* **62**: 329–62, 41 figs.   (1944) **Beaumont, J. de**

Die Faltenwespen Mitteleuropas (Hym. Diploptera). *Abh. dt. Akad. Wiss. Berl.* 1961(2): 248 pp, 71 figs.   (1961) **Blüthgen, P.**

Les guêpes sociales (Hymenoptera Vespidae) d'Europe occidentale et septentrionale. *Faune de l'Europe et du Bassin Méditerranéen* **6**: viii + 181 pp, 41 figs, 3 pls.   (1972) **Guiglia, D.**

Die sozialen Faltwespen Mitteleuropas. Parey, Berlin and Hamburg. 180 pp, 82 figs.   (1967) **Kemper, H. and Döhring, E.**

Social wasps in Norway (Hymenoptera, Vespidae). *Norsk ent. Tidsskr.* **12**(5–8): 195–218, 24 figs, 2 pls.   (1964) **Løken, A.**

*Wasps. An account of the biology and natural history of solitary and social wasps with
particular reference to those of the British Isles.* Sidgwick and Jackson, London.
xvi + 408 pp, 131 figs, 29 pls.   (1973) **Spradbury, J. P.**

Greek wasps of the family Eumenidae, with a key to the European genera.
Entomologist's Gaz. **31**: 39–59, 20 figs.   (1980) **Guichard, K. M.**

SPHECOIDEA

Hymenoptera: Sphecidae. *Insecta helvetica* **3**: 168 pp, 551 figs.
                          (1964) **Beaumont, J. de**

The Sphecidae (Hymenoptera) of Fennoscandia and Denmark. *Fauna ent.
scand.* **4**(1): 1–224, 272 figs; **4**(2): 225–452, 192 figs.
                          (1975, 1976) **Lomholdt, O.**

Sphecid wasps of the world—a generic revision. University of California Press, Los
Angeles and London. ix + 695 pp, 190 figs, 1 pl.

(1976)   **Bohart, R. M. and Menke, A. S.**

Beiträge zur Insekten-Fauna der DDR: Hymenoptera–Sphecidae. *Beitr. Ent.*
20(7–8): 615–812, 387 figs.   (1970) **Oehlke, J.**

Solitary wasps. *Naturalists' handbook.* Cambridge University Press, Cambridge.
Vol. 3, 65 pp.   (1983) **Yeo, P. F. and Corbet, S. A.**

APOIDEA

Bumblebees. Davis-Poynter, London. xii + 352 pp, 210 figs, 27 maps, 56 pls.
                          (1975) **Alford, D. V.**
Covers the British species of *Bombus* and *Psithyrus* with keys and distribution maps.

The world of the honeybee. Collins, London. New Naturalist, no. 29, xii + 226 pp,
42 pls.   (1974) **Butler, C. G.**

Die Bienen des Genus *Halictus* Latr. s.l. im Grossraum von Linz (Hymenoptera, Apidae) 1–3. *Naturk. Jb. Stadt Linz* **1969**: 133–83, 36 figs; **1970**: 19–
82, 28 figs; **1971**: 63–156, 18 figs.   (1969–71) **Ebmer, A. W.**

Systematik der Mittel- und Nordeuropäischen *Bombus* und *Psithyrus* (Hym.,
Apidae). *Ent. Meddr* **38**: 257–302, 18 figs.
                          (1970) **Faester, K. and Hammer, K.**

Colletes halophila Verhoeff (Hym., Apidae) and its *Epeolus* parasite at
Swanscombe in Kent, with a key to the British species of *Colletes* Latreille.
Entomologist's Gaz. **25**(3): 195–9.   (1974) **Guichard, K. M.**

Studies on Scandinavian bumble bees (Hymenoptera, Apidae). *Norsk ent.
Tidsskr.* **20**(1): 218 pp, 99 figs.   (1973) **Løken, A.**

Naturgeschichte der Urbienen. Fischer, Budapest. 214 pp, 60 pls.
                          (1935) **Néhelÿ, L.**
Revision of *Hylaeus* of middle Europe.

Die palaearktischen *Colletes*-Arten. *Pr. nauk. Wyd. Tow. Nauk. Lwow* **3**: 1–532, 40 figs, 28 pls.   (1936) **Noskiewicz, J.**

The British species of *Andrena* and *Nomada*. *Trans. ent. Soc. Lond.* **52**: 218–316, 5 pls.   (1919) **Perkins, R. C. L.**

The British species of *Halictus* and *Sphecodes*. *Entomologist's mon. Mag.* **58**: 46–52, 94–101, 167–74.   (1922) **Perkins, R. C. L.**

The British species of *Megachile* with descriptions of some new varieties from Ireland, and of a species new to Britain in F. Smith's collection. *Entomologist's mon. Mag.* **61**: 95–101.   (1925) **Perkins, R. C. L.**

A note on some British species of *Halictus*. *Entomologist's mon. Mag.* **71**: 104–6.
                                           (1935) **Perkins, R. C. L.**

A study of the British species of *Epeolus* Latr. and their races, with a key to the species of *Colletes* (Hymen., Apidae). *Trans. Soc. Br. Ent.* **4**: 89–130, 21 figs, 29 tables.   (1937) **Richards, O. W.**

Die Westpaläarktischen Arten der Bienenfamilie Melittidae (Hymenoptera). *Polskie Pismo Ent.* **43**(1): 97–126, 33 figs.   (1973) **Warncke, K.**

Recent additions to the British bee-fauna, with comments and corrections. *Entomologist's mon. Mag.* **104**: 60–4.   (1968) **Yarrow, I. H. H.**

Hoplitis claviventris (Thomson 1872) (=*Osmia leucomelana auctt. nec* Kirby) and the identity of *Apis leucomelana* Kirby 1802 (Hymenoptera, Megachilidae). *Entomologist* **103**: 62–9.   (1970) **Yarrow, I. H. H.**

Bees. Hymenoptera Apoidea. *Handbk Ident. Br. Insects* **6**: (in press).
                                           (1987) **Else, G. R.**

Scandinavian species of the genus *Psithyrus* Lepeletier. *Ent. scand.* (Suppl.)**23**: 45 pp, 35 figs.   (1984) **Løken, A.**

TARDIGRADA

How to begin the study of tardigrades. *Countryside* (N.S.) **18**(8): 1–11, 31 figs.
                                           (1958) **Le Gros, A. E.**

La vie des tardigrades. Gallimard, Paris. 132 pp, 40 pls.   (1928) **May, R. M.**

Über die Phylogenie und das System der Tardigraden. *Hereditas* **11**: 207–66, 29 figs.   (1928) **Thulin, G.**

Schema per una nuova sistemazione della famiglie e dei generi degli Eutardigarada. *Boll. Accad. gioenia Sci. nat.* (IV) **10**: 181–93.   (1969) **Pilato, G.**

Tardigrada. *Tierreich* **66**: xvii + 340 pp, 306 figs.   (1936) **Marcus, E.**

A review of the systematics and ecology of marine Tardigrada. *Smithson. Contr. Zool.* **76**: 109–17, 8 figs. (1971) **Renaud-Mornant, J. and Pollock, L. W.**

On some British marine Tardigrada, including two new species of *Batillipes*. *J. mar. biol. Ass. UK* **51**: 93–103, 3 figs.   (1971) **Pollock, L. W.**

Clare Island Survey. Arctiscoida. *Proc. R. Ir. Acad.* **31**(37): 1–16, 3 pls.
(1911) **Murray, J.**

Some notes on the Tardigrada of the Mulet Peninsula, including four additions to the Irish fauna and a key to the Irish species. *Ir. Nat. J.* **18**: 165–77, 1 fig.
(1975) **Morgan, C. I.**

Studies on the British tardigrade fauna; some zoogeographical and ecological notes. *J. nat. Hist.* **10**: 607–32, 6 figs.
(1976) **Morgan, C. I.**

British tardigrades. *Synopses Br. Fauna* (N.S.) **9**: vi + 132 pp, 28 figs, 5 maps.
(1976) **Morgan, C. I. and King, P. E.**

Scottish Tardigrada; a review of our present knowledge. *Ann. Scot. nat. Hist.* **78**: 88–95, 4 figs.
(1911) **Murray, J.**

Tardigrades. *Faune Fr.* **24**: 1–96, 98 figs.
(1932) **Cuénot, L.**

The Tardigrada of the Netherlands. A review of records from literature and a revision of the Loman collection. *Zool. Meded., Leiden* **38**: 195–206, 10 figs.
(1963) **Land, J. van der**

Danmarks Tardigrader. *Natur. Mus., Aarhus* **13**(4): 1–18, 19 figs.
(1969) **Hallas, T. E.**

Nordische Tardigraden. *Zool. Anz.* **27**: 168–72, 2 figs.   (1903) **Richters, F.**

Beiträge zur Kenntnis der Tardigradenfauna Schwedens. *Ark. Zool.* **7**(16): 1–60, 31 figs.
(1911) **Thulin, G.**

MOLLUSCA

About 1200 British species.

General

Handbuch der systematischen Weichtierkunde. G. Fischer, Jena. 2 vols, 1154 pp, 897 figs. [Reprinted (1963) by Asher, Amsterdam.]

Part 1 (1929), Loricata, pp. 1–22, figs 1–12; Gastropoda I: Prosobranchia, pp. 23–376, figs 13–470.

Part 2 (1931), Gastropoda II: Opisthobranchia, pp. 377–461, figs 471–553; Gastropoda III: Pulmonata, pp. 461–734, figs 554–782.

Part 3 (1934), Scaphopoda, pp. 779–82, figs 784–7; Bivalvia, pp. 782–948, figs 788–867; Cephalopoda, pp. 948–95, figs 868–93.

Part 4 (1935), Vergleichende Morphologie; Phylogenie; Geographische Verbreitung, pp. 1023–153, figs 894–7.

(1929–35) **Thiele, J.**

Handbuch der Paläozoologie. Borntraeger, Berlin.
(1938–60) **Schindewolf, O. H.**, ed.

Vol. 6 part 1, Gastropoda: Allgemeiner Teil und Prosobranchia. xii + 1639 pp, 4211 figs.   (1938–44; Reprinted 1960–62) **Wenz, W.**
Vol. 6 part 2, Gastropoda: Euthyneura. xii + 834 pp, 2515 figs.
(1959–60) **Zilch, A.**

Treatise on invertebrate paleontology. Geological Society of America, Kansas.
Part I (1960), Mollusca 1. Mollusca—general features, Scaphopoda, Amphineura, Monoplacophora, Gastropoda—general features, Archaeogastropoda and some Caenogastropoda and Opisthobranchia. xiii + 351 pp, 216 figs.
Part N (1969–71), Mollusca 6. Bivalvia. 3 vols. xxxvii, ii, iv + 1224 pp, 763 figs.   (1960–71) **Moore, R. C.**, ed.

A history of British Mollusca and their shells. van Voorst, London. 4 vols, 59 pls of Mollusca + 132 pls of shells.   (1848–53) **Forbes, E. and Hanley, S.**

British conchology. van Voorst, London. 5 vols, 40 pls of Mollusca + 102 pls of shells.   (1862–69) **Jeffreys, J. G.**

British shells. Warne, London. xii + 196 pp, 2 figs, 80 pls.
(1968) **McMillan, N. F.**

Mollusques. *La Faune de la France illustrée* **9**: 20–143, numerous figs.
(1930) **Perrier, R.**

Gastropoda Prosobranchia et Pulmonata. *Fauna Nederl.* **7**: 1–387, 300 figs.
(1933) **Jutting, T. van Benthem**

Lamellibranchia. *Tierwelt N.-u. Ostsee* **9d**: 1–96, 41 figs.   (1926) **Haas, F.**

Lamellibranchia. *Fauna Nederl.* **12**: 1–477, 144 figs.
(1943) **Jutting, T. van Benthem**

Marine

GENERAL

Molluscs: Caudofoveata, Solenogastres, Polyplacophora and Scaphopoda. *Synopses Br. Fauna* (NS) **37**: 123 pp, 27 figs, 3 pls, maps.
(1987) **Jones, A. M. and Baxter, J. M.**

Aculifera. *Tierwelt N.-u. Ostsee* **9a**: 1–64, 58 figs.
(1929) **Nierstrasz, H. F. and Hoffman, H.**

Concordance to the field card for British marine Mollusca. Conchological Society of Great Britain and Ireland, London. 66 pp.   (1973) **Turk, S. M.**

Sea area atlas of the marine Molluscs of Britain and Ireland. Nature Conservancy Council, Peterborough. 53 pp, 746 maps.   (1982) **Seaward, D. R.**

Nederlandse Zeemollusken. Wereldbibliotheek, Amsterdam. 231 pp, 44 figs, 16 pls.   (1942) **Kaas, P. and Ten Broek, A. N. C.**

Gastropoda Opisthobranchia; Amphineura et Scaphopoda. *Fauna Nederl.* **8**:
  1–106. Contains:
  Opisthobranchia: pp. 9–73, figs 1–29, 2 pls.
                                (1933) **Jutting, T. van Benthem and Engel, H.**
  Amphineura: pp. 75–87, figs 30–6 and Scaphopoda: pp. 89–94, figs 37–8.
                                        (1933) **Jutting, T. van Benthem**
An annotated check-list of marine molluscs of the Norwegian coast and
  adjacent waters. *Sarsia* **71**: 73–145.   (1986) **Høisæter, T.**
Mollusca regionis Arcticae Norvegiae. Christiania. xvi + 466 pp, 52 pls, map.
                                                  (1878) **Sars, G. O.**

CAUDOFOVEATA

Mollusca: Caudofoveata. *Mar. Invert. Scand.* **4**: 1–55, 59 figs.
                                      (1975) **Salvini-Plawen, L. von**

POLYPLACOPHORA

Catalogue of living chitons. Backhuys, Rotterdam. 144 pp, 1 fig.
                                      (1980) **Kaas, P. and Belle, R. A. van**
Monograph of living chitons. Brill, Leiden. Vol. 1, Order Neoloricata:
  Lepidopleurina. 240 pp, 95 figs, 45 maps. Vol. 2, Suborder Ischnochitonina.
  Ischnochitonidae: Schizoplacinae, Callochitoninae and Lepidochitoninae.
  198 pp, 76 figs, 40 maps.   (1985) **Kaas, P. and Belle, R. A. van**
A key for use in the identification of British chitons. *Proc. malac. Soc. Lond.* **29**:
  241–8, 6 figs.   (1953) **Matthews, G.**
The identification of British chitons. *Pap. Students conch. Soc. Gt Br. Ireland* **9**: 3
  pp, 1 pl.   (1967) **Matthews, G.**
A guide to the shell characters of British chitons. *Conchologists' Newsl.* No. 93:
  259–64, 5 figs.   (1985) **Palmer, C. P.**
Skallus, Søtænder, Blæksprutter. *Danm. Fauna* **65**: 25–52, figs 13–29.
                                                  (1959) **Muus, B. J.**
Scandinavian species of *Leptochiton* Gray, 1847. *Sarsia* **66**: 217–29, 10 figs.
                                                  (1981) **Kaas, P.**
The genus *Lepidochitona* Gray, 1821, in the northeastern Atlantic Ocean, the
  Mediterranean Sea and the Black Sea. *Zool. Verh. Leiden* **185**: 1–43, 128 figs,
  3 maps.   (1981) **Kaas, P. and Belle, R. A. van**
The genus *Acanthochitona* Gray, 1821, in the north-eastern Atlantic Ocean and
  in the Mediterranean Sea, with designations of neotypes of *A. fascicularis* (L.,
  1767) and of *A. crinita* (Pennant, 1777). *Bull. Mus. natn. Hist. nat., Paris* (4) **7**
  (A) No. 3: 579–609, 92 figs.   (1985) **Kaas, P.**

[Shell-bearing mollusks (Loricata) of the seas of the USSR.] *Opred. Faune SSSR* **45**: 107 pp, 53 figs, 11 pls [Russian]. English edition (1965) by Israel Program for Scientific Translations, Jerusalem. (1952) **Yakovleva, A. M.**

GASTROPODA

Iconographie der schalentragenden europäischen Meeresconchylien. Cassel. 4 vols, 126 pls. (1887–1908) **Kobelt, W.**
A facsimile reprint of this rare and important work is proposed for publication in 1986–87 by Carlo Settepassi of Rome.

Die europäischen Meeres-Gehäuseschnecken (Prosobranchia) vom Eismeer bis Kapverden, Mittelmeer und Schwarzes Meer, 2nd edn. G. Fischer, Stuttgart. xii + 539 pp, 112 pls (8 col.). (1982) **Nordsieck, F.**

Key to the British marine gastropods. *Inf. Ser. R. Scott. Mus.* (Nat. Hist.) **2**: viii + 44 pp, 4 figs. (1974) **Smith, S. M.**

The prosobranch molluscs of Britain and Denmark. Part 1 (1976). Pleurotomariacea, Fissurellacea and Patellacea. *J. moll. Stud.* Suppl. **1**: ii, 1–37, figs 1–25. Part 2 (1977). Trochacea. *J. moll. Stud.* Suppl. **3**: 39–100, figs 26–74. Part 3 (1978). Neritacea, Viviparacea, terrestrial and freshwater Littorinacea and Rissoacea. *J. moll. Stud.* Suppl. **5**: 101–52, figs 101–30. Part 4 (1978). Marine Rissoacea. *J. moll. Stud.* Suppl. **6**: 153–241, figs 131–95. Part 5 (1980). Marine Littorinacea. *J. moll Stud.* Suppl. **7**: 243–84, figs 196–214. Part 6 (1981). Cerithiacea, Strombacea, Hipponicacea, Calyptraeacea, Lamellariacea, Cypraeacea, Naticacea, Tonnacea, Heteropoda. *J. moll. Stud.* Suppl. **9**: 285–363, figs 215–56. Part 7 (1982). 'Heterogastropoda' (Cerithiopsacea, Triforacea, Epitoniacea, Eulimacea). *J. moll. Stud.* Suppl. **11**: 363–434, figs 257–309. Part 8 (1985). Neogastropoda. *J. moll. Stud.* Suppl. **15**: 435–556, figs 310–76. (1976–85) **Fretter, V. and Graham, A.**
Part 9 (1986). Pyramidellacea. *J. moll. Stud.* Suppl. **16**: 557–649, figs 377–451. (1986) **Fretter, V., Graham, A., and Andrews, E. B.**

Revision of the northeast Atlantic bathyal and abyssal Aclididae, Eulimidae, Epitoniidae. *Boll. malac.* Suppl. **2**: 297–576, figs 724–1267.
(1986) **Bouchet, P. and Warén, A.**

Revision of the northeast Atlantic bathyal and abyssal Neogastropoda excluding Turridae. *Boll. malac.* Suppl. **1**: 121–296, figs 282–723.
(1985) **Bouchet, P. and Warén, A.**

Revision of the north-east Atlantic bathyal and abyssal Turridae. *J. moll. Stud.* Suppl. **8**: 1–119, figs 1–281. (1980) **Bouchet, P. and Warén, A.**

On the systematics of recent *Rissoa* of the subgenus *Turboella* Gray, 1847, from the Mediterranean and European Atlantic coasts. *Basteria* **40**: 21–73, 7 figs, 9 pls. (1976) **Verduin, A.**

Revision des *Skeneopsis* (Gastropoda, Rissoacea) européennes et nord-africaines, avec description d'une espèce nouvelle. *Boll. malac.* **18**: 225–34, 8 figs. (1982) **Gofas, S.**

Revision of the east Atlantic and Mediterranean Caecidae. *Basteria* **41**: 7–19, 25 figs.   (1977) **Aartsen, J. J. van**

On taxonomy and variability of Recent European species of the genus *Bittium* Leach. *Basteria* **46**: 93–120, 31 figs.   (1982) **Verduin, A.**

European Pyramidellidae. I. *Chrysallida*. *Conchiglie* **13**: 49–64, 4 pls.
                                                      (1977) **Aartsen, J. J. van**

European Pyramidellidae. II. *Turbonilla*. *Boll. malac.* **17**: 61–88, 2 figs, 6 pls.
                                                      (1977) **Aartsen, J. J. van**

Pseudothecosomata, Gymnosomata and Heteropoda. Bohn, Scheltema, and Holkema, Utrecht. 484 pp, 166 figs, 81 maps.   (1976) **Spoel, S. van der**

Heteropoda. *Fich. Ident. Zooplancton* **66**: 4 pp, 12 figs.   (1957) **Dales, R. P.**

Heteropoda. *Oceanogr. mar. Biol.* **11**: 237–61, 11 figs.
                                                  (1973) **Thiriot-Quiévreux, C.**

Revision des Trophoninae d'Europe. *Inf. Soc. belg. Malac.* **9**: 1–70, 24 figs, 6 pls.
                                                        (1981) **Houart, R.**

Les Terebridae de l'Atlantique oriental. *Boll. malac.* **18**: 185–216, 53 figs.
                                                        (1982) **Bouchet, P.**

Prosobranchia. Veliger larvae of Taenioglossa and Stenoglossa. *Fich. Ident. Zooplancton* **129–132**: 26 pp, 35 figs.
                                      .   (1970) **Fretter, V. and Pilkington, M. C.**

British prosobranch Mollusca: their functional anatomy and ecology. Ray Society, London. 755 pp, 317 figs.   (1962) **Fretter, V. and Graham, A.**

British prosobranchs. *Synopses Br. Fauna* (N.S.) **2**: 1–114, 119 figs.
                                                        (1971) **Graham, A.**

Die Schnecken (Gastropoda Prosobranchia) der deutschen Meeresgebiete und brackigen Küstengewässer. *Helgoländer wiss. Meeresunters.* **13**: 1–61, 7 figs, 20 pls.   (1966) **Ziegelmeier, E.**

Die Schnecken (Gastropoda Prosobranchia) der deutschen Meeresgebiete und brackigen Küstengewässer, revised edn. Biologische Anstalt Helgoland, Hamburg. 66 pp, 7 figs, 20 pls.   (1973) **Ziegelmeier, E.**

Skeldýrafána Íslands. II. Sæsniglar með skel (Gastropoda Prosobranchia and Tectibranchia). Prentsmiðjan Leiftur, Reykjavík. 167 pp, 156 figs.
                                                        (1962) **Óskarsson, I.**

Framgälade snäckor från svenska västkusten. Naturhistoriska Museet, Göteborg. 74 pp, 284 figs.   (1976) **Hubendick, B. and Warén, A.**
Republication in book form of seven papers on Småsnäckor vid svenska västkusten from the *Årstryck* of Göteborgs Naturhistoriska Museum, 1969–75.

Mollusca Prosobranchia. Forgjellesnegler nordiske marine arter. Universitetsforlaget, Oslo. 55 pp, 60 figs.   (1975) **Sneli, J. A.**

Revision of the Rissoidae from the Norwegian North Atlantic Expedition 1876–78. *Sarsia* **53**: 1–13, 26 figs.   (1973) **Warén, A. H.**

Revision of the Arctic–Atlantic Rissoidae. *Zoologica Scr.* **3**: 121–35, 3 figs, 3 pls.
(1974) **Warén, A.**

An account of the studies on marine prosobranch gastropods in the USSR. *Shell* **16** (178–9): 21–3.   (1984) **Golikov, A. N.**
Russian publications on the prosobranchs of the coasts of the USSR have not been included here as generally the preponderance of the included species are those of the Siberian Arctic and the North Pacific oceans. This bibliography is a very useful guide to an extensive but little-known literature.

Die europäischen Meeresschnecken (Opisthobranchia mit Pyramidellidae; Rissoacea) vom Eismeer bis Kapverden, Mittelmeer und Schwarzes Meer. G. Fischer, Stuttgart. xiii + 327 pp, 41 pls (4 col.).   (1972) **Nordsieck, F.**

An annotated list of the North Atlantic Opisthobranchia (excluding Thecosomata and Gymnosomata). *Ophelia* Suppl. **2**, Appendix: 150–70, 1 fig.
(1985) **Platts, E.**

Euthecosomata, a group with remarkable developmental stages (Gastropoda, Pteropoda). Noorduijn, Gorinchem. 375 pp, 338 figs, 28 maps. (1967) **Spoel, S. van der**

Pteropoda Thecosomata. *Fich. Ident. Zooplancton* **140–2**: 12 pp, 57 figs.
(1972) **Spoel, S. van der**

Opisthobranchia Gymnosomata. *Fich. Ident. Zooplancton* **79–80**: 8 pp, 12 figs.
(1958) **Morton, J. E.**

Opisthobranchia. The veliger larvae of the Nudibranchia. *Fich. Ident. Zooplancton* **106**: 3 pp, 9 figs.   (1965) **Hadfield, M. G.**

Opisthobranchia. *Tierwelt N.-u. Ostsee* **9c**: 1–52, 30 figs. (1926) **Hoffmann, H.**

North Atlantic nudibranchs (Mollusca) seen by Henning Lemche. *Ophelia* Suppl. **2**: 1–150, 4 figs, 60 colour pls.   (1985) **Just, H. and Edmunds, M.**

British opisthobranch molluscs. *Synopses Br. Fauna* (N.S.) **8**: 1–204, 105 figs, 1 pl.   (1976) **Thompson, T. E. and Brown, G. H.**

Biology of opisthobranch molluscs. Ray Society, London. 2 vols.
Vol. 1, General account; Bullomorpha, Aplysiomorpha, Pleurobranchomorpha, Acochlidiacea and Sacoglossa. 206 pp, 106 figs, 20 pls (8 col.).
(1976) **Thompson, T. E.**
Vol. 2, Nudibranchia. 229 pp, 40 figs, 57 pls (35 col.), 72 maps.
(1984) **Thompson, T. E. and Brown, G. H.**

Nudibranchs of the British Isles—a colour guide. Underwater Conservation Society, Ross-on-Wye. 30 pp + 107 colour photographs.
(1979) **Brown, G. H. and Picton, B. E.**

Mollusques opisthobranches. *Faune Fr.* **58**: 1–460, 173 figs, 1 col. pl.
(1954) **Pruvot-Fol, A.**

Northern and Arctic tectibranch gastropods. I. The larval shells. II. A revision of the cephalaspid species. *Biol. Skr.* **5**(3): 1–136.
(1948) **Lemche, H.**

SCAPHOPODA

Scaphopoda. *Tierwelt N.-u. Ostsee* **9c**: 67–80, 12 figs.
(1926) Jutting, T. van Benthem

Skallus, Søtænder, Blæksprutter. *Danm. Fauna* **65**: 53–70, figs 30–43.
(1959) Muus, B. J.

PELECYPODA

Die europäischen Meeresmuscheln (Bivalvia) vom Eismeer bis Kapverden, Mittelmeer und Schwarzes Meer. G. Fischer, Stuttgart. xiii + 256 pp, 27 pls (2 col.).
(1969) Nordsieck, F.

Liste provisoire des mollusques marins de la province celtique et leur synonymes. I. Pélécypodes. *Inf. Soc. belg. Malac.* **8**: 35–56, 8 figs + 63 pp systematic list of synonyms + 48 pp alphabetical list of synonyms.
(1980) Peuchot, R. and Tassin, A.

The identification and classification of lamellibranch larvae. *Hull. Bull. mar. Ecol.* **3**: 73–104, 4 figs, 5 pls. (1950) Rees, C. B.

Espèces européennes récentes de la superfamille des Veneroidea C S Rafinesque, 1815. *Inf. Soc. belg. Malac.* **6**: 45–60, 12 figs.
(1978) Lambiotte, M.

The genus *Ensis* in Europe. *Basteria* **28**: 13–44, 4 pls. (1964) Urk, R. M. van

British bivalve seashells, 2nd edn. HMSO, Edinburgh. 212 pp, 110 figs, 12 pls.
(1976) Tebble, N.

An annotated key to the British species of Cardiidae. *Pap. Students conch. Soc. Gt Br. Ireland* **16**: 8 pp, 2 pls. (1978) Turk, S. M.

Pélécypodes marins de la côte atlantique française. Laboratoire de Zoologie, Brest. 21 pp.
(1968) Glémarec, M.

Die Muscheln (Bivalvia) der deutschen Meeresgebiete. *Helgoländer wiss. Meeresunters.* **6**: 1–51, 3 figs, 13 pls. (1957) Ziegelmeier, E.

Die Muscheln (Bivalvia) der deutschen Meeresgebiete, revised edn. Biologische Anstalt Helgoland, Hamburg. 56 pp, 3 figs, 14 pls. (1962) Ziegelmeier, E.

Saltvandsmuslinger. *Danm. Fauna* **40**: 1–208, 175 figs.
(1934) Jensen, A. S. and Sparck, R.

Skeldýrafána Íslands, 2nd edn. I. Samlokur í sjó (Lamellibranchia). Bókaútgáfan Asór, Reykjavík. Vol. 1, 123 pp, 108 figs. (1964) Óskarsson, I.

[*Key to the genera of Neogene bivalves of the USSR (on punched cards).*] Nauka, Moscow. 40 pp, 4 figs + 213 illustrated punched cards. [Russian].
(1974) Merklin, R. L. and Nevesskaya, L. A.

CEPHALOPODA

A monograph of the Recent Cephalopoda. British Museum (Natural History), London. Vol. 1 (1929), Octopodinae. xi + 595 pp, 89 figs, 7 pls. Vol. 2 (1932), Octopoda (excluding Octopodinae). xi + 359 pp, 79 figs, 6 pls.
(1929 and 1932) **Robson, G. C.**

FAO species catalogue. Cephalopods of the world. An annotated and illustrated catalogue of species of interest to fisheries. *FAO Fish. Synopsis* **125**, Vol. 3, viii + 277 pp, 78 figs, species illustrations and distribution maps.   (1984) **Roper, C. F. E., Sweeney, M. J., and Nauen, C. F.**

A handbook for the identification of cephalopod beaks. Clarendon Press, Oxford. 273 pp, 139 figs.   (1986) **Clarke, M. R.**, ed.

Sepioidea. *Fich. Ident. Zooplancton* **94**: 5 pp, 2 pls. Reprinted (1985) in *Conchologists' Newsl.* **92**: 238–42.   (1963) **Muus, B. J.**

Teuthoidea. *Fich. Ident. Zooplancton* **95–7**: 14 pp, 6 pls. Reprinted (1985) in *Conchologists' Newsl.* **92–3**: 243–9, 265–8.   (1963) **Muus, B. J.**

Octopoda. *Fich. Ident. Zooplancton* **98**: 4 pp, 5 figs. Reprinted (1985) in *Conchologists' Newsl.* **93**: 269–71.   (1963) **Muus, B. J.**

An illustrated key to the families of the order Teuthoidea. *Smithson. contrib. Zool.* **13**: 1–32, 2 figs, 16 pls.
(1969) **Roper, C. F. E., Young, R. E., and Voss, G. L.**

Cephalopoden. *Tierwelt N.-u. Ostsee* **9b**: 479–723, 86 figs.
(1958) **Jaeckel , S. G. A.**

Skallus, Søtænder, Blæksprutter. *Danm. Fauna* **65**: 71–228, figs 44–117.
(1959) **Muus, B. J.**

[*Cephalopods of the seas of the USSR.*] Akademii Nauk SSSR, Moscow. 235 pp, 60 figs. [Russian]. English edition (1965) by Israel Program for Scientific Translations, Jerusalem.   (1963) **Akimushkin, I. I.**

Non-marine

GENERAL

Mollusca. *Tierwelt Mitteleur.* **2**(1): 1–264, 147 figs, 13 pls. (1956) **Ehrmann, P.**
Mollusken. Nachtrag. *Tierwelt Mitteleur.* **2**(1) Ergänzung: 1–294, 9 pls.
(1962) **Zilch, A. and Jaeckel, S. G. A.**

Einiger Bemerkungen zum Ergänzungsband zu Ehrmann's 'Mollusca' in 'Die Tierwelt Mitteleuropas'. *Arch. Molluskenk.* **95**: 49–68, 6 figs.
(1966) **Waldén, H. W.**

Land and freshwater molluscs. Burke: *Young specialist* series, London. 179 pp, 34 + 164 figs, 4 pls.   (1965) **Janus, H.**

Synonymy of British non-marine Mollusca. British Museum (Natural History), London. xxiv + 447 pp.   (1926) **Kennard, A. S. and Woodward, B. B.**

A list of the fresh and brackish-water Mollusca of the British Isles. *J. Conch.,*
Lond. **29**: 26–8.   (1976) **Kerney, M. P.**

Atlas of the non-marine Mollusca of the British Isles. Institute of Terrestrial Ecology,
Cambridge. 8 pp + 200 maps.   (1976) **Kerney, M. P.**, ed.

Mollusques terrestres et fluviatiles. *Faune Fr.* **21, 22**: 897 + xiv pp, 860 figs, 26
pls.   (1930–31) **Germain, L.**

Mollusques I. Mollusques terrestres et dulcicoles. *Faune Belg.* 402 pp, 163 figs,
4 pls.   (1960) **Adam, W.**

Liste commentée des mollusques recents non-marins de Belgique. *Docums*
Trav. Inst. r. Sci. nat. Belg. **17**: 38 pp. [Also published in Flemish as **16** of this
series.]   (1984) **Goethem, J. L. van**

Zoetwatermollusken van Nederland. Nederlandse Jeugdbond voor Natuurstudie,
Amsterdam. 160 pp, 12 figs, 17 pls.
                                    (1965) **Janssen, A. W. and Vogel, E. F. de**

Unsere Schnecken und Muscheln. Stuttgart. 124 pp, 154 figs, 4 pls.
                                                           (1958) **Janus, H.**

Süsswassermollusken. Deutscher Jugendbund für Naturbeobachtung, Hamburg.
73 pp, illustrated with photographs and figs.
                (1980) **Glöer, P., Meier-Brook, C., and Ostermann, O.**

Ferskvandsbløddyr. *Danm. Fauna* **54**: 1–175, 100 figs.
                                                           (1949) **Mandahl-Barth, G.**

GASTROPODA

A field guide to the land snails of Britain and north-west Europe. Collins, London. 288
pp, 167 figs, 24 col. pls, 396 maps.
                            (1979) **Kerney, M. P. and Cameron, R. A. D.**

Die Landschnecken Nord- und Mitteleuropas. Parey, Hamburg. 384 pp, 482 figs, 24
col. pls, 368 maps.
                (1983) **Kerney, M. P., Cameron, R. A. D., and Jungbluth, J. H.**

Recent Lymnaeidae, their variation, morphology, taxonomy, nomenclature
and distribution. *K. svenska VetenskAkad. Handl.* (4) **3**(1): 1–122, 369 figs, 5
pls.   (1951) **Hubendick, B.**

The European freshwater limpets (Ancylidae and Acroloxidae). *Inf. Soc. belg.*
Malac. **1**: 109–28, 39 figs.   (1972) **Hubendick, B.**

Taxonomische Revision paläarktischer Zonitinae. *Arch. Molluskenk.* **86**: 101–
36, 19 figs; **88**: 7–34, 13 figs, 3 pls; **89**: 1–22, 4 figs, 2 pls.
                                                           (1957–60) **Forcart, L.**

Epiphallus anatomy in the *Arion hortensis* species aggregate. *Zoologica Scr.* **15**:
61–8, 5 figs.   (1986) **Backeljau, T. and Beeck, M. van**

Segregates of the *Arion hortensis* complex (Pulmonata: Arionidae), with the description of a new species. *J. Conch., Lond.* **30**: 123–7. (1979) **Davies, S. M.**

British snails, 2nd edn. Clarendon Press, Oxford. 298 pp, 11 figs, 14 pls.
(1969) **Ellis, A. E.**

A nomenclatural list of the land Mollusca of the British Isles. *J. Conch., Lond.* **29**: 21–5. (1976) **Waldén, H. W.**

A key to the British fresh- and brackish-water gastropods with notes on their ecology, 4th edn. *Scient. Publs Freshwat. biol. Ass.* **13**: 1–46, 15 figs.
(1977) **Macan, T. T.**

British slugs (Pulmonata; Testacellidae, Arionidae, Limacidae). *Bull. Br. Mus. nat. Hist.* (Zool.) **6**: 105–226, 19 figs, 1 pl, 20 maps. (1960) **Quick, H. E.**

A field key to the slugs of the British Isles. *Fld Stud.* **5**: 807–24, 2 figs, 4 col. pls.
(1983) **Cameron, R. A. D., Eversham, B., and Jackson, N.**

Atlas provisoire des gastéropodes terrestres de la Belgique. Institut Royal des Sciences Naturelles de Belgique, Bruxelles. 285 pp, 133 figs each with full-page distribution map.
(1986) **Wilde, J. J. de, Marquet, R., and Goethem, J. L. van**

De Landslakken van Nederland, 2nd edn. KNNV, Hoogwoud. 184 pp, 192 figs, 102 maps. (1984) **Gittenberger, E., Backhuys, W., and Ripken, T. E. J.**

Landsnegle. *Danm. Fauna* **10**: 1–221, 181 figs. (1911) **Steenberg, C. M.**
[Additions and corrections.] *Danm. Fauna* **54**: 221–40, 7 figs.
(1949) **Mandahl-Barth, G.**

Ferskvandssneglenes ægkapsler. *Danm. Fauna* **54**: 176–220, 29 figs.
(1949) **Bondesen, P.**

Våra snäckor. Snäckor; söt och bräckt vatten. Bonniers, Stockholm. 100 pp, 172 figs, 4 pls. (1949) **Hubendick, B.**

The land Gastropoda of the vicinity of Stockholm. *Ark. Zool.* **7**: 391–448, 15 figs, 1 pl. (1955) **Waldén, H. W.**

PELECYPODA

British freshwater bivalve Mollusca: keys and notes for the identification of the species. *Synopses Br. Fauna* (N.S.) **11**: 1–109, 39 figs, 15 pls.
(1962) **Ellis, A. E.**

BRYOZOA (ECTOPROCTA)

More than 400 species have been recorded from the Arctic and boreal marine and fresh waters of Europe, of which 286 are known from Britain.

General

Bryozoans. Hutchinson, London. 175 pp, 21 figs.   (1970) **Ryland, J. S.**

Bryozoa as a community component on the north east coast of Britain. In *Living and fossil Bryozoa* (ed. G. P. Larwood), pp. 21–36, 3 figs. Academic Press, London and New York.   (1973) **Moore, P. G.**

A revised key for the identification of intertidal Bryozoa (Polyzoa). *Fld Stud.* **4**: 77–86, 4 figs.   (1974) **Ryland, J. S.**

The marine fauna of the Cullercoats district, 3a, Ectoprocta. *Rep. Dove mar. Lab.* **3**(18): 5–30.   (1975) **Eggleston, D.**

Physiology and ecology of marine Bryozoans. *Adv. Mar. Biol.* **14**: 285–443, 44 figs.   (1976) **Ryland, J. S.**

Biology of Bryozoans. Academic Press, London and New York. xvii + 566 pp, 274 figs.   (1977) **Woollacott, R. M. and Zimmer, R. L.**, eds.
Detailed reviews of aspects of bryozoan biology; a good source of references.

Fouling and settlement

Polyzoa. Catalogue of main marine fouling organisms. OECD, Paris. Vol. 2, 83 pp, 38 figs.   (1965) **Ryland, J. S.**

Polyzoa (Bryozoa) order Cheilostomata, Cyphonautes larvae. *Fich. Ident. Zooplancton* **107**, 5 pp.   (1965) **Ryland, J. S.**

Bryozoa and marine fouling. In *Marine borers, fungi and fouling organisms of wood* (ed. G. Jones and S. K. Eltringham), pp. 137–54. OECD, Paris.
                    (1972) **Ryland, J. S.**

PHYLACTOLAEMATA

25–30 species, nine known from Britain.

Bryozaires, Pt 1. Entoproctes, Phylactolèmes. *Faune Fr.* **60**: 1–398, 151 figs.
                    (1956) **Prenant, M. and Bobin, G.**
Includes some Ctenostomata, see next section, and Entoprocta, see p. 60.

A monograph of the freshwater Bryozoa—Phylactolaemata. *Zool. Verh. Leiden* **93**: 1–159, 13 figs, 18 pls.   (1968) **Lacourt, A. W.**

The freshwater Ectoprocta. In *Living and fossil Bryozoa* (ed. G. P. Larwood), pp. 503–22, 6 figs. Academic Press, London and New York.
                    (1973) **Bushnell, J. M.**

A key to the British and European freshwater Bryozoans. *Sci. Publ. Freshwater*

Biol. Ass. **41**: 1–32, 11 figs.   (1980) **Mundy, S. P.**
Includes freshwater Ctenostomata, see next section.

General features of the Class Phylactolaemata. In *Treatise on invertebrate paleontology* (ed. R. A. Robinson), Part G, Bryozoa Revised, Vol. 1 (ed. R. S. Boardman *et al.*), pp. 287–303, 11 figs. Geological Society of America, New York and University of Kansas, Lawrence, Kansas.   (1983) **Wood, T. S.**

GYMNOLAEMATA (CTENOSTOMATA and CHEILOSTOMATA)

Mosdyr (Bryozoa eller Polyzoa). *Danm. Fauna* **46**: 1–401, 221 figs. [Danish].
                                                        (1940) **Marcus, E.**
Systematic and biological studies on Polyzoa (Bryozoa) from western Norway. *Sarsia* **14**: 1–59, 14 figs.   (1963) **Ryland, J. S.**
Bryozoaires (Pt 2), Chilostomes *Anasca. Faune Fr.* **68**: 1–647, 210 figs.
                                        (1966) **Prenant, M. and Bobin, G.**
Keys to species.

[Bryozoa of the Northern Seas of the USSR.] *Opred Faune SSSR* **76**: 1–584, 404 figs. [Russian]. English translation (1975) by Amerind Publishing, New Delhi. xxiv + 711 pp, 404 figs.   (1962) **Kluge, G. A.**
Keys to species; includes Stenolaemata (next section).

British Anascan Bryozoans. *Synopses Br. Fauna* (N.S.) **10**, vi + 188 pp, 85 figs.
                                    (1977) **Ryland, J. S. and Hayward, P. J.**
Keys to species.

British Ascophoran Bryozoans. *Synopses Br. Fauna* (N.S.) **14**, 312 pp, 129 figs.
                                    (1979) **Hayward, P. J. and Ryland, J. S.**
Keys to species.

Ctenostome Bryozoans. *Synopses Br. Fauna* (N.S.) **33**, 169 pp, 53 figs.
                                                    (1985) **Hayward, P. J.**
Keys to species.

STENOLAEMATA

On the development of *Tubulipora*, and on some British and Northern species of this genus. *Q. Jl micr. Sci.* **41**: 73–158, 3 pls.   (1898) **Harmer, S. F.**
Crisiidae (Polyzoa) from western Norway. *Sarsia* **29**: 269–82, 5 figs.
                                                        (1967) **Ryland, J. S.**
Bryozoans. *Ophelia* **7**(2): 217–56, 41 figs.   (1970) **Nielsen, C.**
Le sous-ordre des Tubuliporina (Bryozoaires Cyclostomes) en Méditerranée, écologie, et systématique. *Mém. Inst. Océanogr. Monaco* **10**: 1–326, 50 figs, 28 pls.   (1976) **Harmelin, J.-G.**
Keys to species, many of which are found in Britain.

Cystid structure and protrusion of the polypide in *Crisia* (Bryozoa, Cyclostomata). *Acta Zool. Stockh.* **60**: 65–88, 24 figs.

(1979) **Nielsen, C. and Pedersen, K. J.**

Cyclostome Bryozoans. *Synopses Br. Fauna* (N.S.) **34**, 147 pp, 48 figs.

(1985) **Hayward, P. J. and Ryland, J. S.**

Keys to species.

PHORONIDEA

British and other phoronids. *Synopses Br. Fauna* (N.S.) **13**, 57 pp, 16 figs.

(1979) **Emig, C. C.**

Phoronidea. *Tierwelt Nord- u. Ostsee.* **7c**: 101–32, 24 figs.   (1932) **Cori, C. J.**

Phoronidea, Phoronidae, Actinotrocha larvae. *Fich. Ident. Zooplancton* **69**, 4 pp.

(1957) **Forneris, L.**

BRACHIOPODA

A monograph of recent Brachiopoda. *Trans. Linn. Soc. Lond.* (2) **4**: 1–248, 30 pls.   (1886–88) **Davidson, T.**

Die anatomie von *Crania anomala* O.F.M. *Untersuchungen über den bau der brachiopoden.* Jena. Vol. I, 65 pp, 7 pls.   (1892) **Blochmann, F.**

Die Brachiopoden der deutschen Tiefsee-Expedition. *Wiss. Ergebn. dt. Tiefsee-Exped. Valdiva, 1898–99* **25**(3): 255–316, 43 figs.   (1940) **Helmcke, J. G.**

Brachiopoda. In *Treatise on invertebrate paleontology* (ed. R. C. Moore), Part H, pp. 524–927, figs. 398–746. University of Kansas Press, Lawrence, Kansas and Geological Society of America, New York.   (1965) **Williams, A.**

Brachiopods in the Linnean collection. *Proc. Linn. Soc. Lond.* **178**: 161–83, 4 pls.

(1967) **Brunton, C. H. C., Cocks, L. R. M., and Dance, S. P.**

Living and fossil brachiopods. Hutchinson, London. 199 pp, 99 figs.

(1970) **Rudwick, M.**

The growth stages of the lophophore of the brachiopods *Platidia davidsoni* (Eudes Deslongchamps) and *P. anomioides* (Philippi), with notes on the feeding mechanism. *J. mar. biol. Ass. UK* **38**: 103–32, figs 1–23.

(1959) **Atkins, D.**

The early growth stages and adult structure of the lophophore of *Macandrevia cranium* (Müller) (Brachiopoda, Dallinidae). *J. mar. bio. Ass. UK* **38**: 335–50, 2 figs.   (1959) **Atkins, D.**

The generic position of the brachiopod *Megerlia echinata* (Fischer and Oehlert). *J. mar. biol. Ass. UK* **41**: 89–94, 4 figs.   (1961) **Atkins, D.**

The growth stages and adult structure of the lophophore of the brachiopods
  Megerlia truncata (L.) and *M. echinata* (Fischer and Oehlert). *J. mar. biol. Ass.*
  UK **41**: 95–111, 16 figs.   (1961) **Atkins, D.**

The lophophore and ciliary feeding mechanism of the brachiopod *Crania*
  anomala (Müller). *J. Mar. biol. Ass. UK* **42**: 469–80, 4 figs.
                                  (1962) **Atkins, D. and Rudwick, M. J. S.**

The Recent brachiopoda of the Mediterranean Sea. *Bull. Inst. Océanogr.*
  Monaco. 112 pp, 12 pls.   (1979) **Logan, A.**
  Includes good illustrations of British species.

Distribution et synécologie des fonds à *Gryphus vitreus* (Brachipoda) en Corse.
  Mar. Biol. Berlin **90**: 139–46, 3 figs.   (1985) **Emig, C. C.**
  Species not known north of the Bay of Biscay.

A new species and genus of Brachiopoda from the Western Approaches and
  the growth stages of the lophophore. *J. mar. biol. Ass. UK* **39**: 71–89, 14 figs, 1
  pl.   (1959) **Atkins, D.**

The brachiopoda of the coast of Ireland. *Proc. R. Ir. Acad.* **37B**(6): 37–46.
                                                                   (1925) **Massy, A. L.**
  Eleven species listed.

British brachiopods. *Synopses Br. Fauna* (N.S.) **17**, 64 pp, 30 figs.
                                  (1979) **Brunton, C. H. C. and Curry, G. B.**
  Contains keys to British species.

Ecology and population structure of the Recent brachiopod *Terebratulina* from
  Scotland. *Palaeontology* **25**: 227–46, 8 figs.   (1982) **Curry, G. B.**

Ecology of the Recent deep-water rhynchonellid brachiopod *Cryptopora* from
  the Rockall trough. *Palaeogeog. Palaeoclimat. Palaeoecol.* **44**: 93–102, 2 pls.
                                                                   (1983) **Curry, G. B.**

Brachiopoda. *Zoology of the Faroes.* Copenhagen. 8 pp.
                                                  (1940) **Wesenberg-Lund, E.**

Brachiopoda. *The Danish Ingolf-Expedition.* Copenhagen. Vol. 4 (12), 17 pp, 3
  figs, 10 charts.   (1941) **Wesenberg-Lund, E.**

Brachiopoda. *The zoology of Iceland.* Copenhagen and Reykjavik. Vol. 4, Pt 67,
  11 pp.   (1938) **Wesenberg-Lund, E.**

Brachiopods from Swedish waters in the Museum of Natural History of
  Gothenburg. *Göteborgs K. Vetensk.-o. vitterhSamh. Handl.* **90**: 8 pp.
                                                  (1941) **Wesenberg-Lund, E.**

CHAETOGNATHA

About 20 species are known from the NE Atlantic.

Sagitta. L.M.B.C. Mem. typ. Br. Mar. Pl. Anim. **28**: 96 pp, 10 figs, 12 pls.
                                                  (1927) **Burfield, S. T.**

Chaetognatha. *Bronn's Kl. Ordn. Tierreichs* **4**(4), **2**(1): 220 pp, 165 figs.
(1938) **Kuhl, W.**

Chaetognatha, 1st revision. *Fich. Ident. Zooplancton* **1**: 6 pp, 18 figs.
(1957) **Fraser, J. H.**

Los quetognatos del atlantico, distribucion y notas esenciales de sistematica. *Trab. Inst. esp. Oceanogr.* **37**: 1–290. (1969) **Alvariño, A.**

Revision der Chaetognathen. *Dt. Südpol. Exped.* **13** (Zool. **5** 1): 1–71, 51 figs.
(1911) **Ritter-Zahony, R. von**

Chaetognatha. *Synopses Br. Fauna* (NS)
(In press) **Pierrot-Bults, A. C. and Chidgey, K. C.**

The Chaetognatha and other zooplankton of the Scottish area and their value as biological indicators of hydrographical conditions. *Mar. Res.* **2**: 52 pp, 4 figs, 3 pls, 21 charts. (1952) **Fraser, J. H.**

Chaetognathes et zooplancton du secteur atlantique marocain. *Revue Trav. Inst. Pêch. marit.* **21** (1–2): 1–356, 104 figs. (1957) **Furnestin, M. L.**

Taxonomy and distribution of certain members of the 'Sagitta serratodentata-group' (Chaetognatha). *Bijdr. Dierk.* **44**(2): 214–34, 11 figs.
(1974) **Pierrot-Bults, A. C.**

Quatre nouveaux chaetognathes atlantiques abyssaux (genre *Heterokrohnia*), description, remarques éthologiques et biogéographique. *Oceanol. Acta.* **9**(4): 469–77, 7 figs. (1986) **Casanova, J.-P.**

Two new species of *Heterokrohnia* (Chaetognatha) from Antarctic waters. *Polar Biol.* **4**: 53–9. (1985) **Kapp, H. and Hagen, W.**

ECHINODERMATA

The Hamlyn guide to the seashore and shallow seas of Britain and Europe. Hamlyn, London, New York, Sydney, Toronto. 320 pp, illustrated.
(1976) **Campbell, A. C. and Nicholls, J.**

Handbook of the echinoderms of the British Isles. Oxford University Press, London. 471 pp, 69 figs. (1927) **Mortensen, T.**

Report on the zoological excursion to Sherkin Island, Co. Cork, Ireland. In *Rep. third zool. Excur. Sherkin Island.* Instituut voor Taxonomische Zoologie, Universiteit van Amsterdam, Amsterdam. Vol. 3, 37 pp. (1979) **Anon.**

A note on the British species of cucumarians, involving the erection of two new nominal genera. *J. mar. Biol. Ass. UK* **50**: 683–7. (1970) **Rowe, F. W. E.**

The marine flora and fauna of the Isles of Scilly: Echinodermata. *J. nat. Hist.* **5**: 233–8. (1971) **Rowe, F. W. E.**

The marine fauna of the Cullercoats district: *Echinodermata. Rep. Dove mar. Lab.* **15**(3): 21–39.   (1966) **Buchanan, J. B.**

Flore et faune fixées sous-marins de Bretagne, 2nd edn. Laboratoire Maritime: Laboratoire d'Oceanographie Biologique, Concarneau, Brest. 96 pp. [French].   (1983) **Castric, A. and Michel, C.**
Includes sedentary benthic fauna; species illustrated with distinguishing characteristics.

The Echinodermata of Guernsey. *Rep. Trans. Soc. guernés.* **18**: 54–7.
                           (1966) **Brehaut, R. N.**

The Echinodermata of the estuarine region of the rivers Rhine, Meuse and Scheldt, with a list of species occurring in the coastal waters of the Netherlands. *Neth. J. Sea Res.* **4**: 59–85, 10 figs.   (1968) **Wolff, W. J.**

Stekelhuidigen—Echinodermata. *Wetensch. Meded. K. med. natuurh. Veren.* No. 105, 18 pp, 33 figs. [Dutch].   (1975) **Wolff, W. J.**
Illustrated keys; figures from Mortensen (1924).

Fauna von Deutschland: ein Bestimmungbuch unserer heimischen Tierwelt. Quelle and Meyer, Heidelberg. 582 pp.   (1982) **Brohmer, P.**

A quantitative investigation of the echinoderm fauna of the central North Sea. *Meddr. Kommn Danm. Fisk.-og Havunders.* (N.S.) **2**(24): 204 pp, 96 figs.
                           (1960) **Ursin, E.**

Pighude (Echinodermen). *Danmarks Fauna* **27**: 274 pp. [Danish].
                           (1924) **Mortensen, T.**

Echinoderma. In *The zoology of the Faroes* (ed. R. Spärck and S. L. Tuxen), Vol. 3(1), pp. 1–20. Andr. Fred. Host and Son, Copenhagen.
                           (1971) **Lieberkind, I.**

Echinodermata: Crinoidea. *Mar. Invert. Scand.* **3**: 1–55, 19 figs.
                           (1970) **Clark, A. M.**

[Sea stars of the Soviet Union.] *Opred. Faune SSSR.* **34**: 199 pp, 23 pls. [Russian]. English translation (1968): Israel program for Scientific Translations, Jerusalem. 183 pp.   (1950) **Dyakonov, A. V.**
Comprehensive on arctic fauna.

[Ophiuroids of the Soviet Union.] *Opred. Faune SSSR.* **55**: 136 pp, 47 figs. [Russian]. English translation (1967): Israel program for Scientific Translations, Jerusalem. 123 pp.   (1954) **Dyakonov, A. M.**
Comprehensive on arctic fauna.

Associates

A survey of the echinoderm associates of the north-east Atlantic area. *Zoologische Verh. Leiden* **156**: 159 pp.
                           (1977) **Barel, C. D. N. and Kramers, P. G. N.**

Larvae

Echinoderm larvae of Port Erin. *Proc. Trans. biol. Lpool biol. Soc.* **28**(22): 467–
98, 9 pls.   (1914) **Chadwick, H. C.**

HEMICHORDATA: ENTEROPNEUSTA

Enteropneusta. *Tierwelt N.-u. Ostsee.* **7a**: 1–12, 7 figs.
(1925) **Horst, C. F. van der**

Observations on the enteropneust, *Protoglossus koehleri* (Caullery and Mesnil).
Proc. zool. Soc. Lond. **127**: 35–8, 19 figs, 1 pl.   (1956) **Burdon-Jones, C.**

A revision of the genus *Saccoglossus* (Enteropneusta) in British waters. *Proc.
zool. Soc. Lond.* **134**: 635–45, 3 figs.
(1960) **Burdon-Jones, C. and Patil, A. M.**

Ptychoderidae, Tornaria larvae. *Fich. ident. Zooplancton* **70**: 6 pp, 16 figs.
(1957) **Burdon-Jones, C.**

Report on the zoological excursion to Sherkin Island, Co. Cork, Ireland. In
Rep. third zool. Excur. Sherkin Island. Instituut voor Taxonomische Zoologie,
Universiteit van Amsterdam, Amsterdam. Vol. 3, 37 pp.   (1979) **Anon.**

The marine flora and fauna of the Isles of Scilly: Enteropneusta, Ascidiacea,
Thaliacea, Larvacea and Cephalochordata. *J. nat. Hist.* **6**: 207–13.
(1972) **Rowe, F. W. E.**

The Tunicata, with an account of the British species. Ray Society, London. 354 pp,
120 figs.   (1950) **Berrill, N. J.**

Manteltiere oder Tunicata. *Tierwelt Dtl.* **17**: 143–63, 16 figs.
(1930) **Buckmann, A.**

Fauna von Deutschland: ein Bestimmungsbuch unserer heimischen Tierwelt. Quelle and
Meyer, Heidelberg. 582 pp.   (1982) **Brohmer, P.**

ASCIDIACEA

Ascidians of European waters. *Catalogue of main marine fouling organisms.* OECD
Publ, Paris. Vol. 4, 34 pp, 22 figs (many in colour).   (1969) **Millar, R. H.**

Ascidians of the British Isles. Marine Conservation Society, Ross-on-Wye. 43 pp,
47 col. photographs.   (1985) **Picton, B. E.**
Identification and other data with colour photograph for each species.

British Ascidians. *Synopses Br. Fauna* **1**: 88 pp, 60 figs.   (1970) **Millar, R. H.**

Flore et faune fixées sous-marins de Bretagne, 2nd edn. Laboratoire Maritime: Laboratoire d'Oceanographie Biologique, Concarneau, Brest. 96 pp.
(1983) **Castric, A. and Michel, C.**

Species illustrated with distinguishing characteristics.

Inventaire des ascidies Didemnidae de Roscoff (Tuniciers). *Cah. Biol. mar.* **24**(4): 377–81.
(1983) **Lafargue, F.**

Révision taxonomique des Didemnidae des côtes de France (Ascidies composées): synthèse des resultats principaux. *Annls Inst. océanogr.* **53**: 1–176.
(1977) **Lafargue, F.**

Tunicata: Ascidiacea. *Mar. Invert. Scand.* **1**: 123 pp, 86 figs.
(1966) **Millar, R. H.**

Tuniciers benthiques récoltés au cours de la campagne Norbi en Mer de Norvège. *Bull. Mus. natn. Hist. nat. Paris (Zool. Biol. Ecol. anim.)* **3**: 563–73.
(1979) **Monniot, C. and Monniot, F.**

Pelagic

A monograph of the order Pyrosomatida (Tunicata, Thaliacea). *J. Plankt. Res.* **3**(4): 603–31.
(1981) **Soest, R. W. M. van**

British pelagic tunicates. *Synopses Br. Fauna* (N.S.) **20**: 57 pp.
(1982) **Fraser, J. H.**

Thaliacea. *Tierwelt N.-u. Ostsee* **12a**: 21–48, 12 figs.   (1927) **Ihle, J. E. W.**

Copelata. *Tierwelt N.-u. Ostsee* **12a**: 1–20, 17 figs.   (1926) **Buckmann, A.**

CEPHALOCHORDATA

On the populations of *Branchiostoma lanceolatum* and their relations with the West African lancelets. *Proc. zool. Soc. Lond.* **127**: 125–40. (1956) **Webb, J. E.**

PISCES

General

The fishes of the British Isles and north west Europe. Macmillan, London. xvii + 613, figs, maps.
(1969) **Wheeler, A.**

Zeevissen. *Wet. Meded. K. ned. natuurh. Veren.* **65**: 1–72, 118 pls.
(1968) **Nijssen, H.**

Die Fische der Nordmark. *Abt. naturw. Ver. Hamburg* (N.F.) Suppl. **3**: 1–432, 145 figs, 1 map. (1960) **Duncker, G.**

Marine and freshwater fishes. *Zoology Faroes* **62**: 1–241.
(1970) **Joensen, J. S. and Tåning, A. V.**

Fiskar och Fiske i Norden. Stockholm. 2 vols, 1064 pp, many figs, 574 pls.
(1942) **Andersson, K. A.**

Freshwater

[*Freshwater fishes of the USSR and adjacent countries.*] [Russian]. English translation by Israel program for scientific translations, Jerusalem. No. 741, vii + 504 pp; no. 742, viii + 496; no. 743, viii + 510 pp. (1962–65k) **Berg, L. S.**

The freshwater fishes of Europe. Vol. 1, pt 1, Petromyzontiformes, Aula, Wiesbaden. 313 pp, 48 figs. (1986) **Holčik, J.,** ed.

Die Süsswasserfische Europas, bis zum Ural und Kaspischen Meer. Parey, Berlin. 250 pp, 44 pls, maps. (1965) **Ladiges, W. and Vogt, D.**

Key to British freshwater fishes. *Scient. Publs Freshwat. biol. Ass.* **27**: 1–139, 64 figs, 1 pl, 55 maps. (1972) **Maitland, P. S.**

The freshwater fishes of Britain and Europe. Collins, London. 222 pp, figs, many pls. (1971) **Muus, B. and Dahlstrøm, P.**

Poissons d'eau douce. *Faune Fr.* **65**: 1–303, 98 figs. (1961) **Spillmann, C. J.**

Marine

[*Fishes of the northern seas of the USSR.*] [Russian]. English translation by Israel program for scientific translations, Jerusalem. No. 836, 566 pp, 300 figs.
(1964) **Andriashev, A. P.**

Fishes of the north-eastern Atlantic and the Mediterranean. UNESCO, Paris. Vol. 1, 510 pp; Vol. II, 511–1007 pp; Vol. III, 1008–1473 pp.
(1984, 1986) **Whitehead, P. J. P.** *et al.*

Guide des poissons marins d'Europe. Delachaux et Niestlé, Lausanne and Paris. 427 pp. (1980) **Bauchot, M. L. and Pras, A.**

AMPHIBIA and REPTILIA

A field guide to the reptiles and amphibians of Britain and Europe. Collins, London. 272 pp, 40 pls, numerous figs, 126 maps.
(1978) **Arnold, E. N., Burton, J. A., and Ovenden, D. W.**

Lurche und Kriechtiere Europas. Neumann, Leipzig. 420 pp, 493 figs, most in colour, 196 maps.

(1985) **Engelmann, W. E., Fritzsche, J., Günther, R., and Obst, F. J.**
In addition to keys, descriptions, and illustrations this work contains lists of species
by country.

Guide des amphibiens et reptiles d'Europe. Delachaux et Niestlé, Paris. 292 pp, 8
figs, 44 pls, 169 maps.   (1983) **Matz, G. and Weber, D.**
Diagnoses, illustrations, and distribution maps.

Kriechtiere und Lurche, 6th edn. Kosmos, Stuttgart. 104 pp, 105 figs, 54 pls.
                                                       (1975) **Mertens, R.**

The British amphibians and reptiles, 5th edn. Collins, London. xiv + 366 pp, 88
figs, 37 pls.   (1973) **Smith, M. A.**

Amphibians and reptiles in Britain, Collins, London. 256 pp, 49 figs, 16 pls.
                                                       (1983) **Frazer, D.**

Guide des reptiles et batraciens de France. Hatier, Paris. 239 pp, 97 figs, 32 pls.
                                                       (1975) **Fretey, J.**

De amfibieën en reptielen van Nederland, België en Luxemburg. Balkema, Rotterdam.
284 pp, 84 figs, 75 pls, 28 maps.   (1981) **Sparreboom, M.**

Zoogeography of the Swedish amphibians and reptiles with notes on their
growth and ecology. *Acta vertebr.* **1**(3): 197–397, 69 figs.
                                                       (1959) **Gislén, T. and Kauri, H.**

AMPHIBIA

Amphibians of Europe—a colour field guide. David and Charles, Newton Abbot.
132 pp, 76 pls, 48 maps.   (1984) **Ballasina, D.**

The tailed amphibians of Europe. David and Charles, Newton Abbot. 180 pp, 5
figs, 17 pls, 16 maps.   (1969) **Steward, J. W.**

Les salamandres d'Europe, d'Asie et d'Afrique du Nord. Lechevalier, Paris. 376 pp,
56 figs, 16 pls, 11 maps.   (1968) **Thorn, R.**

REPTILIA

The reptiles of northern and central Europe. Batsford, London. xi + 268 pp, 44 pls.
                                                       (1979) **Street, D.**

The snakes of Europe. David and Charles, Newton Abbot. 238 pp, 33 figs, 17 pls,
36 maps.   (1971) **Steward, J. W.**

British turtles. Guide for the identification of stranded turtles on British coasts. British
Museum (Natural History), London. viii + 23 pp, 19 figs.
                                                       (1967) **Brongersma, L. D.**

AVES

Atlas of birds of the western Palaearctic. Collins, London. 322 pp, 673 maps, 810 figs. (1982) **Harrison, C.**

Handbook of the birds of Europe, the Middle East, and North Africa. The birds of the western Palaearctic. Oxford University Press, Oxford. Non-passerines in 4 vols: Vol. 1, 722 pp, 108 pls; Vol. 2, 695 pp, 96 pls; Vol. 3, 913 pp, 105 pls; Vol. 4, 960 pp, 98 pls. (1977–85) **Cramp, S. and Simmons, K. E. L.**, eds. Encyclopaedic, all plumages and eggs illustrated in colour. (Passerines, Vols. 5–7, to be published.)

The handbook of British birds. Witherby, London. 5 vols, 1882 pp, 47 pls. (1938) **Witherby, H. F., Jourdain, F. C. R., Ticehurst, N. F., and Tucker, B. W.** Describes all plumages. Digest of other information. Reprinted.

The birds of Great Britain and Europe, with North Africa and the Middle East. Collins, London. 320 pp, 16 pp, maps. (1972) **Heinzel, H., Fitter, R. S. R., and Parslow, J.** Every species and some subspecies figured in colour with brief text and map.

A field guide to the birds of Britain and Europe, 4th edn. Collins, London. 240 pp, 77 pls, 362 maps. (1983) **Peterson, R., Mountfort, G., and Hollom, P. A. D.**

The Hamlyn guide to birds of Britain and Europe. Hamlyn, London. 319 pp. (1970) **Bruun, B. and Singer, A.** Every species figured in colour with brief text and map.

A field guide to the nests, eggs, and nestlings of British and European birds, 2nd edn. Collins, London. 432 pp, 64 pls, figs. (1985) **Harrison, C.** Eggs and nestlings of 145 species illustrated in colour. Keys to nests, eggs, and young. [Available in Danish, Dutch, French, German, and Spanish editions.]

Guide to the young of European precocial birds. Skarv, Tisvildeleje, Denmark. 285 pp, 39 pls, 30 figs. (1977) **Fjeldså, J.**

Flight identification of European raptors. Poyser, Berkhamsted. 184 pp, 80 photos, figs. (1974) **Porter, R. F., Willis, I., Christensen, S., and Nielsen, B. P.**

Seabirds, an identification guide, 2nd edn. Croom Helm, Beckenham, Kent. 448 pp, 88 pls, 312 maps. (1985) **Harrison, P.**

Gulls, a guide to identification. Poyser, Calton. 280 pp, 376 photos, 54 figs, 53 maps. (1982) **Grant, P. J.**

Shorebirds, an identification guide to waders of the world. Croom Helm, Beckenham, Kent. 411 pp, 88 pls, figs. (1986) **Hayman, P., Marchant, J., and Prater, T.**

Identification guide to European passerines, 3rd edn. Published by author, Stockholm. 312 pp, figs. (1984) **Svensson, L.** Guide to the bird in the hand.

MAMMALIA

The mammals of the Palaearctic region: a taxonomic review. British Museum (Natural History), London and Cornell University Press, Ithaca, New York. 314 pp, 104 maps.   (1978) **Corbet, G. B.**
A technical text including keys and information on distribution.

Guide des mammifères d'Europe. Delachaux et Niestlé, Neuchâtel. 263 pp, 135 figs, 32 pls.   (1967) **Brink, F.-H. van den**
A field guide to all mammals. Includes key characters, colour and black and white illustrations, and distribution maps. See also the reference to the English translation of this book which follows.

A field guide to the mammals of Britain and Europe. Collins, London. 221 pp, many figs, 32 pls.   (1967) **Brink, F.-H. van den**
A field guide to all mammals. Includes key characters, colour and black and white illustrations, and distribution maps. This is an English translation, translated and edited by H. Kruuk and H. N. Southern.

The mammals of Britain and Europe. Collins, London. 253 pp, figs, 40 pls.
                      (1980) **Corbet, G. B. and Ovenden, D.**
Includes a checklist, maps, an identification guide, and coloured illustrations of European mammals.

The terrestrial mammals of western Europe. G. T. Foulis and Co. Ltd, London. 264 pp, 22 figs, 15 pls.   (1966) **Corbet, G. B.**

Finding and identifying mammals in Britain. British Museum (Natural History), London. 56 pp, many figs, 4 pls.   (1975) **Corbet, G. B.**
A checklist and illustrated guide with keys and tables to all mammals except the Cetacea.

The handbook of British mammals. Blackwell Scientific Publications, London. xxxii + 520 pp, 184 figs. (1977) **Corbet, G. B. and Southern, H. N.**, eds.
Comprehensive guide to biology of all species, including keys, distribution maps, and illustrations.

Mammals of Britain: their tracks, trails and signs. Blandford Press, London. 298 pp, many figs.   (1974) **Lawrence, M. J. and Brown, R. W.**
A field guide to all mammals except Cetacea but including a checklist to all mammals. Keys are provided for the identification of teeth, skulls, and bones, and tracks, droppings, and other signs are illustrated.

The identification of remains in owl pellets. *Occ. Publs Mamm. Soc.* 8 pp, 32 figs.
                                    (1977) **Yalden, D. W.**
Consists of keys to mammal skulls and bones found in British owl pellets.

CHIROPTERA

The identification of British bats. *Occ. Publs Mamm. Soc.* 14 pp, figs, 1 table.
                                    (1985) **Yalden, D. W.**
Contains keys to external features and to skulls, and notes on field identification.

PINNIPEDIA

British seals. Collins, London. 256 pp, 52 figs, 59 pls.   (1974) **Hewer, H. R.**

CETACEA

British whales, dolphins and porpoises: a guide for the identification and reporting of stranded whales, dolphins and porpoises on the British coasts, 5th edn. British Museum (Natural History), London. x + 34 pp, 31 figs.

(1976) **Fraser, F. C.**

ARTIODACTYLA

Field guide to the British deer, 3rd edn. Blackwell Scientific Publications, Oxford, London, Edinburgh, Boston, Melbourne. 95 pp, 30 figs.

(1982) **Taylor Page, F. J.**, ed.

RODENTIA

Handbuch der Säugetiere Europas. Akademische Verlagsgesellschaft, Wiesbaden. Rodentia 1 (Sciuridae, Castoridae, Gliridae, Muridae). Vol. 1, 476 pp, 85 figs; Rodentia 2 (Cricetidae, Arvicolidae, Zapodidae, Spalacidae, Hystricidae, Capromyidae). Vol. 2(1), 649 pp, 207 figs.

(1978–82) **Niethammer, J. and Krapp, E.**, eds.

Fungi

British species about 12 000.

The Fungi are now usually treated as a Kingdom in their own right independent of the plants (Plantae) and animals (Animalia). Two main divisions are recognized, the Myxomycota for plasmodial types and Eumycota for non-plasmodial (mainly mycelia) types. The lichen-forming fungi are integrated here not as in the previous edition of *Key Works* (see p. 215 and under ordinal names).

General

For details of publications cataloguing new taxa see Hawksworth (1974), this section.

Ainsworth and Bisby's dictionary of the fungi, 7th edn. Commonwealth Mycological Institute, Kew. xii + 445 pp.
(1983) **Hawksworth, D. L., Sutton, B. C., and Ainsworth, G. C.** Notes on genera, major literature, terms, etc.; an essential handbook; includes lichenized fungi.

The fungi. An advanced treatise. Academic Press, New York and London. Vols. IV A–B, 621 + 504 pp.
(1973) **Ainsworth, G. C., Sparrow, F. K., and Sussman, A. S.**, eds. Includes keys to most accepted genera and extensive literature lists; the standard multi-authored reference work **to be used for all groups of non-lichenized fungi treated here.**

The genera of fungi sporulating in pure culture, 3rd edn. J. Cramer, Vaduz. 424 pp.
(1981) **von Arx, J. A.**
Keys to genera.

Morphology and taxonomy of fungi. Constable, London. Reprint (1964) by Hafner, New York. 791 pp.　　　　　　　(1950) **Bessey, E. A.**

Bibliography of systematic mycology. CAB International Mycological Institute, Kew.　　　　　　　(1947 on)
Twice-yearly listing of publications on systematic mycology.

Bulletin of the British Mycological Society. British Mycological Society, London.
(1967 on)
Twice-yearly; includes keys and literature lists from time to time.

The genera of fungi. Wilson, New York. Reprint (1965) by Hafner, New York. iv
+ 496 pp.   (1931) **Clements, F. E. and Shear, C. L.**
Keys to genera; including lichen-forming groups.

Microfungi on land plants. An identification handbook. Croom Helm, London and
Sydney. ix + 818 pp.   (1985) **Ellis, M. B. and Ellis, J. P.**
Arranged by host/substrate; 2125 spp. illustrated by line drawings.

CBS course of mycology, 3rd edn. Centraalbureau voor Schimmelcultures, Baern
& Delft. 136 pp.
(1987) **Gams, W., Aa, H. A. van der, Plaats-Niterink, A. J. van der,
Samson, R. A., and Stalpers, J. A.**
Includes useful literature lists.

*Mycologist's handbook. An introduction to the principles of taxonomy and nomenclature in
the Fungi and lichens.* Commonwealth Mycological Institute, Kew. 231 pp.
(1974) **Hawksworth, D. L.**

Guide to the literature for the identification of British fungi, 4th edn. Part 1.
General, Myxomycota, Mastigomycotina, Zygomycotina, Ascomycotina.
Bull. Br. mycol. Soc. **15**: 36–55. Part 2. Basidiomycotina, Deutermycotina;
Bull. Br. mycol. Soc. **16**: 92–112.   (1982) **Holden, M.**

Mycological papers. CAB International Mycological Institute, Kew. (1925 on)
Includes monographs of many groups of microfungi.

Mycotaxon. Mycotaxon, Ithaca, New York.   (1974 on)
A quarterly journal of systematic mycology.

Persoonia. Rijksherbarium, Leiden.   (1959 on)
A journal of systematic mycology including monographs and revisions of many
groups.

Rabenhorst's Kryptogamen-Flora von Deutschland, Österreich und der Schweiz.
Akademie Verlag, Leipzig. Vol. 1.   (1881–1920)
Detailed multi-authored monograph in German issued in many parts, with keys.
Vol. 8 (1930) deals with lichenicolous fungi (see below) and Vol. 9 (1931–60) with
lichen-forming fungi.

Mycology guidebook. University of Washington Press, Seattle and London. xxiv
+ 703 pp.   (1974) **Stevens, R. B.**
Guide to techniques and methods of study in most groups.

Studies in mycology. Centraalbureau voor Schimmelcutures, Baarn. (1972 on)
Monographs of microfungi (mainly growing in culture).

Transactions of the British Mycological Society. Cambridge University Press,
London and New York.   (1897 on)
The journal of the British Mycological Society; bimonthly.

Habitat keys and lists

Works dealing with particular taxonomic groups in restricted habitats are listed only under their group and not in this section.

ANIMALS AND MAN

Fungal diseases of animals, 2nd edn. Commonwealth Agricultural Bureaux, Farnham Royal. vii + 508 pp.
(1973) **Ainsworth, G. C. and Austwick, P. K.**

Medical mycology, 3rd edn. Lea and Febiger, Philadelphia. ix + 592 pp.
(1977) **Emmons, C. W., Binford, C. H., Utz, J. P., and Kwon-Chung, K. J.**

Laboratory handbook of medical mycology. Academic Press, New York. xiii + 661 pp.
(1980) **McGinnis, M. R.**

Nomenclature of Fungi pathogenic to man and animals, 4th edn. Medical Research Council, London. v + 26 pp.
(1977)

Review of medical and veterinary mycology. CAB International, Wallingford.
(1943 on)
A comprehensive quarterly abstracting journal.

BIODETERIOGENIC FUNGI

See also under Industrial fungi (p. 215).

Biodeterioration abstracts. CAB International, Wallingford.
(1987 on)
Reviews and abstracts; quarterly.

Introduction to biodeterioration. Edward Arnold, London, 136 pp.
(1986) **Allsopp, D. and and Seal, K. J.**

Microbial aspects of the deterioration of materials. Society for Applied Bacteriology Technical Series No. 7. Academic Press, London, New York and San Francisco. xiii + 261 pp. (1975) **Lovelock, D. W. and Gilbert, R. J.**, eds.

BRYOPHYTIC FUNGI

Moosbewohnende Ascomyceten I. Die Pyrenocarpen, den Gametophyten besiedelnden Arten. *Mitt. Bot. StSamml., München* **14**: 1–360.
(1978) **Döbbeler, P.**

Étude systematique et biologique des champignons bryophiles. *Mém. Mus. natn. Hist. Nat., Paris*, n.s., **B10**: 1–288, pls. I–LXXXIV.
(1959) **Racovitza, A.**

COPROPHILOUS FUNGI

Dung Fungi. An illustrated guide to coprophilous fungi in New Zealand. Victoria University Press, Wellington. 88 pp.
(1983) **Bell, A.**
Useful as many species are cosmopolitan.

Keys tc Fungi on dung, 2nd edn. British Mycological Society, London. [Originally published in *Bull. Br. mycol. Soc.* **2–3** (1969) 26 pp.]
(1975) **Richardson, M. J. and Watling, R.**

A key to Hyphomycetes on dung. *Univ. Waterloo Biol. Ser.* **27**: 1–62.
(1983) **Seifert, K. A., Kendrick, B., and Murase, G.**

ENTOMOGENOUS FUNGI

A host catalogue of British entomogenous fungi. *Entomologist's mon. Mag.* **94**: 103–5; **97**: 226–7; **101**: 163–4.
(1958–66) **Leatherdale, D.**

The arthropod hosts of entomogenous fungi in Britain. *Entomophaga* **15**: 419–35.
(1970) **Leatherdale, D.**

Pilzkrankheiten bei Insekten. Parey, Berlin. 444 pp.   (1965) **Müller-Kögler, E.**

A revised list of British entomogenous fungi. *Trans. Br. mycol. Soc.* **31**: 286–304.
(1948) **Petch, C.**

An atlas of entomopathogenic fungi. Springer, Berlin. (In press).
(1987) **Samson, R. A., Evans, H. C., and Latge, J. P.**

FOOD SPOILAGE FUNGI

Fungi and food spoilage. Academic Press, Sydney. 413 pp.
(1985) **Pitt, J. I. and Hocking, A.**

Introduction to food-borne fungi, 2nd edn. Centraalbureau voor Schimmelcultures, Baarn. 248 pp.
(1984) **Samson, R. A., Hoekstra, E. S., and van Oorschot, C. A. N.**

FRESHWATER FUNGI

See Ingold 1975 (p. 269), Sparrow 1960 (p. 224), and under Chytridiomycetes (p. 224) and Oomycetes (p. 224).

FUNGICOLOUS FUNGI

A survey of the fungicolous conidial fungi. In *The biology of conidial fungi* (ed. G. T. Cole and B. Kendrick), Vol. 1, pp. 171–244. Academic Press, New York.
(1981) **Hawksworth, D. L.**

Clé pour la determination des espèces banales de champignons fongicoles. *Revue mycol.* **31**: 393–9.
(1966) **Nicot, J.**

HYPOGEOUS FUNGI

Studies in the hypogean fungi of Norway I. *Endogone* and Tuberales. *Nytt Mag. Bot.* **3**: 35–41.
(1954) **Eckblad, F.-E.**

British hypogeous fungi. *Phil. Trans. R. Soc.* **B237**: 429–546.
(1954) **Hawker, L. E.**

Revised annotated list of British hypogeous fungi. *Trans. Br. mycol. Soc.* **63**: 67–76. **(**1974**) Hawker, L. E.**

Danish hypogeous macromycetes. *Dansk bot. Arkiv.* **16**: 1–84.

(1956**) Lange, M.**

INDUSTRIAL FUNGI

Smith's introduction to industrial mycology, 7th edn. Edward Arnold, London. viii + 398 pp.   **(**1981**) Onions, A. H. S., Allsopp, D., and Eggins, H. O. W.**

LICHEN-FORMING FUNGI

British species 1500.
See under Ascomycotina (pp. 226–44) for treatments of particular groups.

General

The Observer's book of lichens, 2nd edn. F. Warne, London. 188 pp.

(1977**) Alvin, K.**

An introductory guide, well illustrated; many illustrations in colour.

Likenoj de okcidenta Eǔropo. Ilustrita determinlibro. *Bull. Soc. bot. Centre-Ouest, n.sér., num. spéc.* **7**, 894 pp. [Esperanto].

(1985**) Clauzade, G. and Roux, C.**

Keys to species; numerous line drawings.

Lichens. An illustrated guide, 2nd edn. Richmond Publishing, Richmond. xi + 320 pp.   **(**1981**) Dobson, F. S.**

Numerous half-tone illustrations; includes most commoner species.

Introduction to British lichens. Buncle, Arbroath. xxiv + 292 pp.

(1970**) Duncan, U. K.** (assisted by **P. W. James**)

The standard British lichen flora; illustrated by line drawings.

Checklist of British lichen-forming, lichenicolous and allied fungi. *Lichenologist* **12**: 1–115. **(**1980**) Hawksworth, D. L., James, P. W., and Coppins, B. J.**

Also available separately; includes references to individual monographs by genus.

Lichens as pollution monitors. Studies in Biology No. 66. E. Arnold, London. 64 pp.   **(**1976**) Hawksworth, D. L. and Rose, F.**

Includes brief notes with some illustrations of species used in air pollution survey work.

Farne. Moose. Flechten Mittel-, Nord- und Westeuropas. BLV Verlagsgesellschaft, Munich, Vienna, and Zurich. 256 pp.   **(**1980**) Jahns, H. M.**

Includes 655 colour photographs.

Collins guide to the ferns mosses and lichens of Britain and North Europe. Collins, London. 272 pp.   **(**1983**) Jahns, H. M.**

English version of the above with a fully revised text by Launert, E., Eddy, A., Laundon, J. R., and Laundon, R. J.

Lichenologist. Academic Press, London, New York, and San Francisco.

(1958 on**)**

The journal of the British Lichen Society including keys and many papers on systematics; quarterly.

A monograph of the British lichens, 2nd edn. British Museum (Natural History), London. 2 vols.   (1926) **Smith, A. L.**
Detailed descriptions; some plates; now dated and in need of an extensive revision.

Die Strauch- und Laubflechten Mitteleuropas. G. Fischer, Jena. Reprinted by Asher, Amsterdam.   (1928) **Anders, J.**
Useful for its plates.

Macrolichens of Denmark, Finland, Norway and Sweden. Universitetsforlaget, Oslo, Bergen, and Tromso. 184 pp.   (1973) **Dahl, E. and Krog, H.**
Keys, in English, with some illustrations; most British macrolichens. An extremely useful book to be strongly recommended.

Flechten. Eine Einführung. A. Ziemsen, Wittenberg Lutherstadt. 243 pp.
                                    (1982) **Doll, R.**

Flechtenflora von Nordwestdeutschland. G. Fischer, Stuttgart. 411 pp.
                                    (1957) **Erichsen, C. F. E.**
Keys and short descriptions.

Natural history of the Danish lichens. H. Aschehoug, Copenhagen. 10 vols.
                                    (1927–72) **Galløe, O.**
Includes 1397 plates and 7010 figures, many in colour; index in Vol. 10.

Kleine Kryptogamenflora. Vol. 3, *Flechten (Lichenes)*. G. Fischer, Jena. viii + 244 pp.   (1967) **Gams, H.**
Keys; some line drawings.

Flechten. *Krypt.-Fl. Mark Brandenb*. Borntraeger, Berlin-Nikolassee. Vol. 8, x + 898 pp.   (1957) **Hillmann, J. and Grummann, V.**
Keys and detailed descriptions.

Opredelitel' lishainikov SSSR. Nauka, Leningrad. 5 vols.
                                    (1971–78) **Kopaczevskaja, E. G.** *et al.*

Lavflora. Norske busk-og bladlav. Universitetsforlaget, Oslo, Bergen, and Tromso. 312 pp.   (1980) **Krog, H., Østhagen, H., and Tønsberg, T.**
Keys in English available as a supplement.

Lavar. En fälthandbok. Interpublishing, Stockholm. 237 pp. (1982) **Moberg, R.**
Many fine coloured photographs.

Porosty Polskie. Panstwowe Wydawnictwo Naukowe, Warsaw and Cracow. 1177 pp.   (1975) **Nowak, J. and Tobolewski, Z.**

Les Lichens. Étude biologique et flore illustrée. Masson et Cie, Paris. 801 pp.
                                    (1970) **Ozenda, P. and Clauzade, G.**
Keys; numerous excellent line drawings and photographs.

Bestimmungsschlüssel europäischer Flechten. J. Cramer, Lehre. (1)–(71) + 757 pp.
                                    (1969) **Poelt, J.**
Keys to many European lichens; the standard European text. Supplements

Erganzungsheft, Vol. I, 258 pp. (1977), and Vol. II, 390 pp. (1981) by Poelt, J. and Vězda, A.

Rabenhorst's Kryptogamen-Flora von Deutschland, Österreich und der Schweiz. Vol. 9. *Die Flechten*. Akademie Verlag, Leipzig. (1933–60)
Detailed but incomplete multi-authored monograph issued in parts with keys.

The lichens of Sweden and Norway. Swedish Museum of Natural History, Stockholm and Uppsala. 333 pp. (1984) **Santesson, R.**
Annotated checklist.

Lichenographia fennica I–IV. *Acta Soc. Fauna Flora fenn.* **49**(2): 1–274; **53**(1): 1–340; **57**(1): 1–138; **57**(2): 1–531. I and II reprinted (1975) by O. Koeltz, Koenigstein. (1921–27, 1934) **Vainio, E. A.**
Detailed but incomplete monograph covering many traditionally 'difficult' taxonomic group including pyrenocarpous genera, Caliciales, Lecideaceae, etc.

Flechtenflora. Ulmer, Stuttgart. 552 pp. (1980) **Wirth, V.**
Southern Germany; keys; well-illustrated.

Habitat keys

Key to crustose pyrenocarpous lichens on limestone and associated substrata (excluding aquatic and marine habitats). *Bull. Br. Lichen Soc.* **54**: 36–45.
(1984) **Coppins, B. J.**

Zonation of supralittoral lichens on rocky shores around the Dale peninsula, Pembrokeshire (with key for their identification). *Fld Stud.* **3**: 41–67.
(1969) **Ferry, B. W. and Sheard, J. W.**
Also available separately.

Key for the identification of British marine and maritime lichens I–II. *Lichenologist* **7**: 1–52, 73–115. (1975) **Fletcher, A.**

The lichen flora of shaded acid rock crevices and overhangs in Britain. *Lichenologist* **4**: 309–22. (1970) **James, P. W.**

Key to British sterile crustose lichens with *Trentepohlia* as phycobiont. *Lichenologist* **11**: 253–62. (1979) **James, P. W. and Coppins, B. J.**

The taxonomy of sterile crustaceous lichens in Britain. 1. Terricolous species. *Lichenologist* **2**: 57–67. (1962) **Laundon, J. R.**

The taxonomy of sterile crustaceous lichens in Britain. 2. Corticolous and lignicolous species. *Lichenologist* **2**: 101–51. (1963) **Laundon, J. R.**

Bibliography

Guide to the literature for the identification of British lichens. *Bull. Br. mycol. Soc.* **4**: 73–95. (1970) **Hawksworth, D. L.**
Also available separately; now dated and to be used with Hawksworth, James, and Coppins (1980) (p. 215).

A bibliographic guide to the lichen floras of the world. In *Lichen ecology* (ed. M. R. D. Seaward), pp. 437–502. Academic Press, London, New York, and San

Francisco.   (1977) **Hawksworth, D. L.**
Titles arranged by country.

Lichenology in the British Isles 1568–1975. An historical and bibliographic survey.
Richmond Publishing, Richmond, Surrey. x + 231 pp.
                        (1977) **Hawksworth, D. L. and Seaward, M. R. D.**
Includes citations of about 2700 publications indexed by vice-county.

A bibliography of books, pamphlets and articles relating to Irish lichenology, 1727–1970.
Privately printed, Galway. 76 pp.   (1971) **Mitchell, M. E.**
422 publications indexed by vice-county.

Distribution

Distribution patterns shown by epiphytic lichens in the British Isles. In
Lichenology: progress and problems (ed. D. H. Brown, D. L. Hawksworth, and
R. H. Bailey), pp. 249–78. Academic Press, London, New York, and San
Francisco.   (1976) **Coppins, B. J.**

Das ozeanische Element der Strauch- und Laubflechtenflora von
Skandinavien. *Acta phytogeogr. suec.* **7**: i–xii + 1–411.   (1935) **Degelius, G.**

Changes in the British lichen flora. In *The changing flora and fauna of Britain* (ed.
D. L. Hawksworth), pp. 47–78. Academic Press, London and New York.
                    (1974) **Hawksworth, D. L., Coppins, B. J., and Rose, F.**

L'élément eu-oceanique dans la flore lichenique du sud-ouest de l'Irlande.
Revta Biol. **2**: 177–256.   (1961) **Mitchell, M. E.**

Census catalogue of Irish lichens. The Stationery Office, Dublin. 32 pp.
                                            (1984) **Seaward, M. R. D.**

Atlas of the lichens of the British Isles, Vol. 1. Institute of Terrestrial Ecology,
Cambridge.   (1982) **Seaward, M. R. D. and Hitch, C. J. B.**

LICHENICOLOUS FUNGI

Les champignons lichénicoles non lichénisés. Institut de Botanique, Montpellier. 110
pp.   (1976) **Clauzade, G. and Roux, C.**

Notes on British lichenicolous fungi. I–V. *Kew Bull.* **30**: 183–203. *Notes R. bot.
Gdn Edinb.* **36**: 181–97, **38**: 165–83, **40**: 357–97, **43**: 497–519.
                                            (1975–86) **Hawksworth, D. L.**

The lichenicolous Hyphomycetes. *Bull. Br. Mus. nat. Hist., Bot.* **6**: 183–300.
                                            (1979) **Hawksworth, D. L.**

Notes on some fungi occurring on *Peltigera*, with a key to accepted species.
Trans. Br. mycol. Soc. **74**: 363–86.   (1980) **Hawksworth, D. L.**

The lichenicolous Coelomycetes. *Bull. Br. Mus. nat. Hist., Bot.* **9**: 1–98.
                                            (1981) **Hawksworth, D. L.**

A key to the lichen-forming, parasitic, parasymbiotic and saprophytic fungi

occurring on lichens in the British Isles. *Lichenologist* **15**: 1–44.
(1983) Hawksworth, D. L.
Also available separately; spores 141 spp. illustrated.

Die Flechtenparasiten. *Rabenh. Krypt.-Fl.* **8**: i–ix + 1–712.
(1930) Keissler, K. A. von

Synopsis des champignons parasites de lichens. *Bull. Soc. mycol. Fr.* **28**: 177–256; **29**: 33–128; **30**: 135–98, 281–329.   (1912–14) **Vouaux, L.**

MARINE FUNGI

Fungi in oceans and estuaries. J. Cramer, Weinheim. xxii + 668 pp.
(1961) Johnson, T. W. and Sparrow, F. K.
Synoptic plates of higher marine fungi, 3rd edn. J. Cramer, Lehre. 88 pp.
(1971) Kohlmeyer, J. and Kohlmeyer, E.
Marine mycology. The higher fungi. Academic Press, New York, San Francisco, London. xiv + 690 pp.   (1979) **Kohlmeyer, J. and Kohlmeyer, E.**
Monograph with keys, illustrations, and full descriptions.

MYXOMYCETICOLOUS FUNGI

Mouldy myxomycetes. *Bull. Br. mycol. Soc.* **8**: 25–30.   (1974) **Ing, B.**

PLANT PATHOGENIC FUNGI

Collins' guide to the pests, diseases and disorders of garden plants. Collins, London. 515 pp.   (1981) **Buczacki, S. T. and Harris, K. M.**
Coloured plates of symptoms.

Names of British plant diseases and their causes, 5th edn. *Phytopath. Papers* **28**: 76 pp.   (1984) **Doling, D. A., Gwynne, D. C., and Johnston, A.**
Microfungi on land plants. An identification handbook. Croom Helm, London and Sydney. ix + 818 pp.   (1985) **Ellis, M. B. and Ellis, J. P.**
Descriptions of plant pathogenic fungi and bacteria. CAB International Mycological Institute, Kew.   (1964 on)
Loose-leaf detailed descriptions; 40 issued each year.

British parasitic fungi. Cambridge University Press, London. 430 pp.
(1959) Moore, W. C.
Plant pathologist's pocketbook, 2nd edn. Commonwealth Mycological Institute, Kew. 439 pp.   (1983) **Johnston, A. and Booth, C.**, eds.
Mycologie et pathologie forestières I. Mycologie forestière. Masson, Paris.
(1978) Lanier, L., Joly, P., Bondoux, P., and Bellemère, A.
Review of plant pathology. CAB International, Wallingford. [Formerly *Review of applied mycology* 1922–69.]   (1970 on)
A comprehensive abstracting journal; monthly.

PREDACIOUS FUNGI

The nematode destroying fungi. Canadian Biological Publications, Guelph,
   Ontario.   (1977) **Barron, G. L.**

A key to the nematode destroying fungi. *Trans. Br. mycol. Soc.* **47**: 61–74.
                                          (1964) **Cooke, R. C. and Godfrey, B. E. S.**

Parasites (animaux et végétaux) des Helminthes. P. Lechevalier, Paris. 482 pp.
                                                              (1946) **Dolfus, R. P.**

SOIL FUNGI

The genera of Hyphomycetes from soil. Williams and Wilkins, Baltimore, Maryland. Reprinted (1972) by R. E. Krieger Publishing, Huntington, New
   York. 346 pp.   (1968) **Barron, G. L.**

Compendium of soil fungi. Academic Press, London. 2 vols, 859 + 405 pp.
               (1980) **Domsch, K. H., Gams, W., and Anderson, T.-H.**

A manual of soil fungi, 2nd edn. Iowa State College Press, Ames, Iowa. xi + 450
   pp.   (1957) **Gilman, J. C.**

THERMOPHILIC FUNGI

Thermophilic fungi. W. H. Freeman, San Francisco and London. xii + 188 pp.
                                       (1964) **Cooney, D. G. and Emerson, R.**

MYXOMYCOTA (MYCETOZOA)

This taxon comprises plasmodial and aggregative organisms showing both
fungal and protozoan affinities. They have been included at different times in
the plant, protistan, animal, or fungal kingdoms, but, for convenience, are
here treated as a phylum of the Fungi, with the major divisions as classes. A
valuable discussion of the taxonomic position and the phylogeny of the group
is given by L. S. Olive (1975 see below p. 223), who also lists the extensive
literature on the biology of these organisms.

MYXOMYCETES (MYXOGASTRES)

Plasmodial slime fungi. British species 335.

Myxomycetes. In *The fungi. An advanced treatise* (ed. G. C. Ainsworth, F. K.
   Sparrow, and A. S. Sussman), Vol. 4B, pp. 39–60. Academic Press, New

York and London. (1973) **Alexopoulos, C. J.**

Die dänischen schleimpilze. *Friesia* **7**: 149–280. 16 pls.
(1963) **Bjornekaer, K. and Klinge, A. B.**
Keys, descriptions, and distributions of the Danish species.

How to know the true slime molds. Wm C. Brown Co, Dubuque, Iowa. 132 pp, 159 figs. (1981) **Farr, M. L.**
An introduction to the commoner species of North America, most of which are common in Europe.

A revised census catalogue of British myxomycetes, Parts 1, 2 and 3. *Bull. Br. mycol. Soc.* **14**(2): 97–111; **16**(1): 26–35; **19**(2): 109–15.
(1980, 1982, 1985) **Ing, B.**
A check-list with details of British distribution by vice-county.

Corticolous myxomycetes VII: Contribution towards a monograph of *Licea*, five new species. *Mycologia* **69**: 667–84.
(1977) **Keller, H. W. and Brooks, T. E.**
Key to all species known at the time.

Sluzowce Polski. Warsaw. 315 pp. (1960) **Krzemieniewska, H.**
Keys, descriptions; the standard work on Polish species.

Monograph of the Mycetozoa, 3rd edn. British Museum (Natural History), London. xxxii + 296 pp. (1925) **Lister, G.**
A comprehensive account, now much out of date, but still valuable for its beautiful plates.

The myxomycetes. University of Iowa Press, Ames, Iowa. 561 pp, 41 col. pls.
(1969) **Martin, G. W. and Alexopoulos, C. J.**
The standard world monograph, with keys and descriptions. The plates are inferior to those in the preceding entry, Lister (1925), but the taxonomic content is of great critical value.

The genera of myxomycetes. University of Iowa Press, Ames, Iowa. 102 pp, 41 col. pls. (1983) **Martin, G. W., Alexopoulos, C. J., and Farr, M. L.**
An updated set of keys and generic descriptions. The plates are those of the 1969 monograph (preceding entry).

A key to the corticolous myxomycetes, Parts 1, 2 and 3. *Bull. Br. mycol. Soc.* **12**(1): 18–42; **12**(2): 90–107; **13**(1): 42–51. (1978, 1979) **Mitchell, D. W.**
A key, with line drawings, to all the bark-inhabiting species known at the time. Re-issued as a separate publication by the British Mycological Society in 1980.

De Nederlandse Myxomyceten. K.N.V., Zutphen. 440 pp. [Dutch].
(1975) **Nannenga-Bremekamp, N. E.**
Excellent account of the Dutch species, most of which occur in Britain, and very clear illustrations of all the taxa described. Keys and full descriptions and clear diagnostic details. Additions and amendment leaflets were issued in 1979 and 1983. [English edn, in prep.]

Systematische studie van de Trichiales en de Stemonitales (Myxomycetes) van Belgie. *Verh. Konink. Acad. Acad. Wetens. Belg.* **40**(146): 1–66.
(1978) **Rammeloo, J.**
A detailed analysis, with keys and descriptions of two important orders in Belgium.
The myxomycete genus *Echinostelium*. *Mycologia* **72**: **950–87.**
(1980) **Whitney, K. D.**
Keys, descriptions, and photographs of all species known at the time.

The works listed above include the single British species of *Ceratiomyxa*, which may be placed in a separate class, Ceratiomyxomycetes, which has links with the next class.

PROTOSTELIOMYCETES (PROTOSTELIA)

British species 7.

The mycetozoans. Academic Press, New York and London. x + 293 pp.
(1975) **Olive, L. S.**

DICTYOSTELIOMYCETES (DICTYOSTELIA)

Cellular slime moulds. British species 10.

The dictyostelids. Princeton University Press, Princeton. x + 453 pp.
(1984) **Raper, K. B.**
The standard monograph with keys, descriptions, plates, and full biological details.

ACRASIOMYCETES (ACRASEA)

British species 3. Acrasid slime moulds.

Acrasiomycetes. In *The fungi. An advanced treatise* (ed. G. C. Ainsworth, F. K. Sparrow, and A. S. Sussman), Vol. 4B, pp. 9–36. Academic Press, New York and London.
(1973) **Raper, K. B.**
The Dictyostelids. Princeton University Press, Princeton. x + 453 pp.
(1984) **Raper, K. B.**
The standard monograph with keys, descriptions, plates, and full biological details.

LABYRINTHULOMYCETES (LABYRINTHULINA)

Net plasmodial moulds. British species about 6.

On *Chlamydomyxa labyrinthiuloides*. *Q. Jl microsc. Sci.* (N.S.) **15**: 107–30.
(1875) **Archer, W.**

Zoosporic marine fungi (Thraustochytriaceae and Dermocystidiaceae). *A. Rev. Microbiol.* **27**: 13–26.
(1973) **Goldstein, S.**

Fungi in oceans and estuaries. J. Cramer, Weinheim. xxii + 668 pp.
(1961) **Johnson, T. W. and Sparrow, F. K.**

The mycetozoans. Academic Press, New York and London. x + 293 pp.
(1975) **Olive, L. S.**

HYDROMYXOMYCETES

Plasmodial water moulds. British species about 4.

The British freshwater Rhizopoda and Heliozoa. Ray Society, London. Vol. 1.
(1905) **Cash, J. and Hopkinson, J.**

Hydromyxomycetes. In *Die natürlichen Pflanzenfamilien* (eds. A. Engler and K. Prantl). Engelman, Leipzig. 2nd edn, Vol. 2, pp. 304–39. (1928) **Jahn, E.**

This group is best regarded as part of the rhizopod Protozoa.

PLASMODIOPHOROMYCETES (PLASMODIOPHORINA)

Club-root fungi. British species about 10.

The Plasmodiophorales, 2nd edn. (revised). Hafner, New York.
(1968) **Karling, J. S.**
This group is often included as a class of Mastigomycotina (see below).

EUMYCOTA

MASTIGOMYCOTINA AND ZYGOMYCOTINA (PHYCOMYCETES)

General

A list of British Phycomycetes. *Trans. Br. mycol. Soc.* **5**: 301–17.
(1916) **Ramsbottom, J.**

The lower fungi. Phycomycetes. McGraw-Hill, New York and London. xi + 331 pp.
(1930) **Fitzpatrick, H. M.**

Aquatic Phycomycetes, 2nd edn. University of Michigan Press, Ann Arbor. 1187 pp.   (1960) **Sparrow, F. K.**

Mastigomycotina (zoosporic fungi). In *The fungi. An advanced treatise* (ed. G. C. Ainsworth, F. K. Sparrow, and A. S. Sussman), Vol. 4B, pp. 61–73. Academic Press, London and New York.   (1973) **Sparrow, F. K.**

Lower fungi in the laboratory. University of Georgia, Athens, Georgia. 212 pp, many illustrations.   (1978) **Fuller, M. S.**, ed.

CHYTRIDIOMYCETES

The simple holocarpic biflagellate Phycomycetes. New York. x + 123 pp.
                                                       (1942) **Karling, J. S.**

Studies on British chytrids I–XXXI. *Trans. Br. mycol. Soc.* **29–56.**
                                                       (1946–71) **Canter, H. M.**

Annotated list of British aquatic chytrids. *Trans. Br. mycol. Soc.* **36**: 278–303.
                                                       (1953) **Canter, H. M.**

Zoosporic marine fungi (Thraustochytriaceae and Dermocystidiaceae). *A. Rev. Microbiol.* **27**: 13–26.   (1973) **Goldstein, S.**

Synchytrium. Academic Press, New York and London. xvii + 470 pp.
                                                       (1964) **Karling, J. S.**

The genus *Monoblepharis. Trans. Br. mycol. Soc.* **38**: 247–82.
                                                       (1955) **Perrot, P. E. T.**

Chytridiomycetes, Hyphochytridiomycetes. In *The fungi. An advanced treatise* (ed. G. C. Ainsworth, F. K. Sparrow, and A. S. Sussman), Vol. 4B, pp. 85–110. Academic Press, London and New York.   (1973) **Sparrow, F. K.**

OOMYCETES

Oomycetes I. *Flora ČSR* **2B**, 475 pp. Czechoslovak Academy of Sciences, Prague.   (1959) **Cejp, K.**

The Saprolegniaceae with notes on other water molds. University of North Carolina Press, Chapel Hill. 201 pp. Reprinted (1968) by J. Cramer, Lehre.
                                                       (1923) **Coker, W. C.**

Saprolegniales. Leptomitales. In *The fungi. An advanced treatise* (ed. G. C. Ainsworth, F. K. Sparrow, and A. S. Sussman), Vol. 4B, pp. 113–58.
                                                       (1973) **Dick, M. W.**

Beiträge zur einter Monographie der Gattung *Peronospora. Beitr. Kryptfl. schweiz* **5**(4): 1–360.   (1923) **Gäumann, E.**

Studies on Nordic *Peronospora*'s I. Taxonomic revision. *Op. bot. Soc. bot. Lund* **3**(1): 1–271.   (1959) **Gustavsson, A.**

The genus Achlya: morphology and taxonomy. University of Michigan Press, Ann Arbor. 180 pp.   (1956) **Johnson, T. W.**

Tabular key to the species of *Phytophthora* de Bary. *Mycol. Papers* **143**: 1–20.
                    (1978) **Newhook, F. J., Waterhouse, G. M., and Stamps, D. J.**

Monograph of the genus *Pythium*. *Stud. mycol.* **21**: 1–242.
                                        (1981) **Plaats-Niterink, A. J. van der**

A monograph of the genus *Aphanomyces*. *Tech. Bull. Va agric. Exp. Stn* **151**: 1–95.
                                                        (1961) **Scott, W. S.**

The genus *Saprolegnia*. *Nova Hedwigia* **19**: 1–160.   (1970) **Seymour, R. L.**

The genus *Pythium* Pringsheim. *Mycol. Papers* **110**: 1–203.
                                        (1968) **Waterhouse, G. M.**

The genus *Phytophthora* de Bary, 2nd edn. *Mycol. Papers* **122**: 1–59.
                                        (1970) **Waterhouse, G. M.**

Peronosporales. In *The fungi. An advanced treatise* (ed. G. C. Ainsworth, F. K.
    Sparrow, and A. S. Sussman), Vol. 4B, pp. 165–83. Academic Press,
    London and New York.   (1973) **Waterhouse, G. M.**

TRICHOMYCETES

Trichomycetes. In *The fungi. An advanced treatise* (ed. G. C. Ainsworth, F. K.
    Sparrow, and A. S. Sussman), Vol. 4B, pp. 237–43. Academic Press,
    London and New York.   (1973) **Lichtwardt, R. W.**

ZYGOMYCETES

The merosporangiferous Mucorales. *Aliso* **4**: 321–433; **5**: 11–19, 273–322; **6**: 1–
    10. Reprinted as one volume (1967) by J. Cramer, Lehre.
                                        (1959–65) **Benjamin, R. K.**

Observations on Thamnidiaceae (Mucorales) I and II. *Aliso* **8**: 301–51, 391–
    424.   (1975, 1976) **Benny, G. L. and Benjamin, R. K.**

Observations on Thamnidiaceae (Mucorales) III. *Mycotaxon* **22**: 119–48.
                    (1985) **Benny, G. L., Kirk, P. M., and Samson, R. A.**

A key to the species of *Mortierella*. *Persoonia* **9**: 381–91.   (1977) **Gams, W.**

Zygomycetes in culture. University of Georgia, Athens, Georgia. 257 pp, many
    illustrations.   (1979) **O'Donnel, K. L.**

The Endogonaceae in the Pacific North West. *Mycologia, Mem.* **5**: 1–76.
                    (1974) **Gerdemann, J. L. and Trappe, J. M.**

Genera of Mucorales with notes on their synonymy. *Mycologia* **47**: 344–63.
                                        (1955) **Hesseltine, C. W.**

Mucorales. In *The fungi. An advanced treatise* (ed. G. C. Ainsworth, F. K.
    Sparrow, and A. S. Sussman), Vol. 4B, pp. 187–217. Academic Press,
    London and New York.   (1973) **Hesseltine, C. W. and Ellis, J. J.**

Taxonomical studies on the genus *Rhizopus*. *J. gen. appl. Microbiol.* **11**(Suppl.):
    122 pp.   (1965) **Inui, T., Takeda, Y., and Iizuka, N.**

A monograph of the Choanephoraceae. *Mycol. Papers* **152**: 1–62.

(1984) **Kirk, P. M.**

Les Mucorinées de la Suisse. K.-J. Wyss, Berne. 180 pp.   (1908) **Lendner, A.**

Opredelitel' Mukoral'nykh Gribov. Naukova Dumka, Kiev. 303 pp.

(1974) **Mil'ko, A. A.**

Clés des Mucorinées (Mucorales). P. Lechevalier, Paris. 137 + xxxvi pp.

(1939) **Naumov, N. A.**

Taxonomy and identification of mucormycosis-causing fungi. I. Synonymity of *Absidia ramosa* with *Absidia corymbifera*. *Sabouraudia* **12**: 64–74.

(1974) **Nottebrock, H., Schloer, H. J., and Wall, M.**

A study on the variability in *Mucor hiemalis* and related species. *Stud. mycol.* **5**: 1–40.

(1973) **Schipper, M. A. A.**

On *Mucor mucedo*, *Mucor flavus* and related species. *Stud. mycol.* **10**: 1–33.

(1975) **Schipper, M. A. A.**

On *Mucor circinelloides*, *Mucor racemosus* and related species. *Stud. mycol.* **12**: 1–40.

(1976) **Schipper, M. A. A.**

1. On certain species of *Mucor* with a key to all accepted species. 2. On the genera *Rhizomucor* and *Parasitella*. *Stud. mycol.* **17**: 1–71.

(1978) **Schipper, M. A. A.**

A revision of the genus *Rhizopus*. *Stud. mycol.* **25**: 1–34.

(1984) **Schipper, M. A. A. and Stalpers, J. A.**

Entomophthorales. In *The fungi. An advanced treatise* (ed. G. C. Ainsworth, F. K. Sparrow, and A. S. Sussman), Vol. 4B, pp. 219–29. Academic Press, London and New York.

(1973) **Waterhouse, G. M.**

Key to the species *Entomophthora*. *Bull. Br. mycol. Soc.* **9**: 15–41.

(1975) **Waterhouse, G. M.**

Mucorales. J. Cramer, Lehre. 355 pp.

(1969) **Zycha, H., Siepmann, R., and Linnemann, G.**

ASCOMYCOTINA (ASCOMYCETES)

British species 5100. See also literature under Lichen-forming fungi (pp. 215–18).

General

Systema Ascomycetum. CAB International Mycological Institute, Kew and University of Umeå, Sweden.   (1982 on)

Twice yearly. Listing of all 6000+ genera and placements, proposals for changes in classification at genus-level and above, outline annual classification, original papers.

Pilze der Schweiz. Vol. 1. *Ascomycetes.* Edition Mykologia, Lucerne. 310 pp.
(1981) Breitenbach, J. and Kränzlin, F.

The British Ascomycotina. An annotated checklist. Commonwealth Mycological Institute, Kew. 302 pp.
(1985) **Cannon, P. F., Hawksworth, D. L., and Sherwood-Pike, M. A.**
The standard reference work; notes on anamorphs, habitats, distribution, literature, etc.; includes lichen-forming species. 1300 generic and 7300 species names.

Non-lichenized discomycetes recorded in Britain in recent years. *Bull. Br. mycol. Soc.* **14**: 125–56. Supplement *ibid.* **16**: 60–6 (1982).
(1980) Clark, M. C.

Die Gattungen der amerosporen Pyrenomyceten. *Beitr. Kryptog. Flora Schweiz* **11**(1): 1–434. (1954) Arx, J. A. von and Müller, E.

Histoire et classification des Discomycètes d'Europe. Klincksieck, Paris. 223 pp. Reprinted (1968) by Asher, Amsterdam. (1907) Boudier, E.

Keys to the genera of amerospored and didymospored Pyrenomycetes. Commonwealth Mycological Institute, Kew. 53 pp. (1969) Butterfill, G.
English translation of keys in von Arx and Müller (1954) and Müller and von Arx (1962).

British ascomycetes, 3rd edn. J. Cramer, Lehre. xxxii + 456 pp.
(1978) Dennis, R. W. G.
Includes keys to genera, and 71 plates (40 coloured); should be consulted for all non-lichenized groups of Ascomycotina.

On ascomycetes on Diapensiales and Ericales in Fennoscandia I. Discomycetes. *Symb. bot. upsal.* **19**(4): 1–71. (1970) Eriksson, B.

On graminicolous ascomycetes from Fennoscandia I–III. *Ark. Bot.* **6**: 339–79, 381–440, 441–6. (1967) Eriksson, O.

Outline of the ascomycetes—1986. *Systema Ascomycetum* **5**: 185–324.
(1986) Eriksson, O. and Hawksworth, D. L.
This system of orders is used for the arrangement of the literature in this edition of *Key works*.

Discomycetes and Tuberales. In *The fungi. An advanced treatise* (ed. G. C. Ainsworth, F. K. Sparrow, and A. S. Sussman), Vol. 4B, pp. 249–319. Academic Press, London and New York. (1973) Korf, R. P.

Le Discomycètes de Madagascar. Museum National d'Histoire Naturelle, Paris. 465 pp. (1973) Le Gal, M.

Kleine Kryptogamenflora. Vol. 2a. *Ascomyceten.* G. Fischer, Stuttgart. 147 pp.
(1963) Moser, M.
Keys, some line drawings.

Gattungen der didymosporen Pyrenomyceten. *Beitr. Kryptog. Flora Schweiz* **11**(2): 1–922. (1962) Müller, E. and von Arx, J. A.

Phyrenomycetes: Meliolales, Coronophorales, Sphaerales. In *The fungi. An advanced treatise* (ed. G. C. Ainsworth, F. K. Sparrow, and A. S. Sussman), Vol. 4B, pp. 87–132. (1973) **Müller, E. and von Arx, J. A.**

Danish Pyrenomycetes. *Dansk bot. Arkiv.* **17**: 1–491. (1957) **Munk, A.**

Studien über die Morphologie und Systematik der nicht-lichenisierten inoperculaten Discomyceten. *Nova Acta R. Soc. scient. upsal.*, ser. 4 **8**(2): 1–368. (1932) **Nannfeldt, J. A.**

Monographia Discomycetum Bohemiae. Prague. 2 vols, 436 pp.

(1934) **Velenovský, J.**

The second volume comprises 31 plates.

ARTHONIALES

Monographia Arthoniarum Scandinaviae. *K. svenska VetenskAkad. Handl.* **17**(6): 1–69. (1880) **Almquist, S.**

ASCOSPHAERALES

Ascosphaerales. *Friesia* **10**: 1–24. (1972) **Skou, J. P.**
Ascosphaerales and their unique ascomata. *Mycotaxon* **15**: 487–99.

(1982) **Skou, J. P.**

CALICIALES

Sphinctrina in Europe. *Lichenologist* **11**: 109–37.

(1979) **Löfgren, O. and Tibell, L.**

Systematische Übersicht der mitteleuropäischen Arten der Flechtenfamilie Caliciaceae. *Studia bot. čsl.* **5**: 6–40. (1942) **Nádvornik, J.**

Anatomische -taxonomische Untersuchungen an europäischen Arten der Flechtenfamilie Caliciaceae. *Mitt. Staatsinst. allg. Bot. Hamburg* **13**: 111–66.

(1970) **Schmidt, A.**

The genus *Cyphelium* in Europe. *Svensk bot. Tidskr.* **65**: 138–64. (1971) **Tibell, L.**

The Caliciales of boreal North America. *Symb. bot. upsal.* **21**(2): 1–128.

(1975) **Tibell, L.**

The genus *Thelomma. Bot. Notiser* **129**: 221–49. (1976) **Tibell, L.**

Lavordningen Caliciales i Sverige. *Svensk Bot. Tidskr.* **71**: 239–59 (*Calicium*); **72**: 171–88 (*Chaenotheca, Coniocybe*); **74**: 55–69 (other genera).

(1977–80) **Tibell, L.**

The lichen genus *Chaenotheca* in the Northern Hemisphere. *Symb. bot. upsal.* **23**(1): 1–65. (1980) **Tibell, L.**

A re-appraisal of the taxonomy of Caliciales. *Beih. Nova Hedwigia* **79**: 597–713.

(1984) **Tibell, L.**

Rodzina Caliciaceae w Polsce. *Pozn. Tow. Przyj. Nauk, Prace Kom. Biol.* **24**(5): 1–105. (1966) **Tobolewski, Z. R.**

CALOSPHAERIALES

Notes on the Calosphaeriales. *Mycologia* **77**: 549–65. (1985) **Barr, M. E.**

CLAVICIPITALES

The hypocrealean fungi (Ascomycetes, Hypocreales). *Mycologia* **62**: 865–910. (1971) **Rogerson, C. T.**

The genus *Cordyceps* and its allies. *Scient. Rep. Tokyo Bunrika Daig.* **5**(84): 53–260. (1941) **Kobayasi, Y.**

DIAPORTHALES

The Diaporthales in North America. *Mycologia Mem.* **7**: 1–232. (1978) **Barr, M. E.**

Monographie taxonomique des Gnomoniaceae. *Sydowia Beih.* **9**: 1–315. (1983) **Monod, M.**

Gaeumannomyces, *Linocarpon*, *Ophiobolus* and several other genera of scolecospored ascomycetes and *Phialophora* conidial states, with a note on hyphopodia. *Mycotaxon* **11**: 1–129. (1980) **Walker, J.**

The Genus Diaporthe Nitschke and its segregates. University of Michigan Press, Ann Arbor. x + 349 pp. (1933) **Wehmeyer, L. E.**

A revision of Melanconis, Pseudovalsava, Prosthecium and Titania. University of Michigan Press, Ann Arbor. vii + 162 pp. (1941) **Wehmeyer, L. E.**

DIATRYPALES

Studies of British pyrenomycetes III. The British species of the genus *Diatrypella*. *Trans. Br. mycol. Soc.* **33**: 45–73. (1950) **Croxall, N. E.**

Diatrypaceae in the Pacific northwest. *Mycotaxon* **20**: 401–60. (1984) **Glawe, D. A. and Rogers, J. D.**

Recherches sur les pyrénomycètes de l'ordre des Diatrypales sensu Chadefaud. *Bull. Soc. mycol. Fr.* **76**: 305–407. (1961) **Schrantz, J. P.**

DOTHIDEALES

See also above: lichen-forming fungi (pp. 215–18).

Revision of the genera *Sporormia* and *Sporormiella*. *Can. J. Bot.* **50**: 419–77. (1972) **Ahmed, S. I. and Cain, R. F.**

Beiträge zur Kenntnis der Gattung *Mycosphaerella*. *Sydowia* **3**: 22–100. (1959) **von Arx, J. A.**

A re-evaluation of the bitunicate Ascomycetes with keys to families and genera. *Stud. mycol.* **9**: 1–159.   (1975) **von Arx, J. A. and Müller, E.**

Preliminary studies on the Dothideales in temperate North America. *Contr. Univ. Mich. Herb.* **9**: 523–638.   (1972) **Barr, M. E.**

A classification of Loculoascomycetes. *Mycologia* **71**: 935–57.
(1979) **Barr, M. E.**
Keys to families.

On the Pleomassariaceae (Pleosporales) in North America. *Mycotaxon* **15**: 349–93.   (1982) **Barr, M. E.**

The ascomycete connection. *Mycologia* **75**: 1–13.   (1983) **Barr, M. E.**
Keys to orders and families.

Monografia dos fungos Micropeltaceae. *Publs. Inst. Mic. Recife* **56**: 1–520.
(1959) **Batista, A. C.**

The Chaetothyriales. *Sydowia Beih.* **31**: 1–129.
(1962) **Batista, A. C. and Ciferri, R.**

Studies on *Massaria* Sacc. and related genera. *Phytopath. Z.* **41**: 121–213.
(1961) **Bose, S. K.**

Studies in the Lophiostomataceae Sacc. *Mycol. Papers* **120**: 1–55.
(1970) **Chesters, C. G. C. and Bell, A.**

Recherches sur le genre *Didymella* Sacc. *Phytopath. Z.* **28**: 375–414.
(1957) **Corbaz, R.**

Ueber die heterogene Ascomycetengattung Pleospora Rabh.; Vorschlag für eine Aufteilung. ADAG Administration, Zurich. 213 pp.   (1983) **Crivelli, P. G.**

British *Microthyrium* species and similar fungi. *Trans. Br. mycol. Soc.* **67**: 381–94.
(1976) **Ellis, J. P.**

The genera *Trichothyrina* and *Actinopeltis* in Britain. *Trans. Br. mycol. Soc.* **68**: 145–55.   (1977) **Ellis, J. P.**

The genus *Morenoina* in Britain. *Trans. Br. mycol. Soc.* **74**: 297–307.
(1980) **Ellis, J. P.**

The families of bitunicate ascomycetes. *Op. bot.* **60**: 1–220.
(1981) **Eriksson, O.**

The corticolous pyrenolichens of the Great Lakes region. *Mich. Bot.* **12**: 1–68.
(1973) **Harris, R. C.**

A redisposition of the species referred to the ascomycete genus *Microthelia. Bull. Br. Mus. nat. Hist., Bot.* **14**: 43–181.   (1985) **Hawksworth, D. L.**

Études taxonomiques sur les Pléosporacées. *Symb. bot. upsal.* **14**(3): 1–188.
(1957) **Holm, L.**

A study of the Leptopeltidaceae. *Bot. Notiser* **130**: 215–29.
(1977) **Holm, L. and Holm, K.**

Sooty moulds. *Mycologia* **68**: 693–820.   (1976) **Hughes, S. J.**

Über *Phaeosphaeria* Miyake und andere bitunicate Ascomyceten mit mehrfach querseptierten Ascosporen. *Sydowia* **37**: 75–198.   (1984) **Leuchtmann, A.**

Conidial states of British species of *Leptosphaeria*. *Trans. Br. mycol. Soc.* **50**: 85–121.   (1967) **Lucas, M. T. and Webster, J.**

Loculoascomycetes. In *The fungi. An advanced treatise* (ed. G. C. Ainsworth, F. K. Sparrow, and A. S. Sussman), Vol. 4A, pp. 135–219.

(1973) **Luttrell, E. S.**

Keys to genera.

The Tubenfiaceae and similar Loculoascomycetes. *Mycol. Papers* **157**: 1–71.

(1987) **Rossman, A. Y.**

Untersuchungen über die Gattung *Didymosphaeria* Fuck. einige verwandte Gattungen. *Ber. Schweiz bot. Ges.* **68**: 325–86.   (1958) **Scheinpflug, H.**

The genus *Herpotrichia* Fuckel. *Mycol. Papers* **127**: 1–37. (1972) **Sivanesan, A.**

The taxonomy and pathology of *Venturia* species. *Bibltheca mycol.* **59**: 1–139.

(1977) **Sivanesan, A.**

The bitunicate ascomycetes and their anamorphs. J. Cramer, Vaduz. 701 pp.

(1984) **Sivanesan, A.**

Graminicolous pyrenomycetes II–VI. *Trans. Br. mycol. Soc.* **34**: 318–21; **35**: 208–14; **38**: 347–65; **40**: 509–22.   (1952–57) **Webster, J.**

A world monograph of the genus Pleospora and its segregates. University of Michigan Press, Ann Arbor. xi + 451 pp.   (1961) **Wehmeyer, L. E.**

Die Hysteriaceae s. str. und Lophiaceae. *Beitr. Kryptfl. Schweiz* **11**(3): 1–190.

(1962) **Zogg, H.**

ELAPHOMYCETALES

See also under hypogeous fungi (p. 214).

Studies in the hypogean fungi of Norway II. Revision of the genus *Elaphomyces*. *Nytt Mag. Bot.* **9**: 199–210.   (1962) **Eckblad, F.-E.**

ENDOMYCETALES (YEASTS, ETC.)

The genera of yeasts and yeast-like fungi. *Stud. mycol.* **14**: 1–42.
(1977) **von Arx, J. A., Rodrigues de Miranda, L., Smith, M. T., and Yarrow, D.**

A new key to the yeasts. North Holland Publishing, Amsterdam and London. 273 pp.   (1974) **Barnett, J. A. and Pankhurst, R. J.**
Keys based mainly on cultural and physiological characters.

Yeasts: characteristics and identification. Cambridge University Press, Cambridge. ix + 811 pp.   (1983) **Barnett, J. A., Payne, R. W., and Yarrow, D.**

Nematosporaceae (Hemiascomycetidae): Taxonomy, pathogenicity, distribution, and vector relations. US Department of Agriculture, Washington, DC. Technical Bulletin No. 1469. 71 pp.   (1973) **Batra, L. R.**

Endomycetales, basidiomycetous yeasts, and related fungi. In *The fungi. An advanced treatise* (ed. G. C. Ainsworth, F. K. Sparrow, and A. S. Sussman), Vol. 4A, pp. 11–32. Academic Press, London and New York.
(1973) **Kreger-van Rij, N. J. W.**

The yeasts: a taxonomic study, 3rd edn. Elsevier Science Publishers, Amsterdam. 1082 pp. (1984) **Kreger-van Rij, N. J. W.**, ed.
The standard multi-authored work including keys and descriptions.

ERYSIPHALES (POWDERY MILDEWS)

Echte Mehltaupilze (Erysiphaceae). Ein Bestimmungsbuch für die in Europa vorkommenden Arten. G. Fischer, Jena. 436 pp. (1967) **Blumer, S. L.**

Erysiphaceae of Sweden. *Symb. bot. upsal.* **19**(1): 1–117. (1967) **Junell, L.**

Pyrenomycetes: Erysiphales. In *The fungi. An advanced treatise* (ed. G. C. Ainsworth, F. K. Sparrow, and A. S. Sussman), Vol. 4A, pp. 71–86. Academic Press, London and New York. (1973) **Yarwood, C. E.**

EUROTIALES

See also under Food spoilage fungi (p. 214).

Plectomycetes: Eurotiales. In *The fungi. An advanced treatise* (ed. G. C. Ainsworth, F. K. Sparrow, and A. S. Sussman), Vol. 4A, pp. 45–68. Academic Press, London and New York. (1973) **Fennell, D. I.**

Five new genera in the new family Pseudeurotiaceae. *Can. J. Bot.* **48**: 1815–25.
(1970) **Malloch, D. and Cain, R. F.**

The Trichocomaceae (Ascomycetes): synonyms in recent publications. *Can. J. Bot.* **51**: 1647–8. (1974) **Malloch, D. and Cain, R. F.**

A synoptic key to the genus *Eupenicillium* and to sclerotigenic *Penicillium* species. *Can. J. Bot.* **52**: 2231–6. (1974) **Pitt, J. I.**

The genus Penicillium and its teleomorphic states Eupenicillium and Talaromyces. Academic Press, London. 634 pp. (1980)('1979') **Pitt, J. I.**

The genus Aspergillus. Williams and Wilkins, Baltimore. 686 pp.
(1965) **Raper, K. B. and Fennell, D. I.**

The genus *Talaromyces. Stud. mycol.* **2**: 1–65.
(1972) **Stolk, A. C. and Samson, R. A.**

The ascomycete genus *Eupenicillium* and related *Penicillium* anamorphs. *Stud. Mycol.* **23**: 11–149. (1983) **Stolk, A. C. and Samson, R. A.**

GRAPHIDALES

Monographische Bearbeitung einer Flectenfamilien. *Beih. Feddes Repert.* **69**: 1–250. (1932–37) **Lettau, G.**

Taxonomical studies on the family Graphidaceae of Japan. *J. Sci. Hiroshima Univ.* B(2) **11**: 41–126.   (1966) **Nakanishi, M.**

Thelotrema Ach. sect. *Thelotrema* 1. The *T. lepadinum* group. *Lichenologist* **5**: 262–74.   (1972) **Salisbury, G.**

Flechtensystematische Studien XI. Beiträge zur Kenntnis der familie Asterothyriaceae (Discolichenes). *Folia geobot. phytotax bohemoslovaca* **14**: 43–94.   (1979) **Vězda, A.**

GYALECTALES

Monographische Bearbeitung einiger Flechtenfamilien. *Beih. Feddes Repert.* **69**: 1–250.   (1932–37) **Lettau, G.**

Flechtensystematische Studien IV. Die Gattung *Gyalidea* Lett. *Folia geobot. phytotax bohemoslovaca* **1**: 311–40.   (1966) **Vězda, A.**

Flechtensystematische Studien VI. Die Gattung *Sagiolechia* Massal. *Folia geobot. phytotax bohemoslovaca* **2**: 383–96.   (1968) **Vězda, A.**

HALOSPHAERIALES

Taxonomic studies of the Halosphaeriaceae: *Corollospora* Wedermann. *Bot. J. Linn. Soc.* **87**: 193–212.
                                   (1983) **Jones, E. B. G., Johnson, R. G., and Moss, S. T.**

Taxonomic studies of the Halosphaericaceae: *Halosphaeria* Linder. *Bot. Marina* **27**: 129–43.   (1984) **Jones, E. B. G.**

Marine mycology. The higher fungi. Academic Press, New York. xiv + 690 pp.
                                   (1979) **Kohlmeyer, J. and Kohlmeyer, E.**

HELOTIALES

Untersuchungen über Discomyceten der Gruppe *Tapesia-Trichobelonium*. *Nova Hedwigia* **23**: 49–112.   (1972) **Aebi, B.**

Bemerkenswerte Ascomyceten der DDR. VI. Die weiss-sporing Geoglossaceen. *Gleditschia* **10**: 141–71.   (1983) **Benkert, D.**

Sclerotiniaceae Daniae. *Friesia* **3**: 235–330.   (1947) **Buchwald, N. F.**

A revision of the British Hyaloscyphaceae with notes on related European species. *Mycol. Papers* **32**: 1–97.   (1949) **Dennis, R. W. G.**

A revision of the British Helotiaceae in the herbarium of the Royal Botanic Gardens, Kew, with notes on related European species. *Mycol. Papers* **62**: 1–216.   (1956) **Dennis, R. W. G.**

Remarks on the genus *Hymenoscyphus* S.F. Gray. *Persoonia* **3**: 29–80.
                                   (1964) **Dennis, R. W. G.**

Contributions to the Geoglossaceae of Norway. *Nytt Mag. Bot.* **10**: 137–58.
                                   (1963) **Eckblad, F.-E.**

Taxonomic notes on Mollisiaceous fungi I–V. *Fungus* **24**: 1–8; **25**: 1–12; **26**: 28–31, 32–7; **27**: 30–3. (1954–57) **Gremmen, J.**

Die Gattung *Hysterostegiella* v. Höhnel (Ascomycetes Dermateaceae). *Nova Hedwigia* **38**: 669–702. (1983) **Hein, B.**

The Danish species of the genus *Pezicula*. *Dansk bot. Arkiv* **13**(3): 1–26. (1949) **Johansen, G.**

A monographic revision of the genus *Sclerotinia*. *Mycotaxon* **9**: 365–444. (1979) **Kohn, L. M.**

A monograph of the Arachnopezizeae. *Lloydia* **14**: 129–80. (1951) **Korf, R. P.**

Synopsis of the Hyaloscyphaceae. [Scripta Mycologica 1]. Academy of Sciences of the Estonian SSR, Tartu. 115 pp. (1970) **Raitviïr, A.**

Untersuchungen über Pseudopezizoideae sensu Nannfeldt. *Phytopath. Z.* **36**: 213–69. (1959) **Schüepp, H.**

Revise Velenovského druhů rodu *Orbilia* (Discomycetes). *Sb. nár. mus. Praze* **10B**(1): 1–23. (1954) **Svrček, M.**

HYPOCREALES

Internationale Bibliographie der Hypomycetaceae. Jena. Bibliographische Mitteilungen der Universität Jena No. 25, 129 pp. (1976) **Arnold, G. R. W.** Bibliography 595 papers with species index.

Studies of Pyrenomycetes: IV. *Nectria* (Part I). *Mycol. Papers* **73**: 1–115. (1959) **Booth, C.**

Revision of the Hypocreales with cultural observations IV. The genus *Hypocrea* and its allies in Japan. *Bull. natn. Sci. Mus., Tokyo* **15**: 649–751. (1972) **Doi, Y.**

A new species of *Nectriella* with ornamented spores from Iceland, with a key to the lichenicolous species. *Nova Hedwigia* **35**: 755–62. (1983)('1982') **Hawksworth, D. L.**

Clef de détermination des *Nectria* d'Europe. *Bull. Soc. mycol. Fr.* **92**: 335–47. (1976) **Perrin, R.**

British Hypocreales. *Trans. Br. mycol. Soc.* **21**: 243–305. (1939) **Petch, T.**

The hypocrealan fungi. *Mycologia* **62**: 865–910. (1971) **Rogerson, C. T.**

A preliminary account of the taxa described in *Calonectria*. *Mycotaxon* **8**: 485–558. (1979) **Rossman, A. Y.**

The phragmosporous species of *Nectria* and related genera. *Mycol. Papers* **150**: 1–164. (1980) **Rossman, A. Y.**

A revision of the fungi formerly classified as *Nectria* subgenus *Hyphonectria*. *Mem. N.Y. bot. Gdn.* **26**: 1–126. (1976) **Samuels, G. J.**

LABOULBENIALES

Bibliographie des Laboulbeniales. *Bull. mens. Soc. linn. Lyon* **40**: 134–49.

(1971) **Balazuc, J.**

Laboulbeniales de France. *Bull. mens. Soc. linn. Lyon* **42**: 244–56, 280–5; **43**: 12–21, 57–64, 73–9, 252–62, 295–315, 346–68.   (1973–74) **Balazuc, J.**

Laboulbeniomycetes. In *The fungi. An advanced treatise* (ed. G. C. Ainsworth, F. K. Sparrow, and A. S. Sussman), Vol. 4A, pp. 223–46. Academic Press, London and New York.   (1973) **Benjamin, R. K.**

Laboulbeniales (Fungi, Ascomycetes). *Mycologia Mem.* **9**: 627 pp.

(1985) **Tavares, I. I.**

Contribution towards a monograph of the Laboulbeniaceae. Parts I—V *Mem. Am. Acad. Arts Sci.* **12**: 187–429; **13**: 217–469; **14**: 309–426; **15**: 427–580; **16**: 1–435. Reprinted (1971) in 2 vols (with valuable new introduction and supplement by R. K. Benjamin) by J. Cramer, Lehre.

(1896–1931) **Thaxter, R.**

LECANORALES

Lecanorineae

Taxonomic studies on reindeer lichens (*Cladonia*, subgenus *Cladina*). *Suomal. eläin-ja Kasvit. Seur. van. kansvit. Julk.* **32**(1): 1–160.   (1961) **Ahti, T.**

Zur Morphologie und Systematik von *Parmelia*, Untergattung *Hypogymnia*. *Hedwigia* **40**: 171–274.   (1901) **Bitter, G.**

The North America species of the *Lecanora subfusca* group. *Beih. Nova Hedwigia* **79**: 63–185.   (1984) **Brodo, I. M.**

Alectoria and allied genera in North America. *Op. bot. Soc. bot. Lund* **42**: 1–164.

(1977) **Brodo, I. M. and Hawksworth, D. L.**

Les *Acarospora* de l'Europe occidentale et de la région méditerranéene. *Bull. Mus. hist. nat. Marseille* **41**: 41–93.   (1981) **Clauzade, G. and Roux, C.**

A taxonomic study of the lichen genus *Micarea* in Europe. *Bull. Br. Mus. nat. Hist. (Bot.)* **11**: 17–214.   (1981) **Coppins, B. J.**

The genus *Vezdaea* in the British Isles. *Lichenologist* **19**: 167–76.

(1987) **Coppins, B. J.**

A review of *Psilolechia*. *Lichenologist* **19**: 29–42

(1987) **Coppins, B. J. and Purvis, D. W.**

The lichen genera *Cetrelia* and *Platismatia* (Parmeliaceae). *Contr. US natn. Herb.* **34**: 449–558.   (1968) **Culberson, W. L. and Culberson, C. F.**

The lichen genus *Collema* in Europe. *Symb. bot. upsal.* **13**(2): 1–499.

(1954) **Degelius, G.**

A chemosystematic revision of the brown *Parmeliae*. *J. Hattori bot. Lab.* **42**: 1–211.   (1977) **Esslinger, T. L.**

Zur Kenntnis der Flechtengattung *Rhizocarpon* in Bayern. *Ber. Bayer. bot. Ges.* **49**: 59–135. (1978) **Feuerer, T.**

A *Cladonia* primer. *Bull. Br. Lichen Soc.* **55**: 15–18. (1984) **Fletcher, A.**

Karschia. Revision einer Sammelgattung an der Grenze on lichenisierten und nichtlichenisierten Ascomyceten. *Beih. Nova Hedwigia* **62**: 1–248.

(1979) **Hafellner, J.**

Studien in Richtung einer natürlichen Gliederung der Sammelfamilien Lecanoraceae und Lecideaceae. *Beih. Nova Hedwigia* **79**: 241–371.

(1984) **Hafellner, J.**

Die Arten der Gattung *Caloplaca* mit pluriloculVaren Sporen (*Meroplacis, Triophthalmidium, Xanthocarpia*). *J. Hattori bot. Lab.* **46**: 1–41.

(1979) **Hafellner, J. and Poelt, J.**

Die Flechtengattung *Candelariella* Müller Argoviensis. *Suomal. eläin-ja Kasvit. Seur. van. kasvit. Julk.* **27**(3): i–vi + 1–127. (1954) **Hakulinen, R.**

A monograph of *Parmelia* subgenus *Amphigymnia*. *Contr. US natn. Herb.* **36**: 193–258. (1965) **Hale, M. E.**

A monograph of the lichen genus *Pseudoparmelia* Lynge (Parmeliaceae). *Smithson. Contr. Bot.* **31**: 1–62. (1976) **Hale, M. E.**

A monograph of the lichen genus *Parmelina* Hale (Parmeliaceae). *Smithson. Contr. Bot.* **33**: 1–60. (1976) **Hale, M. E.**

Studies on *Parmelia* subgenus *Parmelia*. *Contr. US natn. Herb.* **36**: 121–91.

(1964) **Hale, M. E. and Kurokawa, S.**

Regional studies on *Alectoria* (Lichenes) II. The British species. *Lichenologist* **5**: 181–261. (1972) **Hawksworth, D. L.**

Kritische Bemerkungen über einige Arten der Flechtengattungen *Lecanora* (Ach.), *Lecidea* (Ach). und *Micarea* (Fr.). *Bih. K. svenska VetenskAkad. Handl.* **18**, **3**(3): 1–102. (1892) **Hedlund, T.**

De Nederlandse *Cladonia*'s (Lichenes). *Wet. Meded. K. ned. natuurh. Veren.* **79**: 1–53. (1968) **Hennipman, E.**
Well illustrated.

Revision einiger calciphiler Formenkreis der Flechtengattung *Lecidea*. *Beih. Nova Hedwigia* **24**: 1–155. (1967) **Hertel, H.**

Beiträge zur Kenntnis der Flechtenfamilie Lecideaceae I–VIII. *Herzogia* **1**: 25–39, 321–9; **2**: 37–62, 231–61, 479–515; **3**: 365–406, **5**: 449–63.

(1968–81) **Hertel, H.**

Ein vorläufiger Bestimmungsschlüssel für die kryptothallinen, schwarzfrüchtigen, saxicolen Arten der Sammelgattung *Lecidea* (Lichenes) in der Holarktis. *Decheniana* **127**: 37–78. (1975) **Hertel, H.**

Remarks on the taxonomy of the European and North American species of *Pilophorus* Th. Fr. *Lichenologist* **4**: 199–213. (1970) **Jahns, H. M.**

Key to *Parmelia* in Great Britain. *Bull. Br. Lichen Soc.* **51**: 27–36.

(1982) **James, P. W.**

The lichen family Pannariaceae in Europe. *Op. bot. Soc. bot. Lund* **45**: 1–124.
(1978) **Jørgensen, P. M.**

Studies on some *Leptogium* species of western Europe. *Lichenologist* **15**: 109–25.
(1983) **Jørgensen, P. M. and James, P. W.**

The brown fructicose species of *Cetraria. Op. bot. Soc. bot. Lund* **46**: 1–150.
(1979) **Karnefelt, I.**

The genus *Umbilicaria* in the British Isles. *Lichenologist* **1**: 251–65.
(1961) **Kershaw, K.**

Revision gesteinsbewohnende Sippen der Flechtengattung *Catillaria* Massal. in Europa. *Herzogia* **5**: 209–448. (1981) **Kilias, R.**

Punctelia, a new lichen genus in the Parmeliaceae. *Nord. J. Bot.* **2**: 287–92.
(1982) **Krog, H.**

The genus *Ramalina* in Fennoscandia and the British Isles. *Norw. J. Bot.* **24**: 15–43. (1977) **Krog, H. and James, P. W.**

A conspectus of the lichen genus *Stereocaulon* (Schreber) Hoffm. *J. Hattori bot. Lab.* **43**: 191–355. (1977) **Lamb, I. M.**

Keys to the species of the lichen genus *Stereocaulon* (Schreber) Hoffm. *J. Hattori bot. Lab.* **44**: 209–50. (1978) **Lamb, I. M.**

A monograph of the lichen family Umbilicariaceae in the Western Hemisphere. Office of Naval Research, Washington, DC. Navexos P-831. 281 pp.
(1950) **Llano, G. A.**

Revision of the lichens of the Netherlands I. Parmeliaceae. *Blumea* **6**: i–viii + 1–199 pp. (1948) **Maas Geesteranus, R. A.**

Key to the species of *Lecidea* in Scandinavia and Finland I–II. *Svensk. bot. Tidskr.* **46**: 178–98, 313–23. (1952) **Magnusson, A. H.**

Die lobaten Arten der Flechtengattung *Lecanora* Ach. sensu ampl. in der Holarktis. *Mitt. bot. StSamml., Münch.* **2**: 411–589. (1958) **Poelt, J.**

Studies in *Rhizocarpon* I–II. *Op. bot. Soc. bot. Lund* **2**(1): 1–152, **2**(2): 1–150.
(1956) **Runemark, H.**

Die Flechtengattung *Psora* sensu Zahlbruckner. *Bibltheca lich.* **13**: 1–291.
(1979) **Schneider, G.**

The genus *Leptogium* in North America north of Mexico. *Bryologist* **67**: 245–317.
(1964) **Sierk, H. A.**

A monograph of the lichen family Megalosporaceae. *Bibltheca lich.* **18**: 1–241.
(1983) **Sipman, H. J. M.**

Notes on *Rhizocarpon* in the arctic. *Nova Hedwigia* **14**: 421–81.
(1967) **Thomson, J. W.**

The lichen genus Cladonia in North America. University of Toronto Press, Toronto. 172 pp. (1968, 1967) **Thomson, J. W.**

Physciineae

Beiträge zu einer Lichenenflora der Schweiz II. III. Die Familie Physciaceae. *Ber. schweiz. bot. Ges.* **73**: 389–503. (1963) **Frey, E.**

A monograph of the genus *Anaptychia. Beih. Nova Hedwigia* **6**: 1–115. (1962) **Kurokawa, S.**

Die saxicolen Arten der Flechtengattungen *Rinodina* und *Rinodinella* in der Alten Welt. *J. Hattori bot. Lab.* **55**: 327–493. (1984) **Mayrhofer, H.**

The lichen genus *Physcia* and allied genera in Fennoscandia. *Sym. bot. upsal.* **22**(1): i–vii + 1–108. (1977) **Moberg, R.**

Zur Kenntnis der Flechtengattung *Physconia. Nova Hedwigia* **12**: 107–35. (1966) **Poelt, J.**

The genus *Buellia* de Notaris in the British Isles (excluding section *Diploicza* (Massal.) Stiz.). *Lichenologist* **2**: 225–62. (1964) **Sheard, J. W.**

A revision of the lichen genus *Rinodina* (Ach.) Gray in the British Isles. *Lichenologist* **3**: 328–67. (1967) **Sheard, J. W.**

The lichen genus *Physcia* in North America. *Beih. Nova Hedwigia* **7**: 1–172. (1963) **Thomson, J. W.**

The British *Anaptychiae* and *Physciae. Lichenologist* **1**: 126–44. (1960) **Wade, A. E.**

LICHINALES

Beiträge zur Kenntnis der Anatomie und Systematik der Gloeolichen. *Nova Acta R. Soc. Scient. upsal.*, Ser. 3 **13**(6): 1–118. (1885) **Forssell, K. B. J.**

Eine Revision der Flechtenfamilien Lichinaceae und Ephebaceae. *Symb. bot. upsal.* **18**(1): 1–123. (1963) **Henssen, A.**

MICROASCALES

Revision of *Microascus* with the description of a new species. *Persoonia* **8**: 191–7. (1975) **von Arx, J. A.**

The genus *Microascus. Can. J. Bot.* **39**: 1609–31. (1961) **Barron, G. L., Cain, R. F., and Gilman, J. C.**

Generic concepts in some ascomycetes occurring on dung. In *Taxonomy of fungi* (ed. C. V. Subramanian) Vol. 1, pp. 241–57. University of Madras, Madras. (1978) **Lodha, B. C.**

New concepts in the Microascaceae illustrated by two new species. *Mycologia* **62**: 727–40. (1970) **Malloch, D.**

ONYGENALES

Revision of British Gymnoascaceae. *Mycol. Papers* **96**: 1–56. (1964) **Apinis, A. E.**

A synopsis of the orders and families of Plectomycetes with a key to genera. *Mycotaxon* **12**: 1–91.   (1980) **Benny, G. L. and Kimbrough, J. W.**

Taxonomy of the Onygenales. Arthrodermataceae, Gymnoascaceae, Myxotrichaceae and Onygenaceae. *Mycotaxon* **24**: 1–216.   (1985) **Currah, R. S.**

OPEGRAPHALES

Key to *Opegrapha* in Great Britain. *Bull. Br. Lichen Soc.* **53**: 27–35.
(1983) **Pentecost, A. and Coppins, B. J.**

The genera *Dirina* and *Roccellina* (Roccellaceae). *Op. bot.* **70**: 1–86.
(1983) **Tehler, A.**

OPHIOSTOMATALES

The genus *Ceratocystis* in Ontario. *Can. J. Bot.* **46**: 689–718.
(1968) **Griffin, H. D.**

Taxonomy of the genus *Ceratocystis*. *Lloydia* **19**: 1–58.   (1952) **Hunt, J.**

Taxonomy of the genus *Ceratocystis* in Manitoba. *Can. J. Bot.* **52**: 1675–711.
(1974) **Olchowecki, A. and Reid, J.**
Includes keys to the world's known species.

A monograph of Ceratocystis and Ceratocystiopsis. University of Georgia Press, Athens, Georgia. 176 pp.   (1981) **Upadhyay, H. P.**

OSTROPALES

A guide to the lignicolous species of *Apostemidium*. *Trans. Br. mycol. Soc.* **48**: 639–46.   (1965) **Graddon, W. D.**

The ostropalean fungi. *Mycotaxon* **5**: 1–277.   (1977) **Sherwood, M. A.**
A world monograph; keys; well illustrated.

The ostropalean fungi II. *Mycotaxon* **6**: 215–60.   (1977) **Sherwood, M. A.**

Sykttea, a new genus of odontotremoid lichenicolous fungi. *Trans. Br. mycol. Soc.* **75**: 479–90.
(1980) **Sherwood, M. A., Hawksworth, D. L., and Coppins, B. J.**

PATELLARIALES (LECANIDIALES)

Studies in the Patellariaceae. *Mycologia* **32**: 791–823.   (1940) **Butler, E. T.**
Studies on the Patellariaceae I. *Eutryblidiella sabina* (de Not.) Höhnel. *Can. J. Bot.* **44**: 655–62.   (1966) **Pirozynski, K. A. and Reid, J.**

Genus *Patellaria* from India. *Beih. Nova Hedwigia* **47**: 459–62.
(1974) **Tilak, S. T. and Srinivasulu, B. V.**

PELTIGERALES

Studies on the genus *Nephroma* I. The European and Macronesian species. *Lichenologist* **19**: 215–68.   (1987) **James, P. W. and White, F. J.**

The species of *Peltigera* of North America north of Mexico. *Am. Midl. nat.* **44**: 1–68.   (1950) **Thomson, J. W.**

A key to *Peltigera* in Great Britain. *Bull. Br. Lichen Soc.* **50**: 28–36.
                                                                (1982) **Vitikainen, O.**

The genus *Lobaria* of eastern Asia. *J. Hattori bot. Lab.* **34**: 231–64.
                                                                (1971) **Yoshimura, I.**

PERTUSARIALES

The chemosystematics of the lichen genus *Pertusaria* in North America and Mexico. *Publs Bot. Milwaukee publ. Mus.* **5**: 1–162.   (1980) **Dibben, M. J.**

Die chemotypen der Flechtengattung *Pertusaria* in Europa. *Bibltheca lich.* **19**: 1–297.   (1983) **Hanko, B.**

The corticolous Norwegian Pertusariaceae and Thelotremaceae. *Nyt Mag. Naturvid.* **61**: 139–78.   (1923) **Hoeg, O.**

The lichen genus *Ochrolechia* in North America north of Mexico. *Bryologist* **73**: 93–130.   (1970) **Howard, G. E.**

Taxonomical studies on the family Pertusariaceae of Japan. *J. Sci. Hiroshima Univ.* B(2) **12**: 81–163.   (1968) **Oshio, M.**

PEZIZALES (TUBERALES)

See also under Hypogeous fungi (pp. 214–15).

A world monograph of the genera *Ascobolus* and *Saccobolus*. *Persoonia*, Suppl. **1**: 1–260.   (1967) **Brummelen, J. van**

The genus *Helvella* in Europe. *Dansk. bot. Ark.* **25**: 1–172. (1966) **Dissing, H.**

Untersuchungen an moosparasitischen Pezizales aus der Verwandtschaft von *Octospora. Nova Hedwigia* **31**: 817–64.   (1980)('1979') **Döbbeler, P.**

The genera of the operculate Discomycetes. *Nytt Mag. Bot.* **15**: 1–191.
                                                                (1968) **Eckblad, F.-E.**

A new taxonomy for *Monascus* species based on cultural and microscopical characters. *Austr. J. Bot.* **31**: 51–61.
                                          (1983) **Hawksworth, D. L. and Pitt, J. I.**

Synoptic key to the genera of the Pezizales. *Mycologia* **64**: 937–94.
                                                                (1972) **Korf, R.**

Revision du genre *Melastiza* Boud. *Documns mycol.* **11**: 1–45. (1980) **Lasseur, R.**

De fungi van Nederland II. Pezizales I–II. *Wet. Meded. K. ned. natuurh. Vereen* **69**: 1–72; **80**: 1–84.   (1967–69) **Maas Geesteranus, R. A.**

The Australian Pezizales in the herbarium of the Royal Botanic Gardens, Kew. *Verh. K. ned. Akad. Wet., III* **57**(3): 1–295.   (1968) **Rifai, M. A.**

Ceské druhy podceledi Lachneoideae (cel. Pezizaceae). *Sb. nár. mus. Praze*

4B(6): 1–95.   (1948) **Svrček, M.**

PHYLLACHORALES (POLYSTIGMATALES)

Die Arten der Gattung *Colletotrichum* Cda. *Phytopath. Z.* **29**: 413–68.
                                              (1957) **von Arx, J. A.**
Includes treatment of teleomorphs.

Studies on graminicolous species of *Phyllachora* Nke. in Fckl. V. A taxonomic
   monograph. *Austr. J. Bot.* **15**: 217–375.   (1967) **Parbery, D. G.**

A monographic study of Indian species of *Phyllachora. Monogr. Univ. Agric. Sci.
   Hebbal, Bangalore* **4**: i–v, 100 pp.
             (1978) **Kamat, M. N., Seshadri, V. S., and Pande, A. A.**
Includes key to 88 spp.

PROTOMYCETALES

Protomycetales and Taphrinales. In *The fungi. An advanced treatise* (ed. G. C.
   Ainsworth, F. K. Sparrow, and A. S. Sussman), Vol. 4A, pp. 33–41.
                                              (1973) **Kramer, C. L.**

A taxonomic revision of the Protomycetales. *Mycotaxon* **3**: 1–50.
                             (1975) **Reddy, M. S. and Kramer, C. L.**

PYRENULALES

See also under Lichen-forming fungi (pp. 215–18).

On the Massariaceae in North America. *Mycotaxon* **9**: 17–37.
                                              (1979) **Barr, M. E.**
The corticolous pyrenolichens of the Great Lakes region. *Mich. Bot.* **12**: 3–68.
                                              (1973) **Harris, R. C.**
Type stydies of *Massaria* from the Wehmeyer collection. *Can. J. Bot.* **54**: 1568–
   98.   (1975) **Shoemaker, R. A. and Le Clair, P. M.**

RHYTISMATALES

The Rhytismatales of the Indian subcontinent. *Mycol. Papers* **155**: 1–123.
                             (1986) **Cannon, P. F. and Minter, D. W.**
The Hypodermataceae of conifers. *Contr. Arnold Arb.* **1**: 1–131.
                                              (1932) **Darker, G. D.**
A revision of the genera of the Hypodermataceae. *Can. J. Bot.* **45**: 1399–444.
                                              (1967) **Darker, G. D.**
Host-genus keys to the Hypodermataceae of conifer leaves. *Mycotaxon* **6**: 481–
   96.   (1978) **Hunt, R. S. and Ziller, W. G.**

Lophoderminium on pines. *Mycol. Papers* **147**: 1–54.   (1981) **Minter, D. W.**

Taxonomic studies in the Phacidiales: the genus *Coccomyces. Occ. Pp. Farlow Herb.* **15**: 1–120.   (1980) **Sherwood, M. A.**

Essai sur la systématique des Phacidiales (Fr.) sensu Nannfeldt (1932). *Beitr. Kryptog.-Flora Schweiz* **9**(2): 1–99.   (1942) **Terrier, C.-E.**

SORDARIALES (CORONOPHORALES)

A monograph of the Chaetomiaceae. US Army, Washington, DC. US Army Research and Development Series No. 2, ix + 125 pp. Reprinted (1969) by J. Cramer, Lehre.   (1963) **Ames, L. M.**

On *Thielavia* and some similar genera of ascomycetes. *Stud. mycol.* **8**: 1–28.
                                                        (1975) **von Arx, J. A.**

A key to the species of *Gelasinospora. Persoonia* **11**: 443–9. (1982) **von Arx, J. A.**

The ascomycete genus *Chaetomium. Beih. Nova Hedwigia* **84**: 1–162.
              (1986) **Arx, J. A. von, Guarro, J., and Figueras, M. J.**

Recherches sur la mycoflore coprophile Centrafricaine. Les genres *Sordaria, Gelasinospora, Bombardia. Bull. Soc. mycol. Fr.* **87**: 461–626.
                                                        (1971) **Cailleux, R.**

Studies of coprophilous Sphaeriales in Ontario. *Univ. Toronto Stud., biol. Ser.* **38**: 1–126.   (1934) **Cain, R. F.**

A re-evaluation of *Melanospora* Corda and similar pyrenomycetes, with a revision of the British species. *Bot. J. Linn. Soc.* **84**: 115–60.
                        (1982) **Cannon, P. F. and Hawksworth, D. L.**

Le genre *Melanospora. Botaniste* **39**: 1–313.   (1955) **Doguet, G.**

Monograph of the *Nitschkiae. Mycologia* **15**: 23–67. (1923) **Fitzpatrick, H. M.**

Ornamentation on the terminal hairs in *Chaetomium* Kunze ex Fr. and some allied genera. *Mycol. Papers* **134**: 1–24.
                              (1973) **Hawksworth, D. L. and Wells, H.**
Includes bibliography on the Chaetomiaceae *s. lat.*

Coniochaeta angustispora sp. nov. from roots in Australia, with a key to the species known in culture. *Austr. J. Bot.* **29**: 377–84.
                              (1980) **Hawksworth, D. L. and Yip, H. Y.**

Nordic Sordariaceae *s. lat. Symb. bot. upsal.* **20**(1): 1–374. (1972) **Lundqvist, N.**

Coniochaeta extramundana, with a synopsis of other *Coniochaeta* species. *Mycologia* **73**: 931–52.   (1981) **Mahoney, D. P. and La Favre, J. S.**

The genus *Thielavia. Mycologia* **65**: 1055–77.
                                      (1973) **Malloch, D. and Cain, R. F.**

Revision of the genus *Podospora. Can J. Bot.* **47**: 1999–2048.
                                      (1969) **Mirza, J. H. and Cain, R. F.**

Stray studies in the Coronophorales (Pyrenomycetes) 4–8. *Svensk bot. Tidskr.* **69**: 289–335.   (1975) **Nannfeldt, J. A.**

Two new genera of Coronophorales with descriptions and key. *Trans. Br. mycol. Soc.* **62**: 35–43. (1974) **Sivanesan, A.**

TAPHRINALES

The genus *Taphrina* in Scotland. *Notes R. bot. Gdn Edinb.* **21**: 165–80.
(1954) **Henderson, D. M.**

Protomycetales and Taphrinales. In *The fungi. An advanced treatise* (ed. G. C. Ainsworth, F. K. Sparrow, and A. S. Sussman), Vol. 4A, pp. 33–41. Academic Pres, London and New York. (1973) **Kramer, C. L.**

A monograph of the genus *Taphrina*. *Univ. Kansas Sci. Bull.* **33**: 1–167. Reprinted (1968) by J. Cramer, Lehre. (1949) **Mix, A. J.**

TELOSCHISTALES

Leproplaca in the British Isles. *Lichenologist* **6**: 102–5. (1974) **Laundon, J. R.**

Caloplaca, sect. Gasparrinia i Nordeuropa. Upssala (Skriv Service) 184 pp, illustrated. [Swedish]. (1972) **Nordin, I.**

Die galappten Arten der Flechtengattung *Caloplaca* in Europa mit besonderer Berücksichtigung Mitteleuropas. *Mitt. bot. StSamml. Münch.* **2**: 11–31.
(1954) **Poelt, J.**

Über einige Arten-gruppen der Flechtengattungen *Caloplaca* und *Fulgensia*. *Mitt. bot. StSamml. Münch.* **5**: 571–607. (1965) **Poelt, J.**

The genus *Caloplaca* Th. Fr. in the British Isles. *Lichenologist* **3**: 1–28.
(1965) **Wade, A. E.**

Schwarzfrüchtige, saxicole Sippen der Gattung *Caloplaca* (Lichenes, Teloschistaceae) in Mitteleuropa, dem Mittelmeergebiet und Vorderasien. *Biblitheca lich.* **3**: 1–186. (1974) **Wunder, H.**

VERRUCARIALES

See also under Lichen-forming fungi (pp. 215–18).

Key to crustose pyrenocarpous lichens on limestone and associated substrata (excluding aquatic and marine habitats). *Bull. Br. Lichen Soc.* **54**: 36–45.
(1984) **Coppins, B. J.**

Amphibious pyrenolichens 1. *Ark. Bot.* **29A**(10): 1–67. (1939) **Santesson, R.** Marine species.

Československé lišejniky čeledi Verrucariaceae. Nakladatelstivi Československá Akademie Ved, Prague. 249 pp. (1954) **Servít, M.**

Pyrenocarpous lichens: 4. Guide to the British species of *Staurothele*. *Lichenologist* **2**: 151–71. (1963) **Swinscow, T. D. V.**

Pyrenocarpous lichens: 13. Freshwater species of *Verrucaria* in the British Isles.
  Lichenologist **4**: 34–54.   (1968) **Swinscow, T. D. V.**

Pyrenocarpous lichens: 15. Key to *Polyblastia* in the British Isles. *Lichenologist* **5**:
  92–113.   (1963) **Swinscow, T. D. V.**

XYLARIALES

Anthostomella Sacc. (Part I). *Mycol. Papers* **139**: 1–97.   (1975) **Francis, S. M.**

A monograph of the world species of Hypoxylon. University of Georgia Press, Athens,
  Georgia. xii + 158 pp.   (1961) **Miller, J. H.**

Haupt- und Nehenfruchtformen europäischer *Hypoxylon*-Arten (Xylariacae,
  Sphaeiales) und verwandter Pilze. *Mycologia Helvetica* **1**: 501–627.
                                   (1986) **Petrini, L. E. and Müller, E.**

Key to the British species of *Hypoxylon*. *Bull. Br. mycol. Soc.* **11**: 45–7.
                                             (1977) **Whalley, A. J. C.**

Numerical taxonomy of *Hypoxylon* II. A key for the identification of British
  species of *Hypoxylon*. *Trans. Br. mycol. Soc.* **61**: 455–9.
                         (1973) **Whalley, A. J. S. and Greenhalgh, G. N.**

BASIDIOMYCOTINA (BASIDIOMYCETES)

UREDINIOMYCETES (UREDINALES)

Rusts.

Rost- und Brandpilze auf Kulturpflanzen. G. Fischer, Jena. 379 pp.
                                               (1963) **Blumer, S.**

Illustrated genera of rust fungi, revised edn. American Phytopathological Society,
  St Paul. v + 152 pp.   (1983) **Cummins, C. B. and Hiratsuka, Y.**

The rust fungi of cereals, grasses and bamboos. Springer, Berlin. xv + 570 pp.
                                           (1971) **Cummins, C. B.**

Die Rostpilze Mitteleuropas. *Beitr. Kryptog Flora Schweiz* **12**: 1–1407.
                                             (1959) **Gäumann, E.**

A revised taxonomic account of Gymnosporangium. Pennsylvania State University,
  University Park and London. 134 pp.   (1973) **Kern, F. D.**

Uredinales. In *The fungi. An advanced treatise* (ed. G. C. Ainsworth, F. K.
  Sparrow, and A. S. Sussman), Vol. 4B, pp. 247–79. Academic Press,
  London and New York.   (1973) **Laundon, G. F.**

British rust fungi. Cambridge University Press, London.
                             (1966) **Wilson, M. and Henderson, D. M.**
  The standard British work.

British rust fungi: additions and corrections. *Notes R. bot. Gdn Edinb.* **37**: 475–
  501.   (1979) **Henderson, D. M. and Bennell, A. P.**

USTILAGINOMYCETES (USTILAGINALES)

Rost- und Brandpilze auf Kulturpflanzen. G. Fischer, Jena. 379 pp.
(1963) **Blumer, S.**

Ustilaginales of the British Isles. *Mycol. Papers* **154**: 1–96.
(1984) **Mordue, J. E. M. and Ainsworth, G. C.**
The standard British work.

Ustilaginales. In *The fungi. An advanced treatise* (ed. G. C. Ainsworth, F. K. Sparrow, and A. S. Sussman), Vol. 4B, pp. 281–300. Academic Press, London and New York.   (1973) **Durán, R.**

The genus Tilletia. Washington State University, Pullman. v + 138 pp.
(1961) **Durán, R. and Fischer, W. G.**

Ustilaginales of Sweden. *Symb. bot. upsal.* **16**(2): 1–175. (1959) **Lindeberg, B.**

Anthracoidea (Ustilaginales) on Nordic Cyperaceae–Caricoideae, a concluding synopsis. *Symb. bot. upsal.* **22**(3): 1–41.   (1979) **Nannfeldt, J. A.**

Carpathian Ustilaginales. *Symb. bot. upsal.* **24**(2): 1–309.   (1985) **Vánky, K.**

Brandpilze Mitteleuropas. *Cryptog. Helvetica* **16**: 7–277.   (1985) **Zogg, H.**

The Ustilaginales of the world. Pennsylvania State College, School of Agriculture, Philadelphia. xi + 410 pp.   (1953) **Zundel, G. L.**

HYMENOMYCETES

GENERAL

The young specialist looks at fungi. Burke Publishing, London. 240 pp.
(1969) **Haas, H.**

Les champignons d'Europe, 2nd edn. Boubée et Cie, Paris. 680 pp.
(1969) **Heim, R.**

Collins' guide to mushrooms and toadstools, 2nd edn. Collins, London. 257 pp, col. pls.   (1965) **Lange, M. and Hora, F. B.**

Handbuch für Pilzfreunde. G. Fischer, Jena. 5 vols, many col. pls.
(1958–70) **Michael, E. and Hennig, B.**

Die Röhrlinge und Blätterpilze (Agaricales). In *Kleine Kryptogamenflora*, 5th edn (ed. H. Gams), Vol. 26(2), pp. 1–533. G. Fischer, Stuttgart. English translation (1983) to 4th edn by Phillips, R. (*Keys to agarics and boleti*).
(1983) **Moser, M.**

Die Nichtblätterpilze, Gallertpilze und Bauchpilze (Aphyllophorales, Heterobasidiomycetes). In *Kleine Kryptogamenflora* (ed. H. Gams), Vol. 2b(1), pp. 1–626. G. Fischer, Stuttgart and New York.   (1984) **Jülich, W.**

New or interesting records of British Hymenomycetes I–VI. *Trans. Br. mycol. Soc.* **38**: 387–99; **41**: 419–45; **48**: 513–37; **55**: 413–41; *Persoonia* **7**: 293–303; *Beih. Nova Hedwigia* **51**: 199–206.   (1955–75) **Reid, D. A.**

An annotated list of some fungi from the Channel Islands, mostly from Jersey. *Trans. Br. mycol. Soc.* **84**: 709–14.   (1985) **Reid, D. A.**

Further records of Jersey Fungi. *Trans. Br. mycol. Soc.* **84**: 715–22.
(1985) **Reid, D. A.**

Mushrooms and other fungi. Hamlyn, London. 333 pp.
(1974) **Rinaldi, A. and Tyndalo, V.**
Well illustrated in colour; valuable for all macromycetes; first published in Italian in 1972.

Nouvelle atlas des champignons. Bordas, Paris. 4 vols. (1956–67) **Romagnesi, H.**
Also issued in three small volumes in 1962.

Keys to genera of the higher fungi. University of Michigan Press, Ann Arbor. iv + 120 pp.   (1964) **Smith, A. H. and Shaffer, R. L.**

Common British fungi. Gawthorn, London. 290 pp., 111 col. pls.; second (revised) edn. Saiga Publishing, Hindhead.
(1950, 1981) **Wakefield, E. M. and Dennis, R. W. G.**

Champignons du nord et du midi. Vol. 1, Meilleur comestibles et principeux veneneux; Vol. 2, Meilleurs comestibles; Vol. 3, Boletales et Aphyllophorales; Vol. 4, Aphyllophorales, Hydnaceae, Gasteromycetes, Ascomycètes; Vol. 5, Russulales; Vol. 6, Lactaires et Pholiotes; Vols 7 & 8, Cortinaires. Soc. Myco. Pyrenees Medit. Perpignan.
(1971–83) **Marchand, A.**
Excellent description and colour photos of macrofungi.

Der grosse Pilzeführer, Vols. 1–4. BLV Verlagsgesellschaft, Munich, Berne, and Vienna.   (1979–84) **Cetto, B.**
Descriptions of 3176 macrofungi with 1694 colour photos.

700 Pilze in Farbfotos. A. T. Verlag Aarau, Stuttgart. 686 pp.
(1980) **Dähncke, R. M. and Dähncke, S. M.**

Svampar. En fälthandbook. Interpublishing, Stockholm. 718 pp.
(1984) **Ryman, S. and Holmasen, I.**

Identification of the larger fungi. Hulton Educational, Amersham. 218 pp.
(1973) **Watling, R.**

Agaricales

GENERAL

Gattungschlüssel für Blätte- und Röhrenpilze nach mikroscopischen Merkmalen. *Beih. Z. Pilzkde* **1**: 1–42.   (1976) **Bresinsky, A.**
Keys to genera based on microscopic characters.

Illustrations of British fungi (Hymenomycetes). Williams and Norgate, London. 8 vols.   (1881–91) **Cooke, M. C.**
Valuable for its 1198 coloured plates; updated names are listed in *Trans. Br. mycol. Soc.* **20**: 33–95 (1936); and *Bull. Soc. mycol. Fr.* **80**: 349–56 (1964).

New check list of British agarics and boleti. *Trans. Br. mycol. Soc.* **43**, Suppl. 225 pp. Reprinted (1974) by J. Cramer, Lehre.

(1960) **Dennis, R. W. G., Orton, P. D., and Hora, F. B.**

British fungus flora: agarics and boleti. Introduction. HMSO, Edinburgh. 58 pp.

(1968) **Henderson, D. M., Orton, P. D., and Watling, R.**

Keys to genera.

New check list of British Agarics and Boleti. Part IV. Validations, new species and critical notes. *Trans. Br. mycol. Soc.* **43**: 440–59.   (1960) **Hora, F. B.**

Icones Selectae Fungorum. P. Lechevalier, Paris. 6 vols.

(1924–37) **Konrad, P. and Maublanc, A.**

Valuable for its 500 coloured plates.

Les Hyménomycetès agaricoides (Agaricales, Tricholomatales, Pluteales, Russulales). Étude générale et classification. *Bull. Soc. linn. Lyon*, numero spéciale.   (1980) **Kühner, R.**

Flore analytique des Champignons supérieurs. Masson et Cie, Paris. 557 pp.

(1953) **Kühner, R. and Romagnesi, H.**

The standard European flora essential to all serious students of Agaricales; five supplements published in *Bull. Soc. mycol. Fr.* **69**: 361–88 (1953) (*Lactarius*); **71**: 169–201 (1955) (*Inocybe*); **72**: 181–249 (1956) (*Pluteus, Volvariella*); *Bull. Soc. nat. Oyonnax* **1954**: 73–131 (Pleurotaceae, Marasmiaceae, Tricholomataceae); **1955**: 3–95 (*Inocybe*); **1956**: 3–94 (Naucoriaceae, Coprinaceae, Lepiotaceae); *Bull. mens. Soc. linn. Lyon* **1955**: 39–53 (*Cortinarius*); *Revue mycol.* **19**: 3–46 (1954); **20**: 196–230 (1955) (*Rhodophyllus*).

Flora agaricana Danica. Copenhagen. 5 vols.   (1935–40) **Lange, J. E.**

Includes 500 fine coloured plates; in English.

New check-list of British Agarics and Boleti. Part III. Notes on genera and species in the list. *Trans. Br. mycol. Soc.* **43**: 159–439.   (1960) **Orton, P. D.**

Includes keys to *Crepidotus, Flammulaster, Hygrophorus, Naucoria, Nolanea*, and *Pluteus*.

Notes on British Agarics II–VIII. *Notes R. bot. Gdn Edinb.* **26**: 43–65; **29**: 75–127; **32**: 135–50; **35**: 147–54; **38**: 315–30; **41**: 565–624. *Kew Bull.* **31**: 709–21.

(1965–84) **Orton, P. D.**

Mushrooms. Spring Books, London. 220 pp.   (1958) **Pilát, A. and Ušák, O.**

Mushrooms and other fungi. P. Nevill, London. 160 pp.

(1961) **Pilát, A. and Ušák, O.**

Coloured illustrations [icones] of rare and interesting fungi. Parts 1–5. *Nova Hedwigia, Suppls..*   (1966–72) **Reid, D. A.**

Die Blätterpilze (Agaricaceae). T. O. Weigel, Leipzig. 2 vols.

(1910–15) **Ricken, A.**

The Agaricales in modern taxonomy, 4th edn. Koeltz Scientific Books, Koenigstein. 981 pp.   (1986) **Singer, R.**

Specialist use only.

Agaricales and related secotioid Gasteromycetes. In *The fungi. An advanced treatise* (ed. G. C. Ainsworth, F. K. Sparrow, and A. S. Sussman), Vol. 4B, pp. 421–50. Academic Press, London and New York. (1973) **Smith, A. H.**

Keys to the families and genera of agarics known to contain species with brown or ochre spore prints. *Bull. Br. mycol. Soc.* **1**: 65–80.   (1967) **Watling, R.**

A literature guide for identifying mushrooms. Mad River Press, Eureka, California. 121 pp.   (1980) **Watling, R. and Watling, A. E.**

MONOGRAPHS

Note. Up-to-date keys to British species of treated genera are included in the series of papers by D. N. Pegler and T. W. K. Young on basidiospore form in *Kew Bull.* **26**: 499–537 (1972) (*Inocybe*); **27**: 483–500 (1972) (*Galerina* and *Kuehneromyces*); **27**: 311–23 (1972) (*Crepidotus*); **28**: 365–79 (1973) (Leucopaxilleae); **29**: 659–67 (1974) (*Lepista* and *Ripartites*); **30**: 19–32 (1975) (*Clitopilus, Rhodocybe* and *Rhodotus*); **30**: 225–40 (1975) (*Naucoria, Phaeogalera,* and *Simocybe*).

Agaricus (Psalliota)

Agaricus [Psalliota] studies I–VI. *Annls hist.-nat. Mus. natn. hung.* **53**: 187–95; **61**: 151–7; **63**: 78–82; **66**: 17–85; **67**: 37–40; **68**: 45–9.   (1961–76) **Bohus, G.**

Clé monographique du genre *Agaricus. Docums Mycol.* **15**(fasc. 60): 1–37.
(1985) **Bon, M.**

Agaricus L: Fr (*Psalliota* Fr). Fungi Europaei. Vol. 1, Saronno (Biella Giovanna): 559 pp, 70 pl.   (1984) **Cappelli, A.**

Les Psalliotes. Atlas mycologique No. 1. P. Lechevalier, Paris.
(1964) **Essette, H.**

Essai d'une clé de determination des genres *Agaricus* et *Micropsalliota. Sydowia* **30**: 6–37.   (1977) **Heinemann, P.**

Agaricales de la zone alpine. Genre *Agaricus. Trav. scient. Parc natn. Vanoise* **5**: 131–47.   (1974) **Kühner, R.**

Danish *Psalliota* species. *Friesia* **4**: 1–60, 135–220.   (1950–52) **Möller, F. H.**

The Bohemian species of the genus *Agaricus. Acta Mus. nat. Prague* **7B**(1): 1–42.
(1951) **Pilát, A.**

Agrocybe

British fungus flora. Part 3. Bolbitiaceae: Agrocybe, Bolbitius, and Conocybe. HMSO, Edinburgh. pp. 7–30.   (1982) **Watling, R.**

Amanita

Het geslacht *Amanita* in Nederland I–II. *Coolia* **9**: 40–4, 57–9. (1962) **Bas, C.**
Morphology and subdivision of *Amanita* and a monograph of its section *Lepidella. Persoonia* **5**: 285–579.   (1969) **Bas, C.**

Les Amanitées. *Naturalistes belg.* **45**: 1–22.   (1964) **Heinemann, P.**

Clé des principales Amanites de la flore française. *Revue mycol.* **32**: 162–75.
(1967) **Joly, P.**

Amanites de sud-ouest de la France. Centre d'Études et de Recherches Scientifiques, Biarritz. 73 pp.   (1960) **Parrot, A. G.**

Armillaria

Species of *Armillaria* in southern England. *Pl. Pathology* **31**: 9–12.
(1982) **Rishbeth, J.**

A pilot scheme. *Bull. Br. mycol. Soc.* **10**: 43–4.   (1976) **Watling, R.**

Le genre *Armillariella* Karst. *Docums Mycol.* **6**(22, 23): 297–8. (1976) **Bon, M.**

Bolbitius

British fungus flora. Part 3. Bolbitiaceae: Agrocybe, Bolbitius, and Conocybe. HMSO, Edinburgh. pp. 31–8.   (1982) **Watling, R.**

Boletus s. lat.

Les bolets. Études mycologiques, No. 1. P. Lechevalier, Paris. 168 pp.
(1962) **Blum, J.**

Les Boletinées. *Naturalistes belg.* **42**: 333–63.   (1961) **Heinemann, P.**

Les bolets. P. Lechevalier, Paris. 148 pp.   (1968) **Leclair, A. and Essette, H.**

Hríbovite Huby. Slovenskj Akademia, Bratislava. 207 pp, 103 col. pls.
(1974) **Pilát, A. and Dermek, A.**

Die Röhrlinge. J. Klinkhardt, Bad Heilbrunn. 2 vols.   (1965–67) **Singer, R.**

British fungus flora. Part 1. Boletaceae: Gomphidiaceae: Paxillaceae. HMSO, Edinburgh. 125 pp.   (1970) **Watling, R.**

Monograph of *Chroogomphus* (Gomphidiaceae). *Mycologia* **56**: 526–49.
(1964) **Miller, O. K.**

A new *Chroogomphus* with loculate hymenium and a revised key to the section *Floccigomphus.* *Mycologia* **62**: 831–6.   (1970) **Miller, O. K.**

Cantharellus s. lat.

A monograph of cantharelloid fungi. Oxford University Press, London. 255 pp.
(1966) **Corner, E. J. H.**

Chanterelles et craterelles. *Revue mycol.* **35**: 280–6.   (1970) **Perreau, J.**

Clitocybe

The genus *Clitocybe* (Agaricales) in Fennoscandia. *Karstenia* **10**: 5–168.
(1969) **Harmaja, H.**

Tricholomataceae de France et d'Europe occidentale 6. *Clitocybe.* *Docums Mycol.* **13**(fasc. 51): 5–39.   (1983) **Bon, M.**

Kompendium der Blatterplize *Clitocybe*. *Beih. Z. Mykol.* **5**: 1–68.
(1984) **Clemençon, H.**

North American species of *Clitocybe* I–II. *Beih. Nova Hedwigia* **72**, **81**.
(1982–85) **Bigelow, H. E.**

Collybia

The genus *Collybia* (Agaricales) in the northeastern United States and adjacent Canada. *Mycologia Mem.* **8**: 148 pp.   (1983) **Halling, R. E.**

Conocybe

The genus *Conocybe* subgenus *Pholiotina* I. The European annulate species. *Persoonia* **6**: 119–65.   (1970) **Kits van Waveren, E.**

Le genre Galera. Encyclopédie mycologiques No. 7. P. Lechevalier, Paris. 240 pp.   (1935) **Kühner, R.**

The genus *Conocybe* subgen. *Pholiotina* II. Some European exannulate species and North American annulate species. *Persoonia* **6**: 313–39.
(1971) **Watling, R.**

British fungus flora. Part 3. Bolbitiaceae: Agrocybe, Bolbitius and Conocybe. HMSO, Edinburgh. pp. 39–102.   (1982) **Watling, R.**

Coprinus

The '*stercorarius* group' of the genus *Coprinus*. *Persoonia* **5**: 131–76.
(1968) **Kits van Waveren, E.**

Species concept in the genus *Coprinus*. *Dansk bot. Archiv* **14**: 1–164.
(1952) **Lange, M.**

The *Coprinus ephemerus* group. *Mycologia* **45**: 747–80.
(1953) **Lange, M. and Smith, A. H.**

Recherches sur les Coprines. *Bull. Soc. mycol. Fr.* **71**: 5–18. (1955) **Locquin, M.**

Observations on the genus *Coprinus*. *Trans. Br. mycol. Soc.* **40**: 263–76.
(1957) **Orton, P. D.**

Notes on British agarics IV. *Notes R. bot. Gdn Edinb.* **32**: 135–50.
(1972) **Orton, P. D.**

Notes on some British agarics I–III. *Notes R. bot. Gdn Edinb.* **28**: 39–56, **31**: 259–364; **32**: 127–33.   (1967–72) **Watling, R.**

British fungus flora. Part 2. Coprinaceae: Coprinus. HMSO, Edinburgh. 149 pp.
(1979) **Orton, P. D. and Watling, R.**

Cortinarius

Die Gattung Phlegmacium. J. Klinkhardt, Bad Heilbrunn. 442 pp.
(1960) **Moser, M.**

Cortinarius I–II. *Naturalist, Hull,* Suppl. 149 pp.   (1955, 1958) **Orton, P. D.**

Crepidotus

North American species of Crepidotus. Hafner, New York. 168 pp.
(1965) **Hesler, L. R. and Smith, A. H.**

Monographie des espèces Européenes du genre Crepidotus Fr. Atlas de Champignons
de l'Europe No. 6. Prague. 84 pp. (1948) **Pilát, A.**

Cystoderma

Observations sur le genre *Cystoderma. Bull. Soc. mycol. Fr.* **89**: 5–34.
(1973) **Heinemann, P. and Thoen, D.**

A monograph of the genus *Cystoderma. Pap. Mich Acad. Sci.* **30**: 71–124.
(1945) **Smith, A. H. and Singer, R.**

Les Cystodermes. *Naturalistes belg.* **48**: 285–97. (1967) **Thoen, D.**

Dermoloma

Notes on British Agarics. VII. *Dermoloma. Notes R. Bot. Gdn Edinb.* **38**: 321–9.
(1980) **Orton, P. D.**

Entoloma

Introduction to the taxonomy of the genus *Entoloma* sensu lato (Agaricales)
Persoonia **11**: 121–51. (1980) **Orton, P. D.**

Entoloma subgenera *Entoloma* and *Allocybe* in the Netherlands and adjacent
regions with a reconnaissance of their remaining taxa in Europe. *Persoonia*
11: 153–256. (1981) **Noordeloos, M. E.**

Flammulaster

The genus *Flammulaster. Notes R. bot. Gdn Edinb.* **28**: 65–72. (1967) **Watling, R.**

Galerina s. lat.

Le genre Galera. Encyclopédie mycologique No. 7. P. Lechevalier, Paris. 240 pp.
(1935) **Kühner, R.**

Agaricales de la zone alpine. Genre *Galerina* Earle. *Bull. Soc. mycol. Fr.* **88**, 41–
118. (1972) **Kühner, R.**

Agaricales de la zone alpine. Genre *Galera* Earle et *Phaeogalera* gen. nov. *Bull.
Soc. mycol. Fr.* **88**: 119–53. (1972) **Kühner, R.**

A monograph of the genus Galerina. Hafner, New York and London. 384 pp.
(1964) **Smith, A. H. and Singer, R.**

Gerronema

Die Gattung *Gerronema. Nova Hedwigia* **7**: 52–92. (1964) **Singer, R.**

Gloiocephala

The genus *Gloiocephala* Massee in Europe. *Persoonia* **2**: 77–89. (1961) **Bas, C.**

Gomphidius

The genus *Gomphidius* with a revised description of the Gomphidiaceae and a
key to the genera. *Mycologia* **63**: 1129–63. (1971) **Miller, O. K.**

Hebeloma

Hebeloma studies I. *Annals hist.-nat. Mus. natn. hung.* **64**: 71–8. (1972) **Bohus, G.**

Contribution à l'étude de genre *Hebeloma* (Fr.) Kummer; partie spéciale. *Bull. mens. Soc. linn. Lyon* **39**: Suppl. **6**: 132 pp. (1970) **Bruchet, G.**

Beiträge zur Kenntnis der Gattung *Hebeloma. Z. Pilzk.* **36**: 61–75.
(1970) **Moser, M.**

Hygrophorus s. lat.

Hygrophores du centre-est de la France étudiés au salon du Museum 1971. *Bull. mens. Soc. linn. Lyon* **43**: 333–43. (1974) **Bon, M.**

Clé monographique des Hygrophoraceae Roze. *Docums mycol.* **7**: 1–24.
(1976) **Bon, M.**

North American species of Hygrophorus. University of Tennessee Press, Knoxville. 416 pp. (1963) **Hesler, L. R. and Smith, A. H.**

Agaricales de la zone alpine. Genre *Hygocybe* (Fr.) Kummer. *Bull. Soc. mycol. Fr.* **92**: 455–515. (1976) **Kühner, R.**

Beitrag zur Kenntnis verschiedener Hygrophoreen. *Z. Pilzk.* **33**: 1–21.
(1967) **Moser, M.**

Kompendium der Blätterpilze. *Beih. Z. Mykol.* **4**: 39–61.
(1982) **Clemençon, H.**

Hypholoma

Observations sur le genre *Hypholoma. Bull. Soc. mycol. Fr.* **52**: 9–30.
(1936) **Kühner, R.**

The North American species of *Naematoloma. Mycologia* **43**: 467–521.
(1951) **Smith, A. H.**

Inocybe

Le genre Inocybe. Encyclopédie mycologique No. 1. P. Lechevalier, Paris. 430 pp. (1931) **Heim, R.**

The genus *Inocybe. Naturalist, Hull* **1954**: 117–40. (1954) **Pearson, A. A.**

Über einige Risspilze Sudbayerns. I–III. *Z. Pilzkde* **37**: 19–32; **39**: 191–202; **42**: 15–32. (1971–78) **Stangl, J.**

Beiträge zur Kenntniss seltenerer *Inocybe*-Arten. *Česká mykol.* **25**: 1–9; **27**: 11–25; **29**: 65–78. (1971–75) **Stangl, J. and Veselský, J.**

Inocybe. Bres. Iconogr. Mycol. **29** (suppl. 3) Trento. 2 vols. incl. col. pl.
(1980) **Alessio, C. L.**

A revision of the genus *Inocybe* in Europe, I. Subgenus *Inosperma* and the smooth-spored species of the subgenus *Inocybe. Persoonia* Suppl. **3**.
(1986) **Kuyper, T. W.**

Laccaria

Tricholomataceae de France et d'Europe occidentale 6. *Laccaria*. *Docums Mycol.* **13**(fasc. 51): 46–51.   (1983) **Bon, M.**

Kompendium der Blätterpilze VI. *Laccaria*. *Zeitschr. Mykol.* **50**: 3–11.
                            (1984) **Clemençon, H.**

Lactarius

Notes on occurrence in Hungary of *Lactarius* species with regard to their range in Europe. *Annls hist.-nat. Mus. natn. hung.* **51**: 171–96.   (1959) **Babos, M.**

Lactaires et Russules au salon du champignons de 1965. *Revue mycol. Suppl.* **31**: 85–106.   (1966) **Blum, J.**

Les Lactaires. Études mycologiques No. 3. P. Lechevalier, Paris. 371 pp, 94 figs, 16 col. pls.   (1976) **Blum, J.**

Les Lactaires à lait rouge. *Revue Mycol.* **15**: 65–79.
                            (1950) **Heim, R. and Leclair, A.**

Les Lactaires, 2nd edn. *Naturalistes belg.* **41**: 133–56.   (1960) **Heinemann, P.**

Die Milchlinge. J. Klinkhardt, Bad Heilbrunn. 248 pp.   (1956) **Neuhoff, W.**

The genus *Lactarius*. *Naturalist, Hull* **1950**: 81–91.   (1950) **Pearson, A. A.**

Clé monographique du genre *Lactarius*. *Docums Mycol.* **10**(fasc. 40): 1–85.
                            (1980) **Bon, M.**

Leccinum

Rauhstielröhrlinge. Die Gattung Leccinum in Europa. Coburg. 76 pp.
                    (1978) **Engel, H., Dermek, A. and Watling, R.**

Lentinellus

The genus *Lentinellus*. *Mycologia* **63**: 333–69.
                    (1971) **Miller, O. K. and Stewart, L.**

Lentinus

Atlas de Champignons de l'Europe. Vol. 5 *Lentinus*. Prague. 46 pp. (1946) **Pilát, A.**

The agaric genera *Lentinus*, *Panus* and *Pleurotus*. *Beih. Nova Hedwigia* **69**: 1–169.
                            (1981) **Corner, E. J. H.**

The genus *Lentinus*. A world monograph. *Kew Bull. Addit. Ser.* **10**: 1–281.
                            (1984) **Pegler, D. N.**

Lepiota (including *Leucocoprinus*)

Studies on Hungarian *Lepiota* species I–IV. *Annls nat.-hist. Mus natn. hung.* **50**: 87–92; **53**: 195–9; **61**: 157–64; **66**: 65–75.   (1958–74) **Babos, M.**

Lépiotes des dunes Vendéennes. *Bull. Soc. mycol. Fr.* **88**: 15–28.
                            (1972) **Bon, M. and Boiffard, J.**

Lépiotes de Vendée et de la Côte Atlantique française I. *Bull. Soc. mycol. Fr.* **90**:
287–306.   (1974) **Bon, M. and Boiffard, J.**

Observations sur les *Lepioteae* Fayod. *Persoonia* **1**: 325–9.
                                                (1960) **Huijsman, H. S. C.**

Les espèces françaises du genre *Leucocoprinus*. Première partie: sectio *Procerae*.
Revue mycol. **16**: 213–34; **17**: 47–54.   (1951–52) **Locquin, M.**

Quelques Lépiotes nouvelles ou critiques. *Friesia* **5**: 292–6.
                                                (1956) **Locquin, M.**

Clé monographique des Lépiotes d'Europe. *Docums Mycol.* **11**(fasc. 43): 1–75.
                                                (1981) **Bon, M.**

Lepista

The status of *Lepista*, a new section of *Clitocybe*. *Brittonia* **21**: 144–77.
                                                (1969) **Bigelow, H. E. and Smith, A. H.**

Tricholomataceae de France et d'Europe occidentale 6. *Clitocybe*. *Docums
Mycol.* **13**(fasc. 51): 39–46.   (1983) **Bon, M.**

Leptonia

The genus *Leptonia* on the Pacific Coast of the United States. *Bibltheca mycol.* **55**:
1–286 pp.   (1977) **Largent, D. L.**

Studies in the rhodophylloid fungi II. *Alboleptonia*, a new genus. *Mycologia* **62**:
437–52.   (1970) **Largent, D. L. and Benedict, R. G.**

Entoloma subgenus *Leptonia* in northwestern Europe 1, section *Leptonia*. *Persoonia* **11**: 451–71.   (1982) **Noordeloos, M. E.**

Leucopaxillus

A monograph of the genus *Leucopaxillus* Boursier. *Pap. Mich. Acad. Sci.* **28**: 85–
132.   (1943) **Singer, R. and Smith, A. H.**

Marasmius

The genus *Marasmius* in the northeastern United States and adjacent Canada.
Mycotaxon **4**: 1–144.   (1976) **Gilliam, M. S.**

Melanoleuca

Essai sur le genre *Melanoleuca* Patouillard emend. *Bull. Soc. mycol.* **64**: 141–65.
                                                (1948) **Metród, G.**

Étude systematique sur les *Melanoleuca* d'Europe et clé des espèces observees
en Catalogne. *Cavanillesia* **7**: 122–32.   (1935) **Singer, R.**

Clé monographique du genre *Melanoleuca*. *Docums Mycol.* **9**(fasc. 33): 45–78.
                                                (1979) **Bon, M.**

Mycena

Le genre Mycena. Encyclopedie mycologique No. 10. P. Lechevalier, Paris. 710
pp.   (1938) **Kühner, R.**

Mycena. Naturalist, Hull **1955**: 41–63. (1955) **Pearson, A. A.**

North American species of Mycena. University of Michigan Press, Ann Arbor. 522 pp. (1947) **Smith, A. H.**

Studies on *Mycena*. 15. A tentative subdivision of the genus *Mycena* in the Northern Hemisphere. *Persoonia* **11**: 93–120.

(1980) **Maas Geesteranus, M. A.**

Conspectus of the Mycenas of the Northern Hemisphere. 1. Sacchariferae, Basipedes, Bulbosae, Clavulares, Exiguae, Longisetae. *Proc. K. Ned. Akad. Wet.*, ser. C, **86**: 401–21; Viscipelles, Amictae, and Supinae, **87**: 131–47; Filepedes **87**: 413–47; Mycena **88**: 339–69; Luculentae, Pterigenae, Carolinenses, Monticola **89**: 83–100; Polyadelphia, Saetulipedes **89**: 159–82; Cinerellae **89**: 183–201; Intermediae, Rubromarginatae **89**: 279–310.

(1983–86) **Maas Geesteranus, M. A.**

Naucoria

Les *Naucoria* du groupe *Centunculus* (*Ramicola* Velen.). *Bull. Soc. mycol. Fr.* **78**: 337–58. (1963) **Romagnesi, H.**

A revision of the British species of *Naucoria* sensu lato. *Trans. Br. mycol. Soc.* **82**: 191–237. (1984) **Reid, D. A.**

Nolanea

Entoloma subgenus *Nolanea* in the Netherlands and adjacent regions with a reconnaissance of its remaining taxa in Europe. *Persoonia* **10**: 427–534.

(1980) **Noordeloos, M. E.**

Entoloma subgenus *Nolanea*—additions. *Persoonia* **11**: 257.

(1981) **Noordeloos, M. E.**

Omphalina

Omphalina in North America. *Mycologia* **62**: 1–32. (1970) **Bigelow, H. E.**
Atlas des champignons de l'Europe. Vo. 4 *Omphalia* (*Fr.*) *Quél.* Prague. 152 pp.

(1936) **Cejp, K.**

Lichenicolous agarics. *Bull. Br. Lichen Soc.* **49**: 28–31. (1981) **Watling, R.**

Oudemansiella

Studien zur Gattung *Oudemansiella* Speg., Schleim- und Sammetrüblinge. *Z. Pilzkde* **19**: 4–11. (1955) **Moser, M.**

Classification of *Oudemansiella* (Basidiomycota: Tricholomataceae) with special reference to spore structure. *Trans. Br. mycol. Soc.* **87**: 583–602.

(1987) **Pegler, D. N. and Young, T. W. K.**

Panaeolus

The genus *Panaeolus* in Britain. *Naturalist, Hull* **1957**: 77–88.

(1957) **Hora, F. B.**

Le genre *Panaeolus. Revue Mycol.*, Mém. sér. **10**: 1–273.   (1969) **O'lah, G. N.**

Panellus

The genus *Panellus* in North America. *Mich. Bot.* **9**: 18–30. (1970) **Miller, O. K.**

Phaeocollybia

Contribution towards a monograph of *Phaeocollybia. Brittonia* **9**: 195–217.
(1957) **Smith, A. H.**

Further additions towards a monograph of *Phaeocollybia. Sydowia* **29**: 28–70.
(1977) **Horak, E.**

Phaeomarasmius

Versuch einer Zusammenstellung der Arten der Gattung *Phaeomarasmius. Schweiz. Z. Pilzkde* **34**: 53–65.   (1956) **Singer, E.**

Pholiota

The North American species of Pholiota. Hafner, New York. 402 pp.
(1968) **Smith, A. H. and Hesler, L. R.**

Pleurotus

Atlas des champignons de l'Europe. Vol. 2. *Pleurotus Fries.* Prague. 193 pp.
(1935) **Pilát, A.**

Die Gattung *Pleurotus. Biblthca mycol.* **87**: 1–448.   (1982) **Hilber, O.**

Pluteus

Contribution towards a monograph of the genus *Pluteus. Trans. Br. mycol. Soc.* **39**: 145–232.   (1956) **Singer, R.**

Notulae ad floram agaricinum neerlandicam VIII. *Pluteus* Fr. in West Europe. *Persoonia* **12**: 337–73.   (1985) **Vellinga, E. C. and Schreurs, J.**

British fungus flora. Part 4. Pluteaceae: *Pluteus* & *Volvariella.* HMSO, Edinburgh. 74 pp.   (1986) **Orton, P. D.**

Pouzaromyces

Entoloma subgenus *Pouzaromyces* emend. in Europe. *Persoonia* **10**: 207–43.
(1979) **Noordeloos, M. E.**

Psathyrella

Notes on the genus *Psathyrella* I–III. *Persoonia* **6**: 249–80, 295–312; **7**: 23–54.
(1971–72) **Kits van Waveren, E.**

The North American species of *Psathyrella. Mem. NY bot. Gdn* **24**: 1–633.
(1972) **Smith, A. H.**

The Dutch, French and British species of *Psathyrella. Persoonia* Suppl. **2**: 300 pp.
(1985) **Kits van Waveren, E.**

Psilocybe

Notes on British agarics. III. *Notes R. bot. Gdn Edinb.* **29**: 75–127.
(1969) **Orton, P. D.**

The genus *Psilocybe. Beih. Nova Hedwigia* **74**: 1–439.   (1983) **Guzmán, G.**

Rhodocybe

Agaricales de la zone alpine. Genre *Rhodocybe* R. Maire. *Bull. Soc. mycol. Fr.* **77**, 15–23.   (1971) **Kühner, R. and Lamoure, D.**

A revision of the genus *Rhodocybe. Beih. Nova Hedwigia* **67**: 1–194.
(1981) **Baroni, T. J.**

Ripartites

Observations sur le genre *Ripartites. Persoonia* **1**: 335–9.
(1960) **Huijsman, H. S. C.**

Russula

Les Russules. *Encyclopédie mycologique* No. 32. P. Lechevalier, Paris. 236 pp.
(1962) **Blum, J.**

The spore ornamentation of the Russules. Baillière, Tindall, and Cox, London. 180 pp.   (1930) **Crawshay, R.**

Les Russules, 4th edn. *Naturalistes belg.* **43**: 1–32.   (1962) **Heinemann, P.**

Agaricales de la zone alpine. Genre *Russula* Pers. ex S. F. Gray. *Bull. Soc. Mycol. Fr.* **91**: 313–90.   (1975) **Kühner, R.**

Keys to the British species of *Russula* I–III. *Bull. Br. mycol. Soc.* **2**: 76–109; **3**: 89–120; **4**: 19–46. Also issued separately; additions and corrections in **10**: 69–73 (1976); 3rd edn, British Mycological Society.
(1968–85) **Rayner, R. W.**

Les Russules de l'Europe et d'Afrique du Nord. Bordas, Paris. 998 pp.
(1967) **Romagnesi, H.**

Russula monographie. Klinkhardt, Bad Heilbrunn. 296 pp. (1952) **Schaeffer, J.**

Die Täublinge.—Beiträge zur ihrer Kenntnis und Verbreitung II–III. *Z. Pilzkde* **39**: 175–90; **40**: 145–58.   (1974) **Schwobel, H.**

Squamanita

The genus *Squamanita. Persoonia* **3**: 331–59.   (1965) **Bas, C.**

Stropharia

Notes on British agarics: VI. *Stropharia. Notes R. Bot. Gdn Edinb.* **35**: 148–53.
(1976) **Orton, P. D.**

Tephrocybe

The genus *Tephrocybe* Donk in Britain. *Bull. Br. mycol. Soc.* **18**: 114–20.
(1984) **Orton, P. D.**

Tricholoma

Révision des Tricholomes. *Bull. Soc. mycol. Fr.* **83**: 324–35; **85**: 475–92.
(1967–69) **Bon, M.**

Tricholomes de France d'Europe Occidentale 1–2. *Docums Mycol.* **3**(12): 1–53;
4(14): 55–110; *Encycl. Mycol.* **36**: 324 pp.   (1974, 1984) **Bon, M.**

Musseronflora. Slekten Tricholoma (Fr. ex Fr.) Kummer sensu lato. Universitetforla-
get, Bergen and Tromso. 99 pp.   (1972) **Gulden, G.**

Tricholomopsis

Tricholomopsis in the W. Hemisphere. *Brittonia* **12**: 41–76. (1960) **Smith, A. H.**

Volvariella

The European species of *Volvariella* Spegazzini. *Bull. mens. Soc. linn. Lyon, nu.
spec.* **1974**: 313–26.   (1974) **Orton, P. D.**

British fungus flora. Part 4. Pluteaceae: *Pluteus* & *Volvariella*. HMSO, Edinburgh.
74 pp.   (1986) **Orton, P. D.**

Xeromphalina

A revision of the genus *Xeromphalina. Mycologia* **60**: 156–88.
(1968) **Miller, O. K.**

Aphyllophorales

GENERAL

Hymenomycètes de France. Hetérobasidiés—Homobasidiés gymnocarpes. Sceaux, Paris.
iv + 761 pp. Reprinted (1969) by J. Cramer, Lehre.
(1928) **Bourdot, H. and Galzin, A.**

Danish resupinate fungi. Part II. Homobasidiomycetes. *Dansk. bot. Arkiv* **19**:
57–388.   (1960) **Christiansen, M. P.**

Notes on resupinate Hymenomycetes I–VI. *Reinwardtia* **2**: 425–34; **3**: 363–79;
Fungus **26**: 3–24; **27**: 1–29; **28**: 16–36; *Persoonia* **2**: 217–38.
(1954–62) **Donk, M. A.**

A conspectus of the families of Aphyllophorales. *Persoonia* **3**: 199–324.
(1964) **Donk, M. A.**

Studies in the Hetero- and Homobasidiomycetes-Aphyllophorales in the
Muddus National Park in north Sweden. *Symb. bot. upsal.* **16**(1): 1–172.
(1958) **Eriksson, J.**

Higher taxa of Basidiomycetes. *Bibltheca mycol.* **85**: 1–485.
(1981–82) **Jülich, W.**

The resupinate non-poroid Aphyllophorales of the temperate Northern Hemisphere. North-
Holland, Amsterdam, Oxford, New York.
(1980) **Jülich, W. and Stalpers, J. A.**

Aphyllophorales I: General characteristics; thelephoroid and cupuloid families. In *The fungi. An advanced treatise* (ed. G. C. Ainsworth, F. K. Sparrow, and A. S. Sussman), Vol. 4B, pp. 327–49. Academic Press, London and New York. (1973) **Talbot, P. H.**

Contribución al estudio de los Aphyllophorales espanoles. *Bibltheca mycol.* **74**: 1–464. (1980) **Telleria, M. T.**

Notes on the British Thelephoraceae. *Trans. Br. mycol. Soc.* **5**: 474–81. (1917) **Wakefield, E. M.**

New or rare British Hymenomycetes (Aphyllophorales). *Trans. Br. mycol. Soc.* **35**: 34–65. (1952) **Wakefield, E. M.**

MONOGRAPHS

Clavariaceae

A monograph of Clavaria and allied genera. Oxford University Press, London. xv + 740 pp. (1950) **Corner, E. J. H.**

Supplement to *A monograph of Clavaria and allied genera*. Beih. Nova Hedwigia **33**: 1–299. (1970) **Corner, E. J. H.**

Les Clavaires. *Revue Mycol.* **33**: 396–415. (1969) **Perreau, J.**

Aphyllophorales II: The clavarioid and cantharelloid basidiomycetes. In *The fungi. An advanced treatise* (ed. G. C. Ainsworth, F. K. Sparrow, and A. S. Sussman), Vol. 4B, pp. 351–68. Academic Press, London and New York. (1973) **Petersen, R. H.**

Notes on Clavarioid Fungi 2. Corrections in the genera *Ramariopsis* and Claaria. *Bull. Torrey Bot. Club* **91**: 274–80. (1964) **Petersen, R. H.**

Notes on Clavarioid Fungi 5. Emendation and additions to *Ramariopsis*. *Mycologia* **58**: 201–7. (1966) **Petersen, R. H.**

Notes on Clavarioid Fungi 10. New species and type studies in *Ramariopsis* with a key to species in North America. *Mycologia* **61**: 549–59. (1969) **Petersen, R. H.**

Ramaria subgenus *Lentomaria* with emphasis on North American taxa. *Bibltheca mycol.* **43**: 1–161. (1975) **Petersen, R. H.**

Clavariales. Fungorum rariorum icones coloratae No. 5. Cramer, Lehre. 44 pp. (1971) **Schild, E.**

Corticiaceae

General

The Corticiaceae of North Europe. Fungiflora, Oslo. Vols 2–7 so far published. (1973–84) **Eriksson, J. and Ryvarden, L.**

Notes on Corticiaceae (Basidiomycetes) I–II, X. *Mycotaxon* **5**: 475–80; **7**: 117–24; **14**: 69–74. (1977–82) **Hjørtstam, K. and Larsson, K. H.**

Notes on Corticiaceae (Basidiomycetes) III–V. *Mycotaxon* **7**: 407–10; **9**: 505–19; **10**: 201–9. (1978–79) **Hjørtstam, K. and Ryvarden, L.**

Notes on Corticiaceae (Basidiomycetes) VI. *Mycotaxon* **10**: 265–8.
(1980) **Hjørtstam, K. and Hogholen, E.**

Notes on Corticiaceae (Basidiomycetes) VII–IX, XII–XIII. *Mycotaxon* **11**: 430–4; **13**: 120–3, 124–6; **17**: 577–84.
(1980–84) **Hjørtstam, K.**

Notes on Corticiaceae (Basidiomycetes) XI. *Mycotaxon* **14**: 75–81.
(1982) **Hjørtstam, K. and Stalpers, J. A.**

Einige neue oder unbekannte Corticiaceae (Basidiomycetes). *Willdenowia* **6**: 215–24.
(1971) **Jülich, W.**

Monographie der Athelieae (Corticiaceae, Basidiomycetes). *Willdenowia Beih.* **7**: 1–284.
(1972) **Jülich, W.**

Studien an resupinaten Basidiomyceten I–II. [Continued as Studies in resupinate Basidiomycetes 3–6.] *Persoonia* **7**: 3–8, 381–8; **8**: 291–305, 431–42; **10**: 137–40; 325–36.
(1973–79) **Jülich, W.**

Studies in resupinate Basidiomycetes 7. *Int. J. Mycol. Lich.* **1**: 27–37.
(1982) **Jülich, W.**

The genera of the Hyphodermoideae (Corticiaceae). *Persoonia* **8**: 59–97.
(1974) **Jülich, W.**

Conspectus systemetis Corticiacearum. Academy of Sciences of the Estonian SSR, Tartu. 261 pp.
(1968) **Parmasto, E.**

Notes on British species of *Corticium*. *Trans. Br. mycol. Soc.* **4**: 113–21.
(1913) **Wakefield, E. M.**

Aleurodiscus

The genus *Aleurodiscus* (sensu stricto) in North America. *Can. J. Bot.* **42**: 213–82.
(1964) **Lemke, P. A.**

The genus *Aleurodiscus* (sensu lato) in North America. *Can. J. Bot.* **42**: 723–68.
(1964) **Lemke, P. A.**

Monographie der mitteleuropäischen Aleurodiscineen. *Annls mycol.* **24**: 202–30.
(1926) **Pilát, A.**

Athelia

Monographie der Athelieae (Corticiaceae, Basidiomycetes). *Willdenowia Beih.* **7**: 1–284.
(1972) **Jülich, W.**

A taxonomic analysis of section *Athele* of the genus *Corticium* II. *Mycologia* **53**: 443–50.
(1962) **Liberta, A. E.**

Botryobasidium

Studies in Corticiaceae (*Botryohypochnus* Donk, *Botryobasidium* Donk and *Gloeocystidiellum* Donk). *Svensk bot. Tidskr.* **52**: 1–17.
(1958) **Eriksson, J.**

Studies in the *Botryobasidium vagum* complex. *Friesia* **9**: 1–17.
(1969) **Eriksson, J. and Hjørtstam, K.**

Christiansenia

Three species of *Christiansenia* (Corticiaceae) and the teratological galls on *Collybia dryophila*. *Bot. Notiser* **131**: 167–73.

(1978) **Ginns, J. and Sunhede, S.**

Hyphodontia

Die Gattung *Hyphodontia* J. Erikss. in Mecklenburg. *Feddes Reprium* **90**: 85–101.

(1979) **Doll, R.**

Four new taxa of *Hyphodontia* (Basidiomycetes). *Svensk bot. Tidskr.* **63**: 217–32.

(1969) **Eriksson, J. and Hjørtstam, K.**

Includes key to species.

Studies in the genus *Hyphodontia* (Basidiomycetes) I. *Hyphodontia* John Erikss. sectio *Hyphodontia*. *Mycotaxon* **17**: 550–4. (1983) **Hjørtstam, K.**

Leucogyrophana

Leucogyrophana (Aphyllophorales): identification of species. *Can. J. Bot.* **56**: 1953–73. (1978) **Ginns, J. H.**

Notes on cyanophily of spores with a discussion of the genus *Leucogyrophana* (Corticiaceae). *Persoonia* **8**: 51–8. (1974) **Jülich, W.**

Merulius

Merulius s.s. and s.l., taxonomic disposition and identification of species. *Can. J. Bot.* **54**: 100–67. (1976) **Ginns, J. H.**

Paullicorticium

The genus *Paullicorticium* (Basidiomycetes) in Sweden. *K. Göteborgs Svampklubb* **1971**: 6–13. (1971) **Hjørtstam, K.**

The genus *Paullicorticium* (Thelephoraceae). *Brittonia* **14**: 219–23.

(1962) **Liberta, A. E.**

Pellicularia s. lat.

Studies of *Pellicularia* and associated genera of Hymenomycetes. *Persoonia* **3**: 371–406. (1965) **Talbot, P. H. B.**

Peniophora

Héterobasidiomycètes saprophytes et Homobasidiomycètes résupinés VIII. *Peniophora* Cke à dendrophyses. *Revue Mycol.* **26**: 153–72. (1961) **Boidin, J.**

Le genre *Peniophora* sensu-stricto en France (Basidiomycètes). *Bull. mens. Soc. linn. Lyon* **34**: 161–9, 213–19. (1965) **Boidin, J.**

Peniophora Cke, sect. *Coloratae* Bourd. and Galz. *Symb. bot. upsal.* **10**(5): 1–76.

(1950) **Eriksson, J.**

The genus Peniophora in New York State and adjacent regions. University State College of Forestry, Syracuse, New York. Technical Publication No. 83. 95 pp. (1960) **Slysh, A. R.**

Trechispora

The genus *Trechispora* (Basidiomycetes, Corticiaceae). *Can. J. Bot.* **51**: 1871–
92.   (1973) **Liberta, A. I.**

Contribucion al estudio del genero *Trechispora* (Aphyllophorales, Basidiomycetes) en España Peninsular. *Acta Bot. Malacit.* **6**: 5–12.
                                                                (1980) **Telleria, M. T.**

Tubulicrinis s. lat.

Die Gattung *Tubulicrinis* Donk s.l. (Corticiaceae). *Z. Pilzkde* **31**: 12–48.
                                                                (1966) **Oberwinkler, F.**

Studien of Canadian Thelephoraceae X. Some species of *Peniophora*, section
Tubuliferae. Can. J. Bot. **31**: 760–78.   (1953) **Weresub, L. K.**

Xenasma

A taxonomic analysis of section *Athele* of the genus *Corticium*. 1. Genus *Xenasma*.
Mycologia **52**: 884–914.   (1962) **Liberta, A. E.**

Cyphellaceae

Rectipilus, eine neue Gattung cyphelloider Pilze. *Persoonia* **7**: 389–436.
                                                                (1973) **Agerer, R.**
Includes key to *Henningsomyces* spp.

Flagelloscypha, Studien an cyphelloiden Basidiomyceten. *Sydowia* **27**: 131–265.
                                                                (1975) **Agerer, R.**

Cyphelloid Pilze aus Teneriffa. *Nova Hedwigia* **30**: 295–341. (1978) **Agerer, R.**

Studien zur Sippenstruktur der Gattung *Cyphellopsis* I. Darstellung zweirer
Ausgangssippen. *Z. Mykol.* **46**: 117–207.
                           (1980) **Agerer, R., Prillinger, H.-J. and Noll, H.-P.**

Typusstrudien an Cyphelloiden Pilzen IV. *Lachnella* Fr. s.l. *Mitt. Bot. StSamml.*
München **19**: 163–334.   (1983) **Agerer, R.**

The cyphellaceous fungi. *Sydowia, Beih.* **4**: 1–144.   (1961) **Cooke, W. B.**

Notes on Cyphellaceae I–II. *Persoonia* **1**: 25–110; **2**: 331–48.
                                                                (1959–62) **Donk, M. A.**

Beiträge zur Kenntnis der Thelephoraceen I. Die Cyphellaceen Bonner. *Annls*
mycol. **22**: 204–18.   (1924) **Pilát, A.**

Zweiter Beiträge zur Kenntnis der tschechoslowakischen Cyphellaceen. *Annls*
mycol. **23**: 144–73.   (1925) **Pilát, A.**

Hydnaceae

Flora Polska. Grzyby (Mycota). Tom XIII. *Podstawezaki (Basidiomycetes),*
Bezblaszkowe (Aphyllophorales), Kolczakowate (Hydnaceae), Zabkowcowate (Stec-
cherinaceae). Warsaw–Cracow. 94 pp.   (1981) **Domanski, S.**

Die Stachelbarte (*Hericium, Creolophus*) und ihr Vorkommen in Westfälen. *Westfälische Pilzbriefe* **5**: 90–100. (1965) **Jahn, H.**

Die terrestrischen Stachelpilze Europas. [The terrestrial hydnums of Europe.] *Verh. K. ned. Akad. Wet., naturk.* (3)**65**: 1–127, 40 col. pls.
(1975) **Maas Geesteranus, R. A.**

Polyporaceae s. lat.

The Polyporaceae of the European U.S.S.R. and Caucasia. Program for Scientific Translations, Jerusalem. 896 pp. (1971) **Bondartsev, A. S.**
Originally published in Russian in 1953.

Étude du genre *Tyromyces* sensu lato: Repartition dans les genres *Leptoporus, Spongioporous* et *Tyromyces* sensu stricto. *Bull. Soc. linn. Lyon* **49**: 6–56.
(1980) **David, A.**

Étude monographique du genre *Skeletocutis* (Polyporaceae). *Naturaliste can.* **109**: 235–72. (1982) **David, A.**

Amyloporiella gen. nov. (Polyporaceae). *Trans. Br. mycol. Soc.* **83**: 659–67.
(1984) **David, A. and Tortic, M.**

Fungi, Polyporaceae I (resupinatae), Mucronoporaceae I (resupinatae). US Department of Commerce National Technical Information Service, Springfield, Virginia. 234 pp. (1972) **Domanski, S.**

Fungi, Polyporaceae II (pileatae), Mucronoporaceae II (pileate), Ganodermataceae, Bondarzewiaceae, Boletopsidaceae, Fistulinaceae. US Department of Commerce National Information Service, Springfield, Virginia. 330 pp.
(1973) **Domanski, S., Orlos, H., and Skirgiello, A.**

Check list of European polypores. *Verh. K. ned. Akad. Wet.* (3)**62**: 1–469.
(1974) **Donk, M. A.**

Mitteleuropäische Porlinge. *Westfälische Pilzbriefe* **4**: 1–143. (1963) **Jahn, H.**

Die resupinaten *Phellinus*-Arten in Mitteleuropa mit Hiweisen auf die resupinaten *Inonotus*-Arten. *Westfälische Pilzbriefe* **6**: 37–124. (1967) **Jahn, H.**

Resupinate Porlinge, *Poria* s. lat. in Westfälen und in Nordlichen Deutschland. *Westfälische Pilzbriefe* **8**: 41–68. (1971) **Jahn, H.**

Über *Polyporus brumalis* und verwandte Arten. *Feddes Reprium* **68**: 129–38.
(1963) **Kreisel, H.**

The genus Fomes. University State College of Forestry, Syracuse, New York. Technical Publication No. 80. (1957) **Lowe, J. L.**

The genus Poria. University State College of Forestry, Syracuse, New York. Technical Publication No. 90. 183 pp. (1966) **Lowe, J. L.**

Polyporaceae of North America: The genus *Tyromyces*. *Mycotaxon* **2**: 1–83.
(1975) **Lowe, J. L.**

On Fennoscandian polypores 1–5. *Ann. Bot. fenn.* **8**: 237–44; **9**: 41–59; **11**: 202–15; **12**: 93–122; 6–9 *Karstenia* **17**: 77–86; **18**: 43–8; **20**: 1–15; **22**: 13–16; **25**: 21–40. (1971–85) **Niemela, T.**

The Polyporaceae of the United States, Alaska and Canada. Michigan State University Press, Ann Arbor. 466 pp.   (1953) **Overholts, O. L.**

A survey of the genus *Inonotus*. *Trans. Br. mycol. Soc.* **47**: 175–97.
(1964) **Pegler, D. N.**

The polypores. *Bull. Br. mycol. Soc.* **7**(Suppl.): 43 pp.   (1973) **Pegler, D. N.**

Aphyllophorales IV: Poroid families. In *The fungi. An advanced treatise* (ed. G. C. Ainsworth, F. K. Sparrow, and A. S. Sussman), Vol. 4B, pp. 397–420. Academic Press, London and New York.   (1973) **Pegler, D. N.**
Keys to genera.

Basidiospore form in the British species of *Ganoderma*. *Kew Bull.* **28**: 351–64.
(1973) **Pegler, D. N. and Young, T. W. K.**

Atlas des champignons de l'Europe. Vol. 3 *Polyporaceae*. Prague. 2 vols.
(1936) **Pilát, A.**

The Polyporaceae of north Europe. 2 vols. Fungiflora, Oslo. 214 pp.
(1976–78) **Ryvarden, L.**

Considerations générales sur le genre *Ganoderma* et plus specialement sur les espèces Européenes. *Bull. Soc. r. bot. Belg.* **100**: 189–211.
(1967) **Steyaert, R. L.**

Ganoderma in northern Europe. *Mycologist* **1**: 62–7.   (1986) **Petersen, J. E.**

Stereaceae

Le genre *Amylostereum* (Basidiomycetes). Intercompatibilités partielles entre espèces allopatreques. *Bull. Soc. mycol. Fr.* **100**: 211–36.
(1984) **Boidin, J. and Lanquetin, P.**

Die Schichtpilze (*Stereum* s. lato) *Schweiz Z. Pilzkde* **46**: 65–74. (1968) **Jahn, H.**

Stereum and allied genera of fungi in the Upper Mississippi Valley. US Department of Agriculture, Beltsville, Maryland. USDA Agricultural Monograph No. 24. 74 pp.   (1955) **Lentz, P. L.**

Monographie der europäischen Stereaceen. *Hedwigia* **70**: 10–132.
(1930) **Pilát, A.**

A monograph of the stipitate stereoid fungi. *Beih. Nova Hedwigia* **18**: 1–384.
(1965) **Reid, D. A.**

Thelephoraceae s. lat.

The genera *Serpula* and *Meruliopsis*. *Mycologia* **49**: 197–225.
(1957) **Cooke, W. B.**

A monograph of *Thelephora*. *Beih. Nova Hedwigia* **27**: 1–100.
(1968) **Corner, E. J. H.**

A monograph of the genus *Coniophora* (Aphyllophorales, Basidiomycetes). *Op. Bot.* **61**: 7–61.   (1982) **Ginns, J.**

Tomentelloid fungi of North America. State University College of Forestry,

Syracuse, New York. Technical Publication No. 93. 157 pp.
(1968) **Larsen, M. J.**

The genus *Pseudotomentalla. Nova Hedwigia* **22**: 599–620. (1973) **Larsen, M. J.**

The genus *Tomentellastrum* (Aphyllophorales, Thelephoraceae s.str.). *Nova Hedwigia* **35**: 1–16. (1981) **Larsen, M. J.**

A contribution to the taxonomy of the genus *Tomentella. Mycologia Mem.* **4**: 1–145. (1974) **Larsen, M. J.**

Contribution to the taxonomy of the resupinate Thelephoraceous fungi. *Česká mykol.* **12**: 66–77. (1958) **Svrček, M.**

Tomentelloideae Cechoslovakiae. Genera resupinata familiae Thelephoraceae s. str. *Sydowia* **14**: 170–245. (1960) **Svrček, M.**

Tomentelloideae in the British Isles. *Trans. Br. mycol. Soc.* **53**: 161–206.
(1969) **Wakefield, E. M.**

Exobasidiales

Exobasidium, a taxonomic reassessment applied to the European species. *Symb. bot. upsal.* **23**(2): 1–72. (1981) **Nannfeldt, J. A.**

Key to the North American species of *Exobasidium. Can. J. Bot.* **37**: 641–56.
(1959) **Savile, D. B. O.**

Tremellales s. lat.

Hyménomycètes de France. Héterobasidiés-Homobasidiés gymnocarpes. Sceaux, Paris. iv + 761 pp. Reprinted (1969) by J. Cramer, Lehre.
(1928) **Bourdot, H. and Galzin, A.**

Danish resupinate fungi part I. Ascomycetes and Heterobasidiomycetes. *Dansk bot. Arkiv* **19**: 7–55. (1959) **Christiansen, M. P.**

Check list of European hymenomycetous Heterobasidiae. *Persoonia* **4**: 145–335. (1966) **Donk, M. A.**

Check list of European hymenomycetous Heterobasidiae. Supplement and corrections. *Persoonia* **8**: 35–50. (1974) **Donk, M. A.**

Parasitic Heterobasidiomycetes on other fungi. A key to European taxa. *Int. J. Mycol. Lich.* **1**: 189–203. (1983) **Jülich, W.**

A contribution towards a revision of the genus *Tulasnella. Persoonia* **9**: 49–64.
(1976) **Jülich, W. and Jülich, U.**

The genus *Heterochaetella. Can. J. Bot.* **38**: 559–69. (1960) **Luck-Allen, E. R.**

The genus *Basidiodendron. Can. J. Bot.* **41**: 1025–52. (1963) **Luck-Allen, E. R.**

Some auriculariaceous fungi from the British Isles. *Trans. Br. mycol. Soc.* **48**: 187–92. (1965) **McNabb, R. F. R.**

Phragmobasidiomycetidae: Tremellales, Auriculariales, Saptobasidiales. In *The fungi. An advanced treatise* (ed. G. C. Ainsworth, F. K. Sparrow, and A. S. Sussman), Vol. 4B, pp. 303–17. Academic Press, London and New York.
(1973) **McNabb, R. F. R.**

Revision of the north-central Tremellales. *Stud. nat. Hist. Iowa Univ.* **19**(3): 1–122. Reprinted (1969) by J. Cramer, Lehre. (1952) **Martin, G. W.**

Die europäischen Arten der Gattung *Tremella. Z. Pilzkde* **10**: 70–5. (1931) **Neuhoff, W.**

Die Gallertpilze Schwedens (Tremellaceae, Dacrymycetaceae, Tulasnellaceae, Auriculariaceae). *Ark. Bot.* **28A**(1): 1–57. (1936) **Neuhoff, W.**

Niedere Basidiomyceten aus Sudbayern III. Die Gattung *Sebacina* Tul. s.l. *Ber. bayer. bot. Ges.* **36**: 41–55. (1963) **Oberwinkler, A.**

Übersicht der europäischen Auriculariales und Tremellales unter besonderer Berücksichtigung der tschechoslowakischen Arten. *Acta Mus. nat. Prague* **13b**(4): 1–210. (1957) **Pilát, A.**

A monograph of the British Dacrymycetales. *Trans. Br. mycol. Soc.* **62**: 433–94. (1974) **Reid, D. A.**

Notes on Dacrymycetales. *Trans. Br. mycol. Soc.* **80**: 111–13. (1983) **Reid, D. A.**

A taxonomic review of the Tulasnellaceae. *Annls mycol.* **31**: 181–203. (1933) **Rogers, D. P.**

The genus *Tremella* in Norway. *Nytt Mag. Naturvid.* **15**: 225–39. (1968) **Torkelsen, A. E.**

Studies of some Tremellaceae III. The genus *Bourdotia. Mycologia* **51**: 541–64. (1959) **Wells, K.**

Studies of some Tremellaceae IV. *Exidiopsis. Mycologia* **53**: 317–70. (1961) **Wells, K.**

Studies of some Tremellaceae V. A new genus, *Efibulobasidium. Mycologia* **67**: 147–56. (1974) **Wells, K.**

The species of *Bourdotia* and *Basidiodendron* (Tremellaceae) of the USSR. *Mycologia* **67**: 904–22. (1975) **Wells, K. and Raitviïr, A.**

The species of *Exidiopsis* (Tremellaceae) of the USSR. *Mycologia* **69**: 987–1007. (1977) **Wells, K. and Raitviïr, A.**

Flora Polska. Grzyby (Mycota). Tom VIII. *Podstawczaki (Basidiomycetes) Trzesakowe (Tremellales) Uszakowe (Auriculariales) Czerwcogrzybowe (Septobasidiales).* Warsaw–Cracow. 334 pp. (1977) **Wojewoda, W.**

GASTEROMYCETES

See also under Hypogeous fungi (pp. 214–15).

The bird's nest fungi. University of Toronto Press, Toronto. 198 pp. (1975) **Brodie, H. J.**

Les Gasteromycètes. *Naturalistes belge.* **50**: 225–70. (1969) **Demoulin, V.**

Key to the Gasteromycetes of Great Britain. *Bull. Br. mycol. Soc.* **15**: 37–56. (1981) **Demoulin, V. and Marriott, J. V. R.**

The genus *Geastrum* in Denmark. *Bot. Tidsskr.* **57**: 1–27; **58**: 64–7.
(1961–62) **Dissing, H. and Lange, M.**

Gasteromycetes. In *The fungi. An advanced treatise* (ed. G. C. Ainsworth, F. K. Sparrow, and A. S. Sussman), Vol. 4B, pp. 451–78. Academic Press, London and New York.   (1973) **Dring, D. M.**

The Gasteromycetes in Norway, the epigeaen genera. *Nytt mag. Bot.* **4**: 19–86.
(1955) **Eckblad, F.-E.**

Monografia del genero *Scleroderma*. *Darwiniana* **16**: 233–407.
(1970) **Guzmán, G.**

Hypogeous fungi IV and V. *Trans. Br. mycol. Soc.* **38**: 73–7. (*Rhizopogon*).
(1955) **Hawker, L.**

Die Lycoperdaceae der DDR. *Feddes Reprium* **64**: 1–114. Reprinted (1973) by J. Cramer, Lehre.   (1962) **Kreisel, H.**

Taxonomische-pflanzengeographische Monographie der Gattung *Bovista*. *Beih. Nova Hedwigia* **25**: 1–224.   (1967) **Kreisel, H.**

Observations on Gasteromycetes 1–3. *Trans. Br. mycol. Soc.* **38**: 317–34. (*Geastrum*).   (1955) **Palmer, J. T.**

A chronological catalogue to the literature of the British Gasteromycetes. *Nova Hedwigia* **15**: 65–178.   (1968) **Palmer, J. T.**

Revision of the Lycoperdaceae of the Netherlands. *Blumea* **6**: 480–516.
(1950) **Perdeck, A. C.**

Gasteromycetes. Czechoslovak Academy of Sciences, Prague. Flora ČSR **B1**. 864 pp.   (1958) **Pilát, A.,** ed.

Monographie der Gattung *Hymenogaster*. *Beih. Nova Hedwigia* **2**: 1–114.
(1962) **Soehner, E.**

Keys to the orders, families and genera of the Gasteromycetes. *Mycologia* **41**: 36–58.   (1949) **Zeller, S. M.**

DEUTEROMYCOTINA (FUNGI IMPERFECTI)

Illustrated genera of imperfect fungi, 3rd edn. Burgess Publishing, Minneapolis. 241 pp.   (1972) **Barnett, H. L. and Hunter, B. B.**

Taxonomy of Fungi Imperfecti. University of Toronto Press, Toronto. 309 pp.
(1971) **Kendrick, W. B.,** ed.

BLASTOMYCETES (YEASTS *p.p.*)

See also p. 231 under Ascomycotina–Endomycetales.

Étude sur les Sporobolomycètes. *Annls mycol.* **28**: 1–23.   (1930) **Derx, H. G.**

The genus *Tilletiopsis*. *Mycologia* **42**: 487–96.   (1950) **Nyland, G.**

Studies in the Sporobolomycetaceae in Japan. *Nagaoa* **1**: 26–31.

(1952) **Tuabki, K.**

COELOMYCETES

Die Arten der Gattung *Colletotrichum* Cda. *Phytopath. Z.* **29**: 413–68.

(1957) **von Arx, J. A.**

A revision of the fungi classified as Gloeosporium, 2nd edn. J. Cramer, Lehre. 203 pp.

(1970) **von Arx, J. A.**

Ordinamento artificiale delle specie del genere *Coniothyrium*. *Sydowia* **12**: 258–320. (1959) **Bestagno Biga, M. L., Cifferi, R., and Bestagno, G.**

British stem- and leaf-fungi (Coelomycetes). Cambridge University Press, Cambridge. 2 vols. Reprinted (1967) by J. Cramer, Lehre. (1932) **Grove, W. B.** Still useful although now very dated; includes host indices.

Monograph of Monochaetia and Pestalotia. Harvard University Press, Cambridge, Massachusetts. 342 pp. (1961) **Guba, E. F.**

Septoria *and Septorioid Fungi on dicotyledons in Norway.* Oslo University Press, Oslo. 110 pp. (1965) **Jorstad, I.**

Septoria *and septorioid fungi on Gramineae in Norway.* Oslo University Press, Oslo. 63 pp. (1967) **Jørstad, I.**

Icones generum coelomycetum I–VII. University of Waterloo Biology Series, Waterloo. (1972–75) **Morgan-Jones, G., Nag Raj, T. R., and Kendrick, W. B.**

Die Gattungen der Pyrenomyceten, Sphaeropsideen und Melanconieen I. *Beih. Feddes Repert.* **42**: 1–551. (1926–27) **Petrak, F. and Sydow, H.**

Coelomycetes I–VI. *Mycol. Papers* **80**: 1–16 (*Pestalotiopsis*); **88**: 1–50 (*Neobarclaya* etc.); **97**: 1–42 (*Seimatosporium* etc.); **123**: 1–46 (*Harknessia* etc.); **138**: 1–224 (*Coryneum*); **141**: 1–253 (Generic names). (1961–77) **Sutton, B. C.**

Coelomycetes. In *The fungi. An advanced treatise* (ed. G. C. Ainsworth, F. K. Sparrow, and A. S. Sussman), Vol. 4B, pp. 513–82. Academic Press, London and New York. (1973) **Sutton, B. C.**

The Coelomycetes. Fungi Imperfecti with pycnidia, acervuli and stromata. Commonwealth Mycological Institute, Kew. 696 pp, 397 figs.

(1980) **Sutton, B. C.**

Studies in *Phyllosticta. Stud. mycol.* **5**: 1–110. (1973) **Aa, H. A. van der**

HYPHOMYCETES

The genera of Hyphomycetes from soil. Williams and Wilkins, Baltimore. Reprinted (1972) by R. E. Krieger Publishing, Huntington, New York. 346 pp.

(1968) **Barron, G. L.**

The genus *Cylindrocarpon. Mycol. Papers* **104**: 1–56. (1966) **Booth, C.**

The genus Fusarium. Commonwealth Mycological Institute, Kew. 237 pp.
(1971) **Booth, C.**

Fusarium. Laboratory guide to the identification of the major species. Commonwealth
Mycological Institute, Kew. 58 pp.   (1977) **Booth, C.**

A monograph of the fungus genus Cercospora. C. Chupp, Ithaca, New York. 667 pp.
(1954) **Chupp, C.**

Genera of Hyphomycetes. University of Alberta Press, Edmonton. 386 pp.
(1980) **Carmichael, J. W., Kendrick, W. B., Conners, I. C., and Sigler,
L.**
Illustrations of all accepted genera.

The genus *Fusarium*—a pictorial atlas. *Mitt. biol. Bundes. Land-Forstw.*, Berlin
209: 1–406.   (1982) **Gerlach, W. and Nirenberg, H.**

The genera *Beauveria, Isaria, Tritirachium* and *Acrodontium*. *Stud. mycol.* **1**: 1–41.
(1972) **de Hoog, G. S.**

The genera *Blastobotrys, Sporothrix, Calicarisporium* and *Calicarisporiella* gen. nov.
Stud. mycol. **7**: 1–84.   (1974) **de Hoog, G. S.**

Taxonomy of the *Dactylaria* complex, IV. *Dactylaria, Neta, Subulispora* and
Scolecobasidium. *Stud. mycol.* **26**: 1–60.   (1985) **de Hoog, G. S.**

A revision of *Geotrichum* and its anamorphs. *Stud. mycol.* **29**: 1–131.
(1986) **Hogg, E. S. de, Smith, M. T., and Guého, E.**

Dematiaceous Hyphomycetes. Commonwealth Mycological Institute, Kew. 608
pp, 419 figs.   (1971) **Ellis, M. B.**
This and the following title together include almost all British brown pigmented
Hyphomycetes.

More dematiaceous Hyphomycetes. Commonwealth Mycological Institute, Kew.
507 pp, 383 figs.   (1976) **Ellis, M. B.**

Cephalosporium-artige Schimmelpilze (Hyphomycetes). G. Fischer, Stuttgart. x +
262 pp.   (1971) **Gams, W.**

Chloridium and some other dematiaceous Hyphomycetes growing on decaying
wood. *Stud. mycol.* **13**: 1–99. (1976) **Gams, W. and Holubova-Jechova, V.**

Guide to aquatic Hyphomycetes. *Scient. Publs Freshwat. biol. Ass.* **30**: 96 pp.
(1975) **Ingold, C. T.**

Hyphomycetes. In *The fungi. An advanced treatise* (ed. G. C. Ainsworth, F. K.
Sparrow, and A. S. Sussman), Vol. 4B, pp. 323–509. Academic Press,
London and New York.   **Kendrick, W. B. and Carmichael, J. W.**
Well illustrated.

Synnematous genera of Fungi Imperfecti. Western Illinois University. Series in
Biological Sciences No. 3, 143 pp.   (1963) **Morris, E. F.**

A monograph of Chalara and allied genera. Wilfrid Laurier University Press,
Waterloo.   (1976) **Nag Raj, T. R. and Kendrick, G.**

Fusarium species—an illustrated manual for identification. Pennsylvania State University Press, Pittsburg and London. 193 pp. (1983) **Nelson, P. E., Toussoun, T. A., and Marasas, W. F. O.**

The genus Penicillium and its teleomorphic states Eupenicillium and Talaromyces. Academic Press, London. 634 pp. (1980) ('1979') **Pitt, J. I.**

The genus Aspergillus. Williams and Wilkins, Baltimore. 686 pp. (1965) **Raper, K. and Fennell, D. I.**

A revision of the genus *Trichoderma. Mycol. Papers* **116**: 1–56. (1969) **Rifai, M. A.**

Paecilomyces and some allied Hyphomycetes. *Stud. mycol.* **6**: 1–119. (1974) **Samson, R. A.**

Advances in Penicillium and Aspergillus systematics. Plenum Press, New York and London. NATO Advanced Study Institute Series **A102**. 483 pp. (1986) ('1985') **Samson, R. A. and Pitt, J. I.**, eds.

Revision of the subsection *Fasciculata* of *Penicillium* and some allied species. *Stud. mycol., Baarn* **11**: 1–47. (1976) **Samson, R. A., Stolk, A. C., and Hadlock, R.**

A monograph of *Stilbella* and some allied Hyphomycetes. *Stud. mycol.* **27**: 1–235. (1985) **Seifert, K. A.**

Hyphomycetes. Indian Council Agricultural Research, New Delhi. 930 pp. (1972) **Subramanian, C. V.**

Hyphomycetes from Manitoba and Saskatchewan. *Mycol. Papers* **132**: 1–143. (1973) **Sutton, B. C.**

List of Hyphomycetes recorded for Britain. *Trans. Br. mycol. Soc.* **25**: 49–126, 427–8. (1941) **Wakefield, E. M. and Bisby, G. R.**

Plantae

Algae

For a comprehensive list of keys to the smaller algae see:

A guide to algal keys (excluding seaweeds). *Br. phycol. J.* **11**: 49–55.

(1976) **George, E. A.**

Freshwater

For further references, see Sarcomastigophora (Mastigophora) under Proto-zoa, pp. 9–11.

Rabenhorst's Kryptogamen-Flora von Deutschland, Osterreich und der Schweiz. Akademie Verlag, Leipzig. Vols. 5, 7, 10–14:

5 1890–97. *Die Characeen* (W. Migula) xii + 765 pp, 149 figs.

7 (3 parts). 1927–66 (unfinished) *Die Kieselalgen* (F. Hustedt) 2581 pp, 3509 figs.

10 (Part 1). 1930. *Silicoflagellatae* (K. Gemeinhardt) (Part 2). 1930. *Coccolithineae* (J. Schiller) 273 pp, 206 figs.

10 (Part 2(i)). 1931–33 *Dinoflagellatae (Peridineae)* (J. Schiller) vi + 617 pp, 631 figs.

10 (Part 2(ii)). 1935–37 *Dinoflagellatae* (J. Schiller) vi + 590 pp, 612 figs.

11 1937–39 *Heterokontae* (A. Pascher) ii + 1092 pp, 912 figs.

12 (4 parts). 1938–40 *Oedogoniales* (K. Gemeinhardt) ix + 453 pp, 539 figs.

13 (Part 1(i)). 1933–37 *Die Desmidiaceen* (W. Krieger) vi + 712 pp, 96 pls.

13 (Part 1(ii)). 1939 *Die Desmidiaceen* (W. Krieger) 117 pp, 45 pls.

13 (2 parts). 1941–44 *Zygnemales* (R. Kolkwitz and H. Krieger) 499 pp, 788 figs.

14 (2 parts). 1930–32 *Cyanophyceae* (L. Geitler) vi + 1196 pp, 780 figs.
(1890–1966)
Descriptions, generic and specific keys, distributions, general biological data; well-illustrated account in German for advanced students.

Die Süsswasser-Flora Deutschlands, Österreichs und der Schweiz. G. Fischer, Jena.

Part 1 (1914) *Flagellatae I* (A. Pascher and E. Lemmermann) iv + 138 pp, 252 figs.

Part 2 (1913) *Flagellatae II* (A. Pascher and E. Lemmermann) iv + 192 pp, 398 figs.

Part 3 (1913) *Dinoflagellatae (Peridineae)* (A. J. Schilling) iv + 66 pp, 69 figs.

Part 4 (1927) *Volvocales-Phytomonadinae. Flagellatae IV.-Chlorophyceae 1* (A. Pascher) vi + 506 pp, 451 figs.

Part 5 (1915) *Chlorophyceae II* (E. Lemmermann, J. Brunnthaler, and A. Pascher) iv + 250 pp, 398 figs.

Part 6 (1914) *Chlorophyceae III* (W. Herring) iv + 250 pp, 34 figs.

Part 7 (1921) *Chlorophyceae IV* (W. Herring) iv + 103 pp, 94 figs.

Part 9 (1913) *Zygnemales* (O. Borge and A. Pascher) iv + 51 pp, 79 figs; 2nd edn* 1932 (V. Czurda) v + 232 pp, 226 figs.

Part 10 (1930) *Bacillariophyta (Diatomeae)* (H. von Schonfeldt) iv + 187 pp; 2nd edn* 1930 (F. Hustedt) viii + 466 pp, 875 figs.

Part 11 (1925) *Heterokontae, Phaeophyta, Rhodophyta, Charophyta* (A. Pascher, J. Schiller, and W. Migula) iv + 250 pp, 211 figs.

Part 12 (1925) *Cyanophyceae* (L. Geitler) [viii + 450 pp, 560 figs]; *Cyanochloridinae–chlorobacteriaceae* (L. Geitler and A. Pascher) [451–463 pp, 14 figs]. viii + 481 pp, 574 figs.
*The two parts of the second edition have the title: *Die Süsswasser-Flora Mitteleuropas.*

A Treatise on the British freshwater algae. Cambridge University Press, Cambridge. xviii + 534 pp, 207 figs. (1927) **West, G. S. and Fritsch, F. E.** Generic keys only, British distribution only, biological and ecological data, descriptions of genera, and a few species.

Das Phytyoplankton der Süsswassers. In *Die Binnengewasser* (ed. A. Thienemann), Vol. 16. Schweizerbart'sche, Stuttgart:

Part 1 (1938) *Blaualgen. Bakterien. Pilze* 342 pp, 66 pls.

Part 2(i) (1941) *Chrysophyceen. Farblose Flagellaten Heterokonten* 365 pp, 107 pls.

Part 3 (1950) *Cryptophyceen, Chloromonadinen, Peridineen* ix + 310 pp, 69 pls.

Part 3 (1968) *Cryptophyceae, Chloromonadophyceae, Dinophyceae* 2nd edn (G. Huber-Pestalozzi and B. Fott) ix + 322 pp, 69 pls.

Part 4 (1955) *Euglenophyceen* ix + 606 pp, 114 pls.

Part 5 (1961) *Chlorophyceae (Grunalgen) Ordnung: Volvocales* xii + 744 pp, 158 pls.   (1938–61) **Huber-Pestalozzi, G.**

Part 6 (1972) *Chlorophyceae (Grunalgen) Ordnung: Tetrasporales* (B. Fott) x + 116 pp, 47 pls.

Part 7(1) (1983) *Chlorophyceae (Grunalgen) Ordnung: Chlorococcales* (J. Komarek and B. Fott) x + 1044 pp, 253 pls.

Part 8(1) (1982) *Conjugatophyceae Zygnematales und Desmidiales (excl. Zygnemataceae)* (K. Forster) viii + 543 pp, 65 pls.
Descriptions and keys to the species; well illustrated. World-wide coverage. Works for the advanced student.

The fresh-water algae of the United States, 2nd edn. MacGraw-Hill, New York and London. 719 pp, 559 figs.   (1950) **Smith, G. M.**
Descriptions and keys to the genera many of which have a world-wide distribution.

Algae of the western Great Lakes area with an illustrated key to the genera of desmids and freshwater diatoms, revised edn. W. C. Brown, Dubuque, Iowa. xiii + 977 pp, 143 pls.   (1962) **Prescott, G. W.**
Descriptions, keys, and illustrations to the species. Most taxa covered have a very wide distribution.

Les algues d'eau douce. Boubée et Cie, Paris. 3 vols.

1 (1966) *Les algues vertes.* 511 pp, 117 pls; 1973. *Les algues vertes*, 2nd edn. 572 pp, 121 pls.

2 (1968) *Les algues jaunes et brunes.* 438 pp, 114 pls.

3 (1970) *Les algues bleues et rouges. Les Eugleniens, Peridiniens et Cryptomonadines.* 512 pp, 137 pls.   (1966–73) **Bourrelly, P.**
Keys to and descriptions of the genera; illustrations of many species. World-wide coverage.

Flora Slodkowodna Polski [*Freshwater Flora of Poland*] Polska Akademia Nauck, Warsaw. [Polish].

2 (1966) *Cyanophyta, Glaucophyta* (K. Starmach) 807 pp, 1101 figs.

3 (1983) *Euglenophyta* (K. Starmach) 594 pp, 1409 figs.

4 (1974) *Cryptophyceae, Dinophyceae, Raphidophyceae* (K. Starmach) 520 pp, 644 figs.

5 (1968) *Chrysophyta I Chrysophyceae* (K. Starmach) 598 pp, 1110 figs.

5 (1980) *Chrysophyta I Chrysophyceae II* (K. Starmach) 775 pp, 1136 figs.

6 (1964) *Chrysophyta II Bacillariophyceae* (J. Sieminska) 610 pp, 993 figs.

7 (1968) *Chrysophyta III Xanthophyceae* (K. Starmach) 394 pp, 511 figs.

10 (1972) *Chlorophyta III. Ulothricales, Ulvales, Prasiolales, Sphaeropleales, Cladophorales, Chaetophorales, Trentepohliales, Siphonales, Dichotomosiphonales* (K. Starmach) 750 pp, 646 figs.

11 (1969) *Oedogoniales* (K. Starmach) 657 pp, 834 figs.

12A (1972) *Zygnemaceae* (J. Z. Kadlubowska) 431 pp, 83 pls.

13 (1964) *Charophyta* (I. Dambska) 126 pp, 56 figs.

14 (1977) *Phaeophyta, Rhodophyta* (K. Starmach) 445 pp, 180 figs.
Keys are in English in Vols 10 (pp. 630–94) and 14 (pp. 377–405); the latter volume
includes also marine algae. Descriptions and illustrations of all the species.

Kleine Kryptogamenflora. In *Makroskopische Algen* (ed. H. Gams). G. Fischer,
Stuttgart. Vol. I(a) (1969) *Makroskopische Süsswasser- und Luftalgen.* 63 pp, 28
pls. Vol. I(b) (1974) *Makroskopische Meeresalgen.* 119 pp, 40 pls.
$$(1969, 1974) \textbf{ Gams, H., ed.}$$
Keys, descriptions, and illustrations of species of freshwater and marine algae.

A manual of the fresh-water algae in North Carolina. *Tech. Bull. N. Carolina
agric. Res. Stn.* **188**: 1–313, 71 pls.
$$(1969) \textbf{ Whitford, L. A. and Schumacher, G. J.}$$
Useful keys and illustrations of freshwater algae. Often with a world-wide
distribution.

Kl'uc na Urcovanie Vytrsnych Rastlin 1 diel *Riasy* [*Key to the Identification of
Cryptogamic Plants Part 1 Algae.*] Slovenske Pedagogicke Nakladatel'stvu,
Bratislava. 37 pp, 382 pls. [Czech]
$$(1975) \textbf{ Hindak, F., Komarek, J., Marvan, P., and Ruzicka, J.}$$
A beginner's guide to freshwater algae. HMSO, London. 47 pp, 109 figs.
$$(1976) \textbf{ Belcher, J. H. and Swale, E. M. F.}$$
A coded list of 1000 freshwater algae of the British Isles. Department of the
Environment, Water Data Unit, Reading. Archive Manual Series No. 2.
335 pp.   (1978) **Whitton, B. A., Holmes, N. T. H., and Sinclair, C.**
Annotated listing with some keys and taxonomic comments.

Süsswasserflora von Mitteleuropa. G. Fischer, Stuttgart and New York.

1 (1985) *Chrysophyceae und Haplophyceae* (K. Starmach) 515 pp, 1051 figs.

3 (1978) *Xanthophyceae* 1 (H. Ettl) xiv + 530 pp, 636 figs.

4 (1980) *Xanthophyceae* 2 (A. Rieth) xiv + 147 pp, 61 figs.

9 (1983) *Chlorophyta* I *Phytomonadina* (H. Ettl) xiv + 807 pp, 1120 figs.

14 (1985) *Chlorophyta* VI *Oedogoniophyceae: Oedogoniales* (T. Mrozinska) 624
pp, 1000 figs.

16 (1984) *Conjugatophyceae* I *Zygnemales* = *Chlorophyta* VIII (J. Z.
Kadlubowska) 532 pp, 798 figs.
Keys, descriptions, and illustrations.

An illustrated guide to river phytoplankton. HMSO, London. 64 pp, 109 figs.
$$(1979) \textbf{ Belcher, [J.] H. and Swale, E. [M. F.]}$$
Useful keys and illustrations for the identification of commoner river algae found in
the British Isles.

A key to common British algae. Institution of Water Engineers and Scientists,
London. 94 pp, 24 pls.   (1980) **Bellinger, E. G.**
Key to the species and some line drawings.

How to know the freshwater algae, 3rd edn. W. C. Brown, Dubuque, Iowa. x + 293
pp, 578 figs.   (1983) **Prescott, G. W.**
A good key with useful illustrations of species of North American algae. The majority

are distributed world-wide with many common in freshwater habitats in the British Isles and in Europe.

Introduction to freshwater algae. Richmond Publishing, Richmond. vii + 247 pp, 48 pls.   (1984) **Pentecost, A.**
Keys; descriptions; line drawings of the commoner British freshwater algae.

Marine

For additional references, see under, Freshwater (pp. 271–5); for specific groups, see the appropriate headings. Most general marine Floras have relevance wider than merely the primary area or country of origin. For listings of general determinative and illustrative literature for benthic marine macro-algae, especially for UK shores, to that time, see:

Seaweeds on shore: data sources for Britain. *Nat. Hist. Book Reviews* **3**: 3–13.
                                                    (1978) **Price, J. H.**

BRITAIN

Seaweeds of the British Isles. British Museum (Natural History), London. Various volumes and authors.
A series of volumes in progress; these will eventually fully replace the long-time standard work by Newton. For volumes published or soon to be so, see the algal group headings on Rhodophyta, Xanthophyceae (=Tribophyceae), Phaeophyceae, Chlorophyta. Six volumes (some with various parts) are currently planned; others may be added.

A field key to the British . . . seaweeds. *Phaeophyta (brown seaweeds). Field Stud.* **5**: 1–44, 22 figs.   (1979) **Hiscock, S.**
Useful illustrated keys to most British brown seaweeds, to genus and/or species levels. Specialist keys for the complex genera *Fucus* and *Elachista*. Also available separately. AIDGAP Scheme.

Rhodophyta (red seaweeds). AIDGAP Scheme, Field Studies Council. 104 + [4] pp, 35 compound figs; many small illustrative sketches in keys; colour plates 1–4 + covers.   (1986) **Hiscock, S.**
Useful illustrated keys to most British red seaweeds. Utilizes essentially characters visible with naked eye or with hand lens. Additional keys to complex genera such as *Porphyra, Polysiphonia, Ceramium.* Glossary. Available from Field Studies Council Centres or from Richmond Publishing Co. Ltd.

Handbook to the British seaweeds. Hulton Press, London. Hulton Group Keys Series.   (in preparation) **Irvine, D. E. G., Norton, T. A., and Tittley, I.**
Keys, associated species descriptions and line drawings.

A photographic guide to some common subtidal seaweeds of the British Isles. Marine Conservation Society, Ross-on-Wye. MCS guide.
                                    (1985) **Maggs, C. A. and Howson, C. M.**

Ring-bound; single-sided A4 reproduced from typed stencils. One side per species—gives description, seasonality, habitat, distribution, similar species, colour photos (with outline silhouettes where helpful). The 'arrangement of species' explanation (pp. 4–5) provides a rudimentary key.

FRANCE

Zeewierengids voor de Belgische en Noordfranse Kust, Vol. I. Uitgave van Belgische Jeugdbond voor Natuurstudie vzw, Gent. [2] + 156 pp. [Flemish].
(1980) **Coppejans, E. and Van Der Ben, D.**
Separate keys to green, brown, and red algal groups. Line drawings. Glossary.

Zeewierengids voor de Belgische en Noordfranse Kust. II Beschrijvingen Groen- en Bruinwieren. *Stentor (Wetenschappelijk tijdschrift van de Belgische Jeugdbond voor Natuurstudie vzw)* **17** (extra nummer): 157–254.
(1982) **Coppejans, E.**
Green and brown algae: descriptions; ecology; biology.

Zeewierengids voor de Belgische en Noordfranse Kust. III Beschrijvingen Roodwieren. *Stentor (Wetenschappelijk tijdschrift van de Belgische Jeugdbond voor Natuurstudie vzw)* **18** (extra nummer): 255–392.   (1982) **Coppejans, E.**
Red algae: descriptions; ecology; biology; additional line drawings.

Les algues des côtes françaises (Manche and Atlantique) . . . Éditions Doin, Paris. 632 + [1] pp.
(1966) **Gayral, P.**
Keys; photographs; species descriptions and drawings; ecology; distribution; glossary.

BELGIUM

See France; the entry under Holland is also relevant.

HOLLAND

Flora van de Nederlande Zeewieren. Koninklijk Nederlandse Natuurhistorische Vereniging, Amsterdam. Uitgave no. 33, 263 pp.
(1983) **Stegenga, H. and Mol, I.**
Keys to divisions, genera, and species (within complex genera); descriptions; line drawings; critical comment; photographs; glossary. Main focus Holland; further application in all adjacent countries with at least some similar shorelines (Belgium, Denmark, UK, France, Germany).

GERMANY

Meeresalgen von Helgoland. Benthische Grun-, Braun- und Rotalgen. *Helgol. wiss. Meeresunters.* **29**: 1–289, Abb. 165.
(1977) **Kornmann, P. and Sahling, P.-H.**
Excellent photographic treatments, plus descriptions. Relevant to all adjacent rocky

areas in other countries. Reprinted (1978) and issued by Biologische Anstalt, Helgoland, Hamburg. Later supplemented in same style (see below). Keys to genera within groups (including down to some species); generic keys to more specialist groups (e.g. Ectocarpales).

Meeresalgen von Helgoland: Ergaenzung. *Helgol. wiss. Meeresunters.* **36**: 1–65.
(1983) Kornmann, P. and Sahling, P.-H.
Supplement to the above; essentially similar treatment.

Meeresalgen Ein Bestimmungsbuch für häufigere Arten der Atlantikküsten, der Nord- und Ostsee. A. Ziemsen Verlag, Wittenburg, Lutherstadt. Die Neue Brehm-Bücherei 489, 152 pp. (1975) **Kremer, B. P.**
Keys to genera in separate groups (browns, greens, reds). Short species descriptions; ecology; geographic distribution; small line drawings; a reasonable beginners' book for German readers.

DENMARK

No modern easily-obtainable Flora, but the relationships between the Danish marine flora and those of adjacent countries can be seen by consulting the following:

Distribution of algae in Danish salt and brackish waters. University of Copenhagen, Copenhagen. 64 pp.
(1985) Christensen, T., Koch, C., and Thomsen, H. A.
Check-list and distribution patterns; no keys or descriptions, but useful for consulting adjacent Floras for aids to determinations. Lists all earlier works on Danish marine algae.

SWEDEN

Die Rhodophyceen der Schwedischen Westküste. *Lunds Univ. Årsskr.*, N.F. Avd. 2, **40**(2): 1–104, Taf. 1–32. (1944) **Kylin, H.**

Die Phaeophyceen der Schwedischen Westküste. *Lunds Univ. Årsskr.*, N.F. Avd. 2, **43**(4): 1–99, Taf. 1–18. (1947) **Kylin, H.**

Die Chlorophyceen der Schwedischen Westküste. *Lunds Univ. Årsskr.*, N.F. Avd. 2, **45**(4): 1–79, 66 figs. (1949) **Kylin, H.**
Keys; descriptions; drawings; photographs; ecology; Swedish distribution.

NORWAY

Norssk Algeflora. Universitetsforlaget, Bergen, Trømsø, and Oslo. 266 pp.
(1977) Rueness, J.
Keys; line drawings; descriptions; ecology; distributions. The 2nd edition is currently in preparation.

Aspects of the marine algal vegetation of North Norway. *Acta Universitatis Gothoburgensis* (*Botanica Gothoburgensia*) **4**: 1–174, 43 figs. (1965) **Jaasund, E.**
No keys, but comparative taxonomic comments; extended descriptions; ecology; distribution; line drawings of many species; usable as work in conjunction with Rueness (1977), this section.

FAROES

As for Denmark, no modern easily-obtainable Flora; details recorded in the listings presented in the work below can be used to trace taxa by employing keys from one of the adjacent cold-water area Floras.

Seaweeds of the Faroes. *Bull. Br. Mus. nat. Hist.* (Bot.) **10**(3): 2 + 109–225.
  The Flora [check-list].   (1982) **Irvine, D. E. G.**
  Sheltered shores [ecological].
              (1982) **Tittley, I., Farnham, W. F., and Gray, P. W. G.**
  Open shores [ecological].   (1982) **Price, J. H. and Farnham, W. F.**
  Lists; ecology; distribution; diagrams; photos; summary tabular data; no keys.

ICELAND

Much recent work, but no modern easily-obtainable Flora: details recorded in the listings in the works below can be used both to trace older works on the area and, in conjunction with adjacent cold-water area Floras (e.g. Norway), to key down species concerned.

Survey of the benthic algal vegetation of the Reydarfjördur as a typical example of the east Icelandic vegetation pattern. *Nova Hedwigia* **37**: 545–640.   (1983) **Munda, I. M.**
  Lists; locality descriptions (including photographs of places and algae); communities; comparisons with adjacent cold-water areas; no full descriptions of taxa; no keys.

Nouvel inventaire des algues marines de l'Islande. *Acta Bot. Islandica* **1**: 5–31.
                                (1972) **Caram, B. and Jónsson, S.**
  Check-list only; no drawings or descriptions.

Addition to the check-list of benthic marine algae from Iceland. *Botanica mar.* **22**: 459–63.   (1979) **Munda, I. M.**
  As above, but with descriptions of morphology and ecology.

SPITSBERGEN

No modern Flora of anything like completeness; use the work listed below in conjunction with cold-water Floras such as that for Norway, Rueness (1977) on p. 277 to key out taxa likely to be detected.

The algal vegetation of Spitsbergen. A survey of the marine algal flora of the outer part of Isfjorden. *Norsk Polarinst. Skr.* **116**: 1–49. (1959) **Svendsen, P.**
Ecology; distribution; taxonomic comments; no detailed comparative descriptions; no keys; few illustrations.

Systematic
CYANOPHYTA (CYANOBACTERIA)

Further important works appear under Freshwater, (pp. 271–5).

Revision of the coccoid Myxophyceae. *Bot. Stud. Butler Univ.* **12**: 1–218, 377 figs.   (1956) **Drouet, F. and Daily, W. A.**
Distributional and habitat data, with keys to genera, species, and, occasionally, forms. Additions and corrections to paper are in *Trans. Am. microsc. Soc.* **76**: 219–22 (1957).

Cyanophyta. Indian Council of Agricultural Research, New Delhi. x + 686 pp, 139 pls.   (1959) **Desikachary, T. V.**

Revision of the classification of the Oscillatoriaceae. *Monogr. Acad. nat Sci. Philad.* **15**: 1–370, 131 figs.   (1968) **Drouet, F.**

Revision of the Nostocaceae with cylindrical trichomes. Hafner Press, New York. xii + 292 pp, 83 figs.   (1973) **Drouet, F.**

Revision of the Nostocaceae with constricted trichomes. *Beih. Nova Hedwigia* **57**: 1–258, 42 figs.   (1978) **Drouet, F.**

Revision of the Stigonemataceae with a summary of the classification of the blue-green algae. *Beih. Nova Hedwigia* **66**: 1–221, 109 figs. (1981) **Drouet, F.**
Includes additions and corrections to the preceding entry, Drouet (1978).

RHODOPHYTA

Additional works appear under general Freshwater and Marine headings.

The freshwater Florideae of Sweden. *Symb. bot. upsal.* **6**(1): 1–135, 3 pls.
                                      (1942) **Israelson, G.**
Keys, biological and ecological information, and descriptions.

Die Rhodophyceen der Schwedischen Westküste. *Lunds Univ. Årsskr.* N.F. Avd. 2. **40**(2): 1–104, 32 pls.   (1944) **Kylin, H.**

Die Gattungen der Rhodophyceen. CWK Gleerups Forlag, Lund. xv + 673 pp, 48 figs.   (1956) **Kylin, H.**
Includes a good deal of information on red algal morphology and systematics. Good descriptions and keys to the families and genera. World-wide coverage of both marine and freshwater genera. Woelkerling has privately distributed a translation of the keys: Woelkerling, W. J. (1975). 'Translation of taxonomic keys in Harold Kylin "Die Gattungen der Rhodophyceen"'. iv + 69 pp [mimeo].

Seaweeds of the British Isles. British Museum (Natural History), London. Includes:

Vol. 1. *Rhodophyta. Part 1. Introduction, Nemaliales, Gigartinales*. xi + 250 pp, 90 figs. (1977) **Dixon, P. S. and Irvine, L. M.**

Vol. 1. *Rhodophyta. Part 2A. Cryptonemiales (sensu stricto) Palmariales, Rhodymeniales*. xii + 115 pp, 28 figs. (1983) **Irvine, L. M.**
Descriptions of species and distribution in the British Isles. Some restricted keys to complex genera and good line drawings. Glossary.

A field key to the British red seaweeds (Rhodophyta). AIDGAP Scheme, Field Studies Council. 104 + [4] pp, 35 compound figs. (1986) **Hiscock, S.**
Useful illustrated key, with good line drawings, to most British red algae. Utilizes essentially characters visible using no more than a hand lens. Subsidiary keys to complex genera. Obtainable from Field Studies Council (see under 'Britain', p. 275).

Studies on marine algae of the British Isles. 10. The genus *Rhodymenia*. *Br. phycol. J.* **12**: 385–425. (1977) **Guiry, M. D.**
Key to British species; comparative gross and cell structure drawings; descriptions; habitat distribution; tables of species comparative characteristics.

The genus *Callithamnion* (Rhodophyta: Ceramiaceae) in the British Isles. *Bull. Brit. Mus. nat. Hist. (Bot.)* **9**: 99–141. (1981) **Dixon, P. S. and Price, J. H.**
Key; descriptions; ecological comments.

Clé des *Polysiphonia* des côtes françaises. *Crypt.: Algologie* **2**: 71–7.
(1981) **Feldmann, J.**
Keys to sections and to species within sections; no illustrations or descriptions.

The taxonomy, morphology and distribution of species of *Scinaia* Biv.-Born. (Nemaliales, Rhodophyta) in north-western Europe. *Nordic J. Bot.* **2**: 517–23. (1982) **Maggs, C. A. and Guiry, M. D.**
Comparisons; descriptions; ecology; distributions; line drawings; photographs.

Studies on the biosystematics and ecology of the epilithic crustose Corallinaceae of the British Isles. *Br. phycol. J.* **8**: 343–407.
(1973) **Adey, W. H. and Adey, P. J.**
Keys (generic; vegetative to species; reproductive to species; anatomical to species); biology (including species descriptions); ecology; line drawings; distribution maps; few photographs.

The genus *Polysiphonia* (Ceramiales, Rhodomelaceae) in Scandinavia. *Giorn. bot. Ital.* **117**: 1–30. (1983) **Kapraun, D. F. and Rueness, J.**
Useful in most of NW Europe; key; line drawings; descriptions; ecology; distribution; few photos.

CHROMOPHYTA

BACILLARIOPHYCEAE

See under Freshwater (pp. 271–5) for other key works covering this family.

Diatomées marines de France. 2ᵉ Partie. *Pseudo-Raphidies*, pp. 1–363. 3ᵉ et dernière
Partie. *Anaraphidies*, pp. 264–491. M. J. Tempère, Paris. 491 pp.
(1897–1908) Peragallo, H. and Peragallo, M.

The planktonic diatoms of northern seas. Ray Society, London. ix + 244 pp, 181 text
figs, 4 pls.   (1930) **Lebour, M. V.**

Marine plankton diatoms of the West Coast of North America. *Bull. Scripps
Inst. Oceanogr. tech. Ser.* **5**: 1–237, 160 figs, 5 pls.   (1943) **Cupp, E. E.**
Reprinted (1977) by O. Koeltz, Koenigstein. Includes keys, descriptions, and good
illustrations.

Die Diatomeen von Schweden und Finnland. *K. svenska VeternskAkad. Handl.*
Ser. 4 **2**(1) (1951): 1–163; **3**(3) (1952): 1–153; **4**(1) (1953): 1–158; **4**(5)
(1954): 1–255; **5**(4) (1955): 1–232.   (1951–55) **Cleve-Euler, A.**
Includes good keys and descriptions.

An introductory account of the smaller algae of British coastal waters. Part V.
Bacillariophyceae (Diatoms). HMSO, London. Fishery Investigations, Series
IV. xxii + 317 pp, 45 pls.   (1964) **Hendey, N. I.**

The diatoms of the United States exclusive of Alaska and Hawaii. *Monogr.
Acad. nat. Sci. Philad.* **13**: Vol. 1 (1966) xi + 688 pp, 64 pls; Vol. 2, Part 1
(1975) ix + 213 pp, 28 pls.   (1966, 1975) **Patrick, R. and Reimer, C. W.**
Freshwater only. Descriptions and figures of all species; keys to both genera and species.

*Marines phytoplankton. Eine Auswahl der Helgolander Planktonalgen (Diatomeen,
Peridineen).* G. Thieme, Stuttgart. vi + 186 pp, 151 figs.
(1974) Drebes, G. von
Covers more common marine phytoplankton; descriptions; illustrations; no keys.

A guide to the morphology of the diatom frustule with a key to the British
freshwater genera. *Scient. Publs Freshwat. biol. Ass.* **44**, 112 pp, 144 figs.
(1981) Barber, H. G. and Haworth, E. Y.
A useful guide to the terminology on diatom shapes, a good key, and a general
introduction to the taxonomic literature.

*Flore des Diatomées Diatomophycées eaux douces et saumatres du Massif Armoricain et des
contrées voisines d'Europe occidentale.* Boubée et Cie, Paris. 444 pp, 169 pls.
(1981) Germain, H.

Naviculaceae. Neue und wenig bekannte Taxa, neue Kombinationen und
Synonyme sowie Bemerkungen zu einigen Gattungen. *Biblthca diatom.* **9**: 1–
230, 43 pls.   (1985) **Krammer, K. and Lange-Bertalot, H.**

The pennate diatoms. [A translation of Hustedt's *Die Kieselalgen*, Vol. 2. Supplements by Norman G. Jensen.] Koeltz Scientific Books, Koenigstein. xvii
+ 918 pp, 870 figs.   (1985) **Krammer, K. and Lange-Bertalot, H.**
The original was published (1959) as Vol. 7 (part 2) of Dr L. Rabenhorst's
Kryptogamen-Flora von Deutschland, Österreich und der Schweiz (see under Freshwater).

CHRYSOPHYCEAE

See other important works under Freshwater (pp. 271–5).

Recherches sur les Chrysophycées. Morphologie, phylogenie, systematique. *Revue algol., Mém., hors-sér.* **1**: 1–412, 11 pls. (1957) **Bourrelly, P.**
Keys and descriptions of genera, and of some species. World-wide coverage.

DINOPHYCEAE

See other important works under Freshwater and Marine as well as under Mastigophora, Sarcomastigophora (free-living Protozoa) on pp. 9–11.

Marines phytoplankton. Eine Auswahl der Helgolander Planktonalgen (Diatomeen, Peridineen). G. Thieme, Stuttgart. vi + 186 pp, 151 figs.
(1974) **Drebes, G. von**
Covers more common marine phytoplankton; descriptions; illustrations; no keys.

Marine dinoflagellates of the British Isles. HMSO, London. 303 pp, 35 figs, 8 pls.
(1982) **Dodge, J. D.**
Good keys and line drawings.

XANTHOPHYCEAE (TRIBOPHYCEAE)

See also under Freshwater (pp. 271–5).

Vaucheriaceae. *N. Am. Flora.*, ser. III **8**: 1–64, 104 figs. (1972) **Blum, J. L.**

De Nederlandse *Vaucheria*-soorten. *Wet. Meded. K. ned. natuurh. Veren.* **120**: 1–32.
(1977) **Simons, J.**

Seaweeds of the British Isles. British Museum (Natural History), London. Vol. 4, Tribophyceae (Xanthophyceae). viii + 36 pp, 8 compound figs.
(1987) **Christensen, T.**
Keys, descriptions, and line drawings.

PHAEOPHYCEAE (FUCOPHYCEAE)

See Marine (pp. 275–9) for other important works.

Phéophycées de France. Paris. xlvii + 431 pp, 10 pls. (1931–39) **Hamel, G.**

Die Phaeophyceen der Schwedischen Westküste. *Lunds Univ. Årsskr.* (N.F.) Avd. 2 **43**(4): 1–99. (1947) **Kylin, H.**

Étude sur les Ectocarpacées de la Manche. *Beih. Nova Hedwigia* **15**: 1–86, 41 figs. (1964) **Cardinal, A.**
Keys to genera; keys to species within genera; descriptions; line drawings, arranged many per figure.

A field key to the British brown seaweeds (Phaeophyta). *Field Stud.* **5**: 1–44, 22 figs. (1979) **Hiscock, S.**
Useful illustrated key to most British brown seaweeds, to genus and/or species level.

Specialist keys for the complex genera *Fucus* and *Elachista*. AIDGAP listed; also available separately.

A taxonomic revision of the european Sphacelariaceae (Sphacelariales, Phaeophyceae). E. J. Brill/Leiden University Press, Leiden. 293 pp, 6 pls, 660 text figs.

(1982) **Prud'homme van Reine, W. F.**

Keys to families; to genera within families; to subgenera and sections in *Sphacelaria*; tables of comparative data; descriptions; ecology; distribution.

Seaweeds of the British Isles. British Museum (Natural History), London. Vol. 3, Part 1, Fucophyceae (Phaeophyceae). [10] + 359 pp, 90 figs, 15 pls.

(1987) **Fletcher, R. L.**

Keys, descriptions, line-drawings, photographs, ecology, distribution around the British Isles and the world and glossary.

EUGLENOPHYTA

See also Freshwater (pp. 271–5).

Euglenoid flagellates. Prentice-Hall, Englewood Cliffs, New Jersey. xiii + 242 pp, 176 figs.   (1967) **Leedale, G.**

CHLOROPHYTA

See other important works under Freshwater (pp. 271–5).

A monograph of the British Desmidiaceae. Ray Society, London. 5 vols, 1198 pp, 167 pls.   (1904–23) **West, W. and West, G. S.***
* Last volume (5) completed by N. Carter following the death of the Wests. Generic and specific keys, some biological data, and both British and foreign distributions; a useful monograph with some good illustrations.

Die Chlorophyceen der Schwedischen Westküste. *Lunds Univ. Årsskr.* (N.F.) Avd. 2 **45**(4): 1–79.   (1949) **Kylin, H.**

A critical survey of European taxa in Ulvales. I. *Capsosiphon, Percursaria, Blidingia, Enteromorpha. Op. bot. Soc. bot. Lund* **8**(3): 1–160, 92 figs.

(1963) **Bliding, C.**

A good survey of most aspects of the biology of some Ulvales. Useful descriptions and photographs but no keys.

Revision of the european species of Cladophora. E. J. Brill, Leiden. vii + 248 pp, 55 pls.   (1963) **van den Hoek, C.**

All aspects of the genus in Europe; key to species and varieties.

Studies in *Cladophora. Acta Hort. gothoburg* **26**(1): 1–147, 125 figs.

(1963) **Söderström, J.**

Concerns the genus in the European Atlantic. Covers most aspects of its biology and includes keys to the species and distribution maps.

A critical survey of European taxa in Ulvales, II. *Ulva, Ulvaria, Monostroma, Kornmannia. Bot. Notiser* **121**: 535–629, 47 figs. (1968) **Bliding, C.**
Same remarks apply as under Bliding (1963), this section; includes addenda to latter and overall systematic conclusions as to families and genera.

Die Desmidiaceen Mitteleuropas. Schweizerbart'sche, Stuttgart. Vol. 1, Part 1 (1977) vi + 291 pp, 44 pls; Vol. 1, Part 2 (1981) ix + 292–736 pp, 72 pls.
(1977, 1981) **Ruzicka, J.**
Keys, illustrations, and good descriptions.

Taxonomic studies on the marine and brackish-water species of *Ulothrix* (Ulotrichales, Chlorophyceae) in western Europe. *Blumea* **24**: 191–299, 16 pls, 19 figs. (1978) **Lokhorst, G. M.**
Keys; descriptions; distribution; ecology; line drawings; photographs.

A key to the commoner desmids of the English Lake District. *Scient. Publs Freshwat. biol. Ass.* **42**, 123 pp, 171 figs. (1980) **Lind, E. M. and Brook, A. J.**

Taxonomic studies on *Urospora* (Acrosiphonales, Chlorophyceae) in western Europe. *Acta Bot. neerl.* **30**: 353–431.
(1981) **Lokhorst, G. M. and Trask, B. J.**
Key; descriptions; distribution; ecology; line drawings; photographs.

The taxonomy of *Ulva* (Chlorophyceae) in the Netherlands. *Br. phycol. J.* **16**: 9–53. (1981) **Koeman, R. P. T. and van den Hoek, C.**
Key to species; drawings (gross and cell structure); photographs (cell structure); ecology; distribution; species descriptions; comparative table of parameters.

The taxonomy of *Enteromorpha* Link, 1820, (Chlorophyceae) in the Netherlands. I. The section Enteromorpha. *Arch. Hydrobiol.* (Suppl.) **63**(3): 279–330. (Algological Studies no. 32).
(1982) **Koeman, R. P. T. and van den Hoek, C.**
For details of content, see parts II and III below.

The taxonomy of *Enteromorpha* Link, 1820, (Chlorophyceae) in the Netherlands. II. The section Proliferae. *Crypt.: Algologie* **3**: 37–70.
(1982) **Koeman, R. P. T. and van den Hoek, C.**
Comparative keying table of the five species recognized; line drawings (gross and anatomical); close-up photographs; descriptions; ecology; distribution; cell sizes.

The taxonomy of *Ulva* (Chlorophyceae) from the coastal region of Roscoff (Brittany, France). *Botanica mar.* **26**: 65–86.
(1983) **Hoeksema, B. W. and van den Hoek, C.**
Key to species; line drawings; descriptions; ecology; distribution; summary table of characteristics.

The taxonomy of *Enteromorpha* Link, 1820 (Chlorophyceae) in the Netherlands. III. The sections Flexuosae and Clathratae and an addition to the section Proliferae. *Crypt.: Algologie* **5**: 21–61.
(1984) **Koeman, R. P. T. and van den Hoek, C.**
Comparative keying table of the four species recognized; line drawings (gross and anatomical); photographs (close-up); descriptions; cell sizes; key to sections of the genus.

Zielenice (Chlorophyta): Edogoniowce (Oedogoniales). In *Flora Polski* [Flora of Poland]. Polska Academia, Instytut Botaniki, Warsaw. 313 pp, 210 figs.
(1984) **Mrozinska, T.**

Morphology and systematics of *Stigeoclonium* Kutz (Chaetophorales). In *Systematics of the green algae* (ed. D. E. G. Irvine and D. M. John), pp. 363–77, 6 figs. Academic Press, London.   (1984) **Francke, J. A. and Simons, J.** Information on the biology of the genus and a useful key to the species. World-wide coverage.

Seaweeds of the British Isles. British Museum (Natural History), London. Vol. 2. Chlorophyta (Chlorophyceae). (In press).   (1988) **Burrows, E. M.** Keys, descriptions, illustrations, and other information on the ecology and distribution of green seaweeds around the British Isles.

CHAROPHYTA

Other important works appear under Freshwater (pp. 271–5).

The British Charophyta. Ray Society, London. Vol. 1 (1920) Nitelleae. 141 pp, 20 pls; Vol. 2 (1924) Chareae. 129 pp, 25 pls.
(1920, 1924) **Grove, J. and Bullock-Webster, F. R.** Descriptions, illustrations, generic and specific keys, British and foreign distributions, biological data fairly full: a useful monograph with good illustrations but now largely superseded by Moore (1986), this section.

Danish Charophyta. Chorological, ecological, and biological investigations. *K. dansk. Vidensk. Selsk. Skr.* **3**: 1–240, 2 pls.   (1944) **Olsen, S.**

British stoneworts (Charophyta). Buncle, Arbroath. 52 pp, 18 figs.
(1950) **Allen, G. O.** A short introduction to the group with keys and descriptions.

Les Charophycées de France et d'Europe occidentale. Bretonne, Rennes. 499 pp.
(1957) **Corillion, R.** Keys and descriptions, but few illustrations.

A revision of the Characeae. J. Cramer, Weinheim. Includes:

Vol. 1. *Monograph of the Characeae*. xxiv + 904 pp.   (1965) **Wood, R. D.**

Vol. 2. *Iconograph of the Characeae*. xv + 7 pp, 395 pls and figs.
(1964) **Wood, R. D. and Imahori, K.** World-wide treatment with descriptions, illustrations, and keys to the species.

Charophytes of Great Britain and Ireland. Botanical Society of the British Isles, London. *BSBI Handbook* no. 5, 140 pp, 25 figs, 17 maps. (1986) **Moore, J. A.** Good keys, descriptions, and illustrations to the representatives of the group found in the British Isles. Distribution maps and mention made of distributions in north-western Europe.

BRYOPHYTA

British species about 980.

General

British mosses and liverworts, 3rd edn. Cambridge University Press, Cambridge. 519 pp, illustrations. (1981) **Watson, E. V.**
Illustrated descriptions of and keys to many British bryophytes.

Distribution maps of British and Irish Bryophytes. *Trans. Br. bryol. Soc.* **4–6** (1963–71); *J. Bryol.* **7** (1972–80). (1963–80) **Smith, A. J. E.**, ed.

Distribution of bryophytes in the British Isles. A census catalogue of their occurrence in vice-counties. British Bryological Society, Cardiff.
(1981) **Corley, M. F. V. and Hill, M. O.**
Full British and Irish distribution by vice-counties up to December 1980: nomenclature corrected: now the British standard check-list. For species added to the British vice-county lists after 1981 see *Bull. Br. bryol. Soc.* (1981 on).

Flora des bryophytes. *Encyclopédie biologique.* P. Lechevalier, Paris. Vol. 4, 702 pp. (1966) **Augier, I.**

Artenliste der moose Macaronesiens. *Cryptogamie, bryol. lichén.* **3**: 282–325.
(1982) **Eggers, J.**

Moosflora. Eugene Ulmer, Stuttgart. 522 pp, illustrated.
(1983) **Frahm, J.-P. and Frey, W.**
Keys, brief descriptions, and illustrations of all German bryophytes. Useful pocket flora.

Kleine Kryptogamenflora. *Die Moos- und Farnpflanzen*, 5th edn. G. Fischer, Stuttgart. Vol. 4, 240 pp. (1973) **Gams, H.**

Beknopte flora van Nederlandse blad- en levermossen. Thieme, Zutphen. 517 pp.
(1982) **Margadant, W. D. and During, H.**

Provisional atlas of the distribution of the bryophytes of the British Isles. British Bryological Society and Institute of Terrestrial Ecology, Cambridge. 105 pp (of maps). (1978) **Smith, A. J. E.**, ed.
Distribution maps of 105 commoner species.

ANTHOCEROTAE (ANTHOCEROTOPSIDA)

Hornworts. British species 3.

ANTHOCEROTACEAE

Anthoceros agrestis, a new name of *A. punctatus* var. *cavernosus* Prosk. 1958 *non* (Nees) Gotsche *et al. J. Bryol.* **10**: 257–61. (1979) **Paton, J. A.**

Phaeoceros laevis (L.) Prosk. subsp. *carolinianus* (Michaux) Prosk. in Britain. *J. Bryol.* **7**: 541–3.   (1983) **Paton, J. A.**

HEPATICAE (HEPATICOPSIDA AND ANTHOCEROTOPSIDA)

Liverworts. British species about 285.

GENERAL

The student's handbook of British hepatics, 2nd edn. Sumfield, Eastbourne. 464 pp. Reprinted (1960) by Wheldon and Wesley, Codicote.
                                    (1926) **Macvicar, S. M.**

Distribution of the European and Macaronesian liverworts (Hepaticophytina). *Bryol. Beitr.* **2**: 1–115.   (1983) **Duell, R.**
Lists every European and Macaronesian country and island in which each liverwort species occurs.

Hepatics of Europe including the Axores: an annotated list of species, with synonyms from the recent literature. *J. Bryol.* **12**: 403–59. (1983) **Grolle, R.**
Taxonomically and nomenclaturally the most up-to-date European list.

Advances in the knowledge of British hepatics since 1926. *Trans. Br. bryol. Soc.* **2**, 1–10.   (1952) **Jones, E. W.**

Atlas van de Nederlandse levermossen. Thieme, Zutphen.   (1980) **Landwehr, J.**
Beautifully illustrated descriptions of all Dutch hepatics.

Hepaticae. *Index to distribution maps of bryophytes.* Svenska Växtageografiska Sällskappet, Uppsala. Vol. 2.   (1980) **Sjödin, Å.**

Die Lebermoose Deutschlands, Öesterreichs und der Schweiz, 3rd edn. *Rabenh. Krypt.-Fl.* **6**(1): 1–756.   (1951) **Müller, K.**
The 1st edn (1906–11) is superior in some respects to the 3rd edn and should also be consulted.

Flora generale de Belgique. Bryophytes. Ministère de l'Agriculture, Brussels. Vol. 1, pp. 1–3.   (1955–57) **Bergen, C. van den**

Hepaticae. *Illustrated moss flora of Fennoscandia.* Gleerups, Lund. Vol. 1, 308 pp.
                                    (1956) **Arnell, S.**

Boreal Hepaticae. A manual of the liverworts of Minnesota and adjacent regions. *Am. Midl. Nat.* **42**: 257–684.   (1953) **Schuster, R. M.**
Well illustrated.

Annotated key to the orders, families and genera of the Hepaticae of America north of Mexico. *Bryologist* **61**: 1–66.   (1958) **Schuster, R. M.**

The Hepaticae and Anthocerotae of North America east of the hundredth meridian. Columbia University Press, New York. 4 vols. (1966–80) **Schuster, R. M.**
An exhaustive and lavishly illustrated work of which one further volume is still to appear.

MONOGRAPHS OF SPECIAL GROUPS

RICCIACEAE

Riccia crystallina L. and *R. cavernosa* Hoffm. in Britain. *Trans. Br. bryol. Soc.* **5**: 222–5.   (1967) **Paton, J. A.**

Observations on *Riccia bifurca* Hoffm. and other species of *Riccia* L. in the British Isles. *J. Bryol.* **11**: 1–6.   (1980) **Paton, J. A.**

METZEGERIACEAE

Metzgeria temperata Kiwah. in the British Isles, and *M. fructicosa* (Dicks.) Evans with sporophytes. *J. Bryol.* **9**: 441–9.   (1977) **Paton, J. A.**

ANEURACEAE

The oil bodies of the genus *Riccardia* Gray. *Trans. Br. bryol. Soc.* **5**: 536–40.
    (1968) **Little, E. R. B.**
Use of oil bodies in distinguishing the polymorphic species of *Riccardia*.

CODONIACEAE

Fossombronia incurva and *Aongstroemia longipes* in Perthshire, new to the British Isles. *Trans. Br. bryol. Soc.* **4**: 767–74.   (1965) **Crundwell, A. C.**

Taxonomic studies on the genus *Fossombronia* Raddi. *J. Bryol.* **7**: 243–52.
    (1973) **Paton, J. A.**
A critical assessment of the status of several British species of *Fossombronia*.

Fossombronia fimbriata sp. nov. *J. Bryol.* **8**: 1–4.   (1974) **Paton, J. A.**

LOPHOZIACEAE

Barbilophozia atlantica (Kaal.) K. Müll. in Britain. *Trans. Br. bryol. Soc.* **4**: 214–20.   (1962) **Fitzgerald, J. W. and Fitzgerald, R. D.**

Beitrag zur Kenntnis von *Barbilophozia*, insbesondere *B. floerkii* und *B. hatcheri*. *Nova Hedwigia* **2**: 555–66.   (1960) **Grolle, R.**

Lophozia capitata (Hook.) K.M. in Britain. *Trans. Br. bryol. Soc.* **1**: 253–6.
    (1950) **Jones, E. W.**

Lophozia opacifolia Culmann in Scotland. *Trans. Br. bryol. Soc.* **3**: 180.
    (1957) **Jones, E. W.**

Lophozia perssonii Buch. and S. Arnell—new to Britain. *Trans. Br. bryol. Soc.* **5**: 439–42.   (1968) **Jones, E. W.**

JUNGERMANNIACEAE

Solenostoma levieri (Steph.) Steph. new to Britain. *Trans. Br. bryol. Soc.* **6**, 50–5.
    (1970) **Paton, J. A.**

PLAGIOCHILACEAE

Azur Synonymik und Verbreitung von *Plagiochila spinulosa* (Dicks.) Dum. und
P. killarniensis Pears. *J. Bryol.* **12**: 215–25.

(1982) **Grolle, R. and Schumacker, R.**

Plagiochila atlantica F. Rose sp. nov.—*P. ambagiosa* auct. *J. Bryol.* **8**, 417–22.

(1975) **Jones, E. W. and Rose, F.**

Plagiochila killarniensis Pears. in the British Isles. *J. Bryol.* **9**: 451–9.

(1977) **Paton, J. A.**

Plagiochila britannica, a new species in the British Isles. *J. Bryol.* **10**. 245–56.

(1979) **Paton, J. A.**

Includes a key to British species of *Plagiochila*.

GEOCALYCACEAE

Lophocolea semiteres (Lehm.) Mitt. and *Telaranea murphyae* sp. nov. established on
Tresco. *Trans. Br. bryol. Soc.* **4**: 775–9.   (1965) **Paton, J. A.**

Lophocolea bispinosa (Hook. f. and Tayl.) Gottsche, Lindenb. and Nees
established in the Isles of Scilly. *J. Bryol.* **8**: 191–6.   (1974) **Paton, J. A.**

The taxonomy of *Lophocolea bidentata* (L.) Dum. and *L. cuspidata* (Nees) Limpr.
J. Bryol. **10**: 49–59.   (1978) **Steel, D. T.**

SCAPANIACEAE

Die Scapanien Nord-europas und Sibiriens. II Systematischer Teil. *Soc. Sci.
Fennica, Commentat. biol.* **3**(1): 1–177.   (1928) **Buch, H.**

Scapania lingulata Buch. in the British Isles. *J. Bryol.* **11**: 399–403.

(1981) **Paton, J. A.**

Scapania paludicola Loeske and K. Müll. discovered in the British Isles. *J. Bryol.*
11: 405–7.   (1981) **Paton, J. A.**

Notes on the genus *Scapania*. I. *S. mucronata* in Britain. *Trans. Br. bryol. Soc.* **4**:
785–9.   (1965) **Perry, A. R.**

LEPIDOZIACEAE

Lophocolea semiteres (Lehm.) Mitt. and *Telaranea murphyae* sp. nov. established on
Tresco. *Trans. Br. bryol. Soc.* **4**: 775–9.   (1965) **Paton, J. A.**

CALYPOGEIACEAE

The genus *Calypogeia* in Britain. *Trans. Br. bryol. Soc.* **4**: 221–9.

(1962) **Paton, J. A.**

HERBERTACEAE

Herbertus borealis, a new species from Scotland and Norway. *Trans. Br. bryol. Soc.*
  6: 41–9.   (1970) **Crundwell, A. C.**

LEJEUNEACEAE

Notes on Lejeuneaceae. II. A quantitative assessment of criteria used in
  distinguishing some British species of *Lejeunea*. *Trans. Br. bryol. Soc.* **2**: 458–
  69.   (1954) **Greig-Smith, P.**

Cololejeunea schaeferi spec. nov., ein verkauntes Lebermoos in Makaronesien. *J.*
  Bryol. **13**: 487–95.   (1985) **Grolle, R.**

Muscinées recoltés dans le nordouest de la Péninsule Ibérique. *Revue bryol.*
  lichén. **7**: 238–48.   (1924) **Buch, H.**

MUSCI (MOSSES)

British species about 690.

GENERAL

The British moss flora. Reeve, London. 3 vols.   (1887–1905) **Braithwaite, R.**
Mosses of Europe and the Azores: an annotated list of species, with synonyms
  from the recent literature. *J. Bryol.* **1**: 609–89.
  (1981) **Corley, M. F. V., Crundwell, A. C., Dull, R., Hill, M. O., and**
                                                              Smith, A. J. E.

Mosses of eastern North America. Columbia University Press, New York.
                                       (1981) **Crum, H. A. and Anderson, L. E.**
  Abundantly illustrated descriptions of species, many of which occur in Europe. No
  keys to families.

The student's handbook of British mosses, 3rd edn. Sumfield, Eastbourne. 580 pp,
  illustrated. Reprinted (1954) by Wheldon and Wesley, Codicote.
                                                              (1924) **Dixon, H. N.**
  Descriptions of and keys to all taxa known in Britain in 1924; nomenclature and
  classification now out of date.

Distribution of the European and Macaronesian mosses (Bryophytina), I.
  Bryol. Beitr. **4**: 1–113.   (1984) **Duell, R.**
  The first of three parts listing every European and Macaronesian country and island
  in which each moss species occurs.

Index to distribution maps of bryophytes 1887–1975. *Vaxtekologiska Studier.*
  Svenska Vaxtageografiska Sällskapet, Uppsala:
  Pt. 1, Musci, **11**, vi + 282 pp;
  Pt. 2, Hepaticae, **12**, vi + 143 pp.   (1980) **Sjödin, Å.**

Atlas van de Nederlandse bladmossen. KNNV, Hoogwoud, 560 pp.
(1966) **Landwehr, J. and Barkman, J. J.**
Beautifully illustrated descriptions of all Dutch mosses.

The moss flora of Britain and Ireland. Cambridge University Press, Cambridge.
705 pp. (1978) **Smith, A. J. E.**
A fully comprehensive modern flora.

Die Laubmoose Europas. *Rabenh. Krypt.-Fl.* **4**(1): 1–836, (2): 1–854, (3): 1–
864, *1–79.* (1890–1904) **Limpricht, K. G.**

Die Laubmoose Europas. *Rabenh. Krypt.-Fl.* **4** (Erganzungsband): 1–960.
Reprinted (1963) by Cramer, Weinheim. (1927) **Monkemeyer, W.**

Conspectus muscorum Europaeorum. Českolovenske Akademie Ved, Prague. 697
pp. (1954) **Podpera, J.**

Musci. *Illustrated moss flora of Fennoscandia.* Gleerups, Lund. Vol. 2, 6 fascicles.
(1954–69) **Nyholm, E.**

Moseflora. Universitetsforlaget, Oslo. 144 pp. (1968) **Lye, K. A.**

Den Danske: moosflora. Gyldendal, Copenhagen. (1976) **Anderson, A. G.** *et al.*

Bryophytes. *Flora générale de Belgique.* Ministère de l'Agriculture, Brussels. Vol.
3, Part 1. (1968) **Sloover, J. De**

Bryophytes. *Flora générale de Belgique.* Ministère de l'Agriculture, Brussels. Vol.
2, Parts 1–3. (1959–64) **Demaret, F. and Castagne, E.**

Moss Flora of North America north of Mexico. Newfane, Vermont. 3 vols.
(1928–40) **Grout, A. J.**

Sphagnopsida (peat mosses)
SPHAGNACEAE

British species 31.

Sphagnopsida. Sphagnaceae. *North American flora.* New York Botanical
Garden, New York. (Ser. 2) pt **11**, 180 pp. (1984) **Crum, H.**
Comprehensive illustrated account with keys.

Handbook of European sphagna. Institute of Terrestrial Ecology, Cambridge, 262
pp. (1986) **Daniels, R. and Eddy, A.**
Comprehensive illustrated account with keys.

An illustrated key to *Sphagnum* mosses. *Trans. Proc. bot. Soc. Edinb.* **29**: 290–301.
(1962) **Duncan, U. K.**

Graphic keys for the identification of *Sphagna. New Phytol.* **37**: 409–20.
(1938) **Fearnsides, M.**

A key for the identification of British *Sphagna* using macroscopic characters.
Bull. Br. bryol. Soc. **27**: 22–31. (1976) **Hill, M. O.**

Key to northern boreal and arctic species of *Sphagnum* based on characteristics
of the stem leaves. *Lindbergia* **8**: 1–29. (1982) **Lange, B.**

Sphagnum subfulvum Sjörs in Ireland compared with the occurrences in Norway.
J. Bryol. **12**: 331–6.   (1983) **Moen, A. and Synnott, D.**

A key to the British species of *Sphagnum. Trans. Br. bryol. Soc.* **2**: 552–60.
                              (1955) **Proctor, M. C. F.**

The identification of *Sphagnum* spores. *Trans. Br. bryol. Soc.* **4**: 209–13.
                              (1962) **Tallis, J. H.**

Untersuchen über die *Sphagnum*-Arten der Gruppe *Subsecunda* in Europa. *Ark. Bot.* **29A**(1): 1–77.   (1937) **Åberg, G.**

Andreacopsida (granite mosses)

British species about 6.

Monographie der Laubmoosgattung *Andreaea*. I. Die costaten Arten. *Willdenowia* **6**: 25–110.   (1970) **Schultze-Motel, W.**

Bryopsida (true mosses)

British species about 655.

POLYTRICHACEAE

Studies on the genus *Atrichum* P. Beauv. *Lindbergia* **1**: 1–33. (1971) **Myholm, E.**

DICRANACEAE

Nerve sections in the Dicranales. I. *Campylopus* and *Dicranodontium. Bull. Br. bryol. Soc.* **28**: 14–15.   (1976) **Corley, M. F. V.**

A progress report on *Campylopus introflexus* (Hedw.) Brid. and *C. polytrichoides* De Not. in Britain and Ireland. *J. Bryol.* **8**: 293–8.
                  (1975) **Richards, P. W. and Smith, A. J. E.**

FISSIDENTACEAE

The *Fissidens viridulus* complex in the British Isles and Europe. *J. Bryol.* **11**: 191–209.   (1980) **Corley, M. F. V.**

Fissidens crassipes Wils. ex B., S. and G., *F. mildeanus* Schimp. and *F. rufulus* B., S. and G. *Trans. Br. bryol. Soc.* **4**: 204–5.
                  (1961) **Smith, A. J. E. and Warburg, E. F.**

ENCALYPTACEAE

Encalypta brevipes and *E. brevicolla*: new records from North America, Iceland, Great Britain and Europe. *J. Bryol.* **11**: 209–12.   (1980) **Horton, D. G.**

POTTIACEAE

Didymodon reedii Robins, a moss new to the British Isles. *J. Bryol.* **13**: 319–21.
                              (1981) **Appleyard, J.**

Barbula tomaculosa, a new species from arable fields in Yorkshire. *J. Bryol.* **11**: 583–9. (1981) **Blockeel, T. J.**

Scopelophila cataractae (Mitt.) Broth. in South Wales, new to Europe. *J. Bryol.* **13**: 323–8. (1985) **Corley, M. F. V. and Perry, R. A.**

A revision of *Weissia*, subgenus *Astomum*. I. The European species. *J. Bryol.* **7**: 7–19. (1972) **Crundwell, A. C. and Nyholm, E.**

Taxonomic revision of *Aloina*, *Aloinella* and *Crossidium* (Musci). *Bryologist* **78**: 245–303. (1975) **Delgadillo, M. C.**

Leptobarbula berica (De Not.) Schimp. in Britain. *J. Bryol.* **13**: 461–70.
(1985) **Hill, M. O. and Whitehouse, H. L. K.**

Bryoerythrophyllum caledonicum, a new moss from Scotland. *J. Bryol.* **12**, 141–57.
(1982) **Long, D. G.**

Tortula solmsii (Schimp.) Limpr. in Devon and Cornwall, newly recorded in the British Isles. *J. Bryol.* **12**: 159–69. (1982) **Long, D. G. and Hill, M. O.**

GRIMMIACEAE

Racomitrium elongatum Frisvoll in Britain and Ireland. *Bull. Br. bryol. Soc.* **43**: 21–5. (1984) **Hill, M. O.**

Monographie der europäischen Grimmiaceen. *Bibltheca bot.* **25**: 1–236.
(1930) **Loeske, L.**

Zur Kenntnis der *gracile*-Formen der Sammelart *Schistidium aporcarpum* (L.) Br. Eur. *Svensk. bot. Tidskr.* **47**: 248–62. (1953) **Poelt, J.**

SELIGERIACEAE

Key to British *Seligeria* species and related genera based on vegetative characters. *Bull. Br. bryol. Soc.* **32**: 26. (1978) **Corley, M. F. V.**

Seligeria parvifolia (Lindb.) Lindb. on Snowdon, newly recorded in the British Isles. *J. Bryol.* **11**: 7–10. (1980) **Hill, M. O.**

FUNARIACEAE

Die Laubmoose Europas. Vol. 2. *Funariaceae*. Berlin. viii + 120 pp, illustrated.
(1914–19) **Loeske, L.**

BRYACEAE

Pholia scotica, a new species of moss from the western highlands of Scotland. *J. Bryol.* **12**: 7–10. (1982) **Crundwell, A. C.**

The European species of the *Bryum erythrocarpum* complex. *Trans. Br. bryol. Soc.* **4**: 597–637. (1964) **Crundwell, A. C. and Nyholm, E.**

On the taxonomy and distribution of *Rhodobrym roseum* and related species (Bryophyta). *Acta bot. fenn.* **96**: 1–22.

(1972) **Iwatsuki, Z. and Koponene, T.**

Die Gattung *Pohlia* Hedw. (Bryales, Bryaceae) in Deutschland und den angrenzenden Gebieten 1. Wenig bekannte und oft übersehene *Pohlia*-arten. *Lindbergia* **8**: 139–47. (1982) **Nordhorn-Richter, G.**

A taxonomic study of *Bryum capillare* Hedw. and related species. *J. Bryol.* **7**: 265–326. (1973) **Syed, H.**

Les espèces du 'complexe *Bryum bicolor*' (Musci). *Bull. Jard. bot. natn. Belg.* **46**: 511–41. (1976) **Wilczek, R. and Demaret, F.**

MNIACEAE

Generic revision of Mniaceae Mitt. (Bryophyta). *Ann. bot. fenn.* **5**: 117–51.

(1968) **Koponen, T.**

The moss genus *Rhizomnium* (Roth.) Kop. with description of *Rhizomnium perssonii*, species nova. *Mem. Soc. Fauna Fl. fenn.* **44**: 33–50.

(1968) **Koponen, T.**

A monograph of *Plagiomnium* sect. *Rosulata* (Mniaceae). *Ann. bot. fenn.* **8**: 305–67. (1971) **Koponen, T.**

ORTHOTRICHACEAE

Die Gattung *Zygodon. Acta hort. bot. Univ. Latv.* **1**: 1–184. (1926) **Malta, N.**
A revision of the genus *Orthotrichum* in North America north of Mexico. *Bibltheca bryoph.* No. 1. J. Cramer, Lehre. (1973) **Vitt, D. H.**

FONTINALACEAE

A monograph of the Fontinalaceae. Martinus Nijhoff, The Hague. 357 pp.

(1960) **Welch, W. H.**

LESKEACEAE

Revision of the genus *Lescuraea* in Europe and N. America. *Bull. Torrey bot. Club* **84**: 281–307. (1957) **Lawton, E.**

AMBLYSTEGIACEAE

Pictus scoticus, a new genus and species of pleurocarpous moss from Scotland. *J. Bryol.* **12**: 1–6. (1982) **Townsend, C. C.**

BRACHYTHECIACAE

Brachythecium appleyardiae sp. nov. in south-west England. *J. Bryol.* **11**: 591–8.

(1981) **McAdam, S. V. and Smith, A. J. E.**

HYPNACEAE

Hypnum uncinulatum Jur. reinstated as an Irish species. *J. Bryol.* **11**: 185–9.
(1980) Ando, H. and Townsend, C. C.

Les *Stereodon* de l'Europe. *Revue bryol. lichén.* **20**: 263–88.
(1950) Doignon, P. and Guillamot, M.

PTERIDOPHYTA
(Ferns and fern allies)

British species about 80.

Welsh ferns, clubmosses, quillworts and horsetails, 6th edn. National Museum of
Wales, Cardiff. 178 pp.
(1978) Hyde, H. A., Wade, A. E., and Harrison, S. G.
Descriptions, keys, British and foreign distribution; cytological data, etc.

Atlas of ferns of the British Isles. Botanical Society of the British Isles and the
British Pteridological Society, London. 101 pp.
(1978) Jermy, A. C., Arnold, H. R., Farrell, L., and Perring, F. H.
Maps of British and Irish distributions, geographical, ecological, and some morphological comments.

*Index to botanical monographs and taxonomic papers relating to phanerogams and vascular
cryptogams found growing wild in the British Isles.* Academic Press, London and
New York. xi + 163 pp.
(1967) Kent, D. H.

Die Farnpflanzen. *Rabenh. Krypt.-Fl.* **3**: 1–906.
(1889) Luerssen, C.
Descriptions, illustrations, generic and specific keys, overseas distribution only
(often inaccurate), scanty biological data: the best book for serious students, but
some British species not included.

The ferns of Britain and Ireland. Cambridge University Press, Cambridge. 447
pp.
(1982) Page, C. N.
Descriptions, illustrations, chart and multi-access keys to main groups, distribution
notes and maps for Britain and Ireland, good field notes; cytological data.

Grasses, ferns, mosses and lichens of Great Britain and Ireland. Pan Books, London.
191 pp.
(1980) Phillips, R.
Photographic illustrations and field notes.

Wayside and woodland ferns, new edn. (by A. B. Jackson) Warne, London. 144
pp.
(1949) Step, E.
Descriptions, illustrations, no keys, British and foreign distribution.

British ferns and mosses. Eyre and Spottiswoode, London. The Kew Series. 231
pp.
(1960) Taylor, P. G.
Descriptions, illustrations, and keys; ecological but no cytological data; includes fern
allies and a few mosses.

SPERMATOPHYTA
(Flowering plants and conifers)

British species about 2000, excluding microspecies of certain genera.

Geographic

Many of the books listed include Charophyta (see also p. 285) and Pteridophyta (see also above). Though some of the works cited are out of date as regards classification and nomenclature, they remain useful.

Flora europaea. Cambridge University Press, Cambridge. 5 vols.
(1964–80) **Tutin, T. G.** *et al.*, eds.
Authoritative, with keys.

Flowers of Europe. A field guide. Oxford University Press, London. 642 pp, 1926 col. photographs, 50 figs.   (1969) **Polunin, O.**

The wild flowers of Britain and northern Europe, 2nd edn. Collins, London. 336 pp, 2900 figs.   (1974) **Fitter, R., Fitter, A., and Blamey, M.**

The alpine flowers of Britain and Europe. Collins, London. 384 pp, coloured illustrations.   (1979) **Grey-Wilson, C. and Blamey, M.**

Atlas florae europaeae. Suomalaisen Kirjallisuudan Kirjapaino Oy, Helsinki. Vols 1–7.   (1972–86) **Jalas, J. and Suominen, J.**
1011 distribution maps of European species. Continuing.

BRITAIN

A new illustrated British flora. L. Hill, London. 2 vols, 1825 figs.
(1961) **Butcher, R. W.**

Further illustrations of British plants, 2nd edn. L. Reeve, London. iv + ii + 476 pp.
(1946) **Butcher, R. W. and Strudwick, F. E.**

Flora of the British Isles, 3rd edn. Cambridge University Press, Cambridge. xxviii + 688 pp, 82 figs.
(1987) **Clapham, A. R., Tutin, T. G., and Moore, D. M.**
The standard British flora.

Flora of the British Isles. (Illustrations by S. J. Roles.) Cambridge University Press, Cambridge. 4 vols, 1910 figs.
(1957–65) **Clapham, A. R., Tutin, T. G., and Warburg, E. F.**

Excursion flora of the British Isles, 3rd edn. Cambridge University Press, Cambridge. xxxv + 499 pp.
(1981) **Clapham, A. R., Tutin, T. G., and Warburg, E. F.**

The identification of flowering plant families. Oliver and Boyd, Edinburgh and

London. 122 pp.   (1965) **Davis, P. H. and Cullen, J.**

Hayward's botanists pocket-book, 19th edn. S. Bell, London. xlv + 310 pp.
Reprinted.   (1948) **Druce, G. C.**

Flowering plants of Wales. National Museum of Wales, Cardiff. ix + 338 pp.
                                                        (1983) **Ellis, R. G.**
Notes on distribution, distribution maps, occasional taxonomic comments.

Illustrations of the British flora, 5th edn. L. Reeve, London. xxvii + 338 pp, 1315
figs. Reprinted.   (1946) **Fitch, W. H. and Smith, W. G.**

The student's flora of the British Isles, 3rd edn. Macmillan, London. xxiii + 563 pp.
Reprinted.   (1937) **Hooker, J. D.**

*Index to botanical monographs and taxonomic papers relating to phanerogams and vascular
cryptogams found growing wild in the British Isles.* Academic Press, London and
New York. xi + 163 pp.   (1967) **Kent, D. H.**

Docks and knotweeds of the British Isles. Botanical Society of the British Isles,
London. BSBI *Handbook* **3**, 205 pp. (1981) **Lousley, J. E. and Kent, D. H.**
Descriptions, illustrations, and keys.

The pocket guide to wild flowers. Collins, London. xii + 340 pp, 1400 figs,
illustrated. Suppl. by D. McClintock (1957) ix + 89 pp. Platt, Kent
(privately printed).   (1956) **McClintock, D. and Fitter, R. S. R.**

The concise British flora in colour, 2nd edn (revised D. H. Kent). Ebury Press and
M. Joseph, London. 254 pp, 1486 figs.   (1969) **Martin, W. K.**

The Cambridge British Flora. Cambridge University Press, Cambridge. Vols 2
and 3 only. Illustrated.   (1914–20) **Moss, C. E.**

The *Taraxacum* flora of the British Isles. *Watsonia* **9** (suppl.): 141 pp.
                                                    (1972) **Richards, A. J.**

The wild flower key—British Isles and N.W. Europe. Warne, London. 480 pp.
                                                        (1981) **Rose, F.**

Drawings of British plants. Bell, London. 31 vols + index, 1317 figs.
                                                (1948–74) **Ross-Craig, S.**

Hybridization and the flora of the British Isles. Academic Press, London, New York,
and San Francisco. xiii + 626 pp.   (1975) **Stace, C. A.**, ed.

Umbellifers of the British Isles. Botanical Society of the British Isles, London.
BSBI *Handbook* **2**, 197 pp.   (1980) **Tutin, T. G.**
Descriptions, illustrations, and keys.

Handbook of the Rubi of Great Britain and Ireland. Cambridge University Press,
Cambridge. xi + 274 pp.   (1958) **Watson, W. C. R.**

An Irish flora, 6th edn (revised). Dundalgan Press, Dundalk. xxxi + 276 pp, 100
figs.   (1977) **Webb, D. A.**

Guide to the identification of some difficult plant groups. Nature Conservancy
Council, Banbury. 145 pp. (1981) **Wigginton, M. J. and Graham, G. G.**

A taxonomic revision of *Euphrasia* in Europe. *Bot. J. Linn. Soc.* **77**: 223–34.
                                                        (1978) **Yeo, P. F.**

British plant list, 2nd edn. Buncle, Arbroath. xi + ii + 154 pp.
(1928) **Druce, G. C.**

List of British vascular plants. British Museum (Natural History) and Botanical Society of the British Isles, London. xvi + 176 pp.   (1958) **Dandy, J. E.**

Nomenclatural changes in the *List of British vascular plants*. *Watsonia* **7**: 157–78.
(1969) **Dandy, J. E.**

Contributions towards a Cybele Hibernica, being outlines of the geographical distribution of plants in Ireland. Second edition founded on the papers of . . . A. G. More. Ponsonby, Dublin, xcvi + 538 pp.   (1898) **Colgan, N. and Scully, R. W.**

The comital flora of the British Isles. T. Buncle, Arbroath. xxxii + 407 pp + map.
(1932) **Druce, G. C.**

Atlas of the British flora. Critical supplement. Nelson, London. viii + 159 pp.
(1968) **Perring, F. H. and Sell, P. D.**

Atlas of the British flora, 3rd edn. EP Publishing, Wakefield. xxvi + 432 pp.
(1983) **Perring, F. H. and Walters, S. M.**, eds.

Irish topographic botany. *Proc. Roy. Ir. Acad.*, 3rd series, **8**, clxxxviii + 410 pp + Supplements published in *Proc. Roy. Ir. Acad.*, **B**. (1901) **Praeger, R. L.**

Topographical botany: Being local and personal records toward showing the distribution of British plants traced through the 112 counties and vice-counties of England, Wales and Scotland, 2nd edn by J. G. Baker and W. W. Newbould. London. xlv + 612 pp. Suppls by A. Bennett (1905) *J. Bot., Lond.* **43**: *Suppl.* 14 pp. and A. Bennett, C. E. Salmon, and J. R. Matthews (1929–30), *J. Bot., Lond.* **67–68**: *Suppl.* 96 pp.   (1883) **Watson, H. C.**

Census catalogue of the flora of Ireland. National Museum of Ireland, Dublin. xvi + 127 pp.   (1972) **Scannell, M. J. P. and Synnott, D. M.** Check-list, with comments on status and vice-county distributions. New edition in press.

FRANCE

Flore complète de France, Suisse et Belgique. Librarie Générale de l'Enseignement . . ., Paris. 12 vols, 12 col. pls. (1911–35) **Bonnier, G. E. M. and Douin, R.**

Flore de France. Asnières (Seine), Paris and Rochefort. 14 vols.
(1893–1913) **Rouy, G. and Foucaud, J.**

Flora descriptive et illustrée de la France. P. Klincksieck, Paris. 3 vols, reprinted, illustrated. Also 6 Suppls by P. Jovet and R. de Vilmorin. Blanchard, Paris.
(1937) **Coste, H.**

Les quatres flores de la France, 2nd edn. Lechevalier, Paris. 2 vols, xlviii + 1105 pp, 304 pl.   (1977) **Fournier, P.**

BELGIUM

Flora de la Belgique, du nord de la France et des regiones voisines . . ., 3rd edn. Jardin Botanique Nationale, Brussels. cviii + 1016 pp. (1983) **Langhe, J.-E.** *et al.*

Flora générale de Belgique. Spermatophytes. Jardin Botanique de l'Etat, Brussels. 4 vols to date by A. Lawalree, illustrated. (1952 on) **Robyns, W.**, ed.

HOLLAND

Flora der Gekweekte, Kruidachtige Gewassen. H. Veenman, Wageningen. 450 pp. (1950) **Boom, B. K. and Ruys, J. D.**

Flora Nederlandica. De Vereeningin, Amsterdam. 7 parts to date, illustrated. (1948 on) **Weevers, Th.** *et al.*, eds.

Flora van Nederland, 20th edn. (ed. W. Heukels). P. Noordhoff, Groningen. 912 pp, 1038 figs. (1983) **Meijden, R. van der** *et al.*

GERMANY

Illustrierten Flora von Mittel-Europa. Lehmann, Munich. 8 vols, illustrated. 2nd edn (1936–71) Hanser, Munich and Berlin. 7 vols. 2nd edn still incomplete, 3rd edn in progress. (1906–31) **Hegi, G.**

Liste der Gefasspflanzen Mitteleuropas, 2nd edn. G. Fischer, Stuttgart. xii + 318 pp. (1973) **Ehrendorfer, F.**

Excursionflora von Deutschland. Spermatophyta, 3rd edn. Volk and Wisson, Berlin. Vols 2–4, illustrated. (1962–63) **Rothmaler, W.**

Flora von Deutschland, 87th edn. O. Schmeil and J. Fitschen, Heidelberg. (1982) **Rauh, W. and Senghas, K.**

Illustrierte Flora Deutschland und angrenzender Gebiete, 23rd edn (ed. K. von Weihe). P. Parey, Berlin and Hamburg. xx + 1607 pp, illustrated. (1972) **Garcke, F. A.**

Synopsis der mitteleuropäischen Flora. Engelmann, Leipzig. 8 vols. (1896–1939) **Ascherson, P. and Graebner, P.**

Prodromus enumerationis specierum plantarum agri- et horticulturae. *Kulturpflanzen Beih.* **2**: i–v + 1–659. (1959) **Mansfield, R.**

Flora von Nord- und Mitteleuropa. G. Fischer, Stuttgart. xii + 1154 pp. (1956) **Hermann, F.**

DENMARK

Den Danske Flora, 20th edn. Gyldendal, Copenhagen. lxiv + 664 pp, 154 figs. (1973) **Rostrup, E.**

ICELAND

Islensk-Ferdaflora, 2nd edn. Almenna Bokaf elagi à, Reykjavik. 428 pp, illustrated. [Icelandic]. (1983) **Löve, A.**

NORWAY AND SWEDEN

Norsk og Svensk Flora, 4th edn. Norske Samlaget, Oslo. 800 pp, 434 figs.
(1963) **Lid, J.**

Nordisk Kärlväxtflora. Almqvist and Wiksell, Stockholm. 2 vols.
(1953–66) **Hylander, N.**

Skanes Flora. Corona, Lund. xxiv + 720 pp.   (1963) **Weimarck, H.**

Monographs of special groups

TREES AND SHRUBS

The trees around us. Weidenfeld and Nicolson, London. 191 pp, illustrated.
(1975) **Barber, P. and Phillips, C. E.**

Nederlandse dendrologie, 9th edn. H. Veenman, Wageningen. 454 pp, 134 figs.
(1975) **Boom, B. K.**

A handbook of Coniferae and Ginkgoaceae, 4th edn (by S. G. Harrison). E. Arnold,
London. xix + 729 pp, illustrated.
(1966) **Dallimore, W. and Jackson, A. B.**

Manuel des conifères. École National des Eaux et Forêts, Nancy. 172 pp,
illustrated.   (1964) **Debazac, E. F.**

Our catkin-bearing plants; an introduction, 2nd edn. Oxford University Press,
London. xii + 61 pp, illustrated.   (1932) **Gilbert-Carter, H.**

*British trees and shrubs including those commonly planted; a systematic introduction to our
conifers and woody dicotyledons.* Clarendon Press, Oxford. xv + 291 pp,
illustrated.   (1936) **Gilbert-Carter, H.**

Trees of Britain. Faber, London. 228 pp.   (1958) **Gurney, R.**

The country life pocket guide to trees in Britain, 5th edn (revised). Country Life,
London. 248 pp, illustrated.   (1966) **Holbrook, A. W.**

Handbuch der Laubgeholze. P. Parey, Berlin and Hamburg. 2 vols.
(1960–62) **Krussmann, G.**

Die Baume Europas. P. Parey, Berlin and Hamburg. 140 pp, illustrated.
(1968) **Krussmann, G.**

Handbuch der Nadelgeholze. P. Parey, Berlin and Hamburg. 366 pp, illustrated.
(1970–71) **Krussmann, G.**

The identification of trees and shrubs, 2nd edn. Dent, London. vii + 375 pp, 128 figs.
(1948) **Makins, F. K.**

British trees and shrubs. Eyre and Spottiswoode, London. Kew Series. 244 pp, 84
figs, 15 col. pls.   (1958) **Meikle, R. D.**

Willows and poplars of Great Britain and Ireland. Botanical Society of the British Isles, London. BSBI *Handbook* **4**, 198 pp. (1984) **Meikle, R. D.**
Good illustrations and descriptions of most taxa.

Conifers, 3rd edn. HMSO, London. *Forestry Commission Booklet* no. 15, 67 pp, illustrated. (1985) **Mitchell, A. F.**

Broadleaves, 2nd edn. HMSO, London. *Forestry Commission Booklet* no. 20, 104 pp, illustrated. (1985) **Mitchell, A. F.**

Conifers in the British Isles. HMSO, London. *Forestry Commission Booklet* no. 33, iii + 322 pp. (1972) **Mitchell, A. F.**
Illustrated with keys and descriptions.

A field guide to the trees of Britain and northern Europe. Collins, London. 415 p, 640 figs, 40 col. pls. (1974) **Mitchell, A. F.**

The Oxford book of trees. Oxford University Press, London. 216 pp, illustrated. (1975) **Nicholson, B. E. and Clapham, A. R.**

Trees and bushes of Europe. Oxford University Press, London. xvi + 208 pp, illustrated. (1976) **Polunin, O. and Everard, B.**

Trees in Britain, Europe and North America. Pan Books, London. 224 pp. (1978) **Phillips, R.**

Trees and shrubs: their identification in summer and winter. W. Heffer, Cambridge. 110 pp, illustrated. (1951) **Prime, C. T. and Deacock, R. J.**

Manual of cultivated trees and shrubs hardy in North America, 2nd edn. Macmillan, New York. xxx + 996 pp + map. (1940) **Rehder, A.**

Elm. Cambridge University Press, Cambridge. xii + 347 pp. (1983) **Richens, R. H.**

Trees: a handbook of forest-botany for the woodlands and the laboratory. Cambridge University Press, Cambridge. 5 vols, illustrated. (1904–09) **Ward, H. M.**

WATER PLANTS

See also under Charophyta (p. 285).

Water plants of the world. A manual for the identification of the genera of the freshwater macrophytes. W. Junk, The Hague. viii + 561 pp, illustrated. (1974) **Cook, C. D. K.** *et al.*

A manual of aquatic plants, 2nd edn. Madison. 405 pp, illustrated. (1972) **Fassett, N. C.**

British water plants. *Fld Stud.* **4**: 243–351. (1975) **Haslam, S. M., Sinker, C. A., and Wolseley, P. L.**
Also available separately; illustrated keys.

Aquatic plants of the United States. Constable, London. x + 374 pp, 154 figs. Reprinted. (1967) **Muenscher, W. C.**

A manual of aquarium plants. Shirley Aquatics, Solihull. 111 pp, illustrated. (1967) **Roe, D.**

Encyclopedia of water plants. T. F. H. Publications, New York. 368 pp, illustrated.   (1967) **Stodola, J.**

Aquarium plants. Blandford, London. 255 pp, illustrated. (English edn prepared by V. Higgins).   (1964) **Wit, H. C. de**

SEDGES AND GRASSES (CYPERACEAE AND GRAMINEAE)

Grassweeds. CIBA-GEIGY, Basle. 2 vols, 142 pp + 2 col. plates, 137 pp + coloured plates.   (1980–81) **Hafliger, E. and Schulz, H.**

Manual of the grasses of the United States, 2nd edn (revised by A. Chase). US Department of Agriculture Publication No. 200. Washington, DC. 1051 pp, illustrated.   (1950) **Hitchcock, A. S.**

Grasses: A guide to their structure, identification, uses and distribution in the British Isles, 3rd edn. Penguin Books, Harmondsworth. 476 pp, illustrated.
                                                          (1984) **Hubbard, C. E.**

British sedges. A handbook to the species of Carex found growing in the British Isles. 2nd edn. Botanical Society of the British Isles, London. BSBI *Handbook* **1**, 268 pp.   (1982) **Jermy, A. C., Chater, A.O., and David, R. W.** Illustrated with distribution maps.

ORCHIDS (ORCHIDACEAE)

Orchideen Europas: Mitteleuropas. Hallwag, Stuttgart. 264 pp, 165 figs.
                                              (1962) **Danesch, O. and Danesch, E.**

Wild orchids of Britain and Europe. Chatto and Windus, London. xii + 256 pp.
                                              (1983) **Davis, P. J. and Huxley, A.**

Orchids of Europe. Blandford Press, London. xiii + 235 pp, 120 figs, 32 col. pls. (English edn by A. J. Huxley.)   (1961) **Duperrex, A.**

Orchids of northern Europe. Penguin Books, Harmondsworth. 146 pp.
                                                          (1979) **Nilsson, S.**

Wild orchids of Britain, 2nd edn. Collins, London. xviii + 366 pp, 72 col. pls.
                                              (1968) **Summerhayes, V. S.**

Europäische und mediterrane Orchideen. Eine Bestimmungsflora mit Berücksichtigung der Ökologie, 2nd edn. Brucke-Verlag Kurt Schmersow, Hildesheim. 243 pp, illustrated.   (1975) **Sundermann, H.**

Bibliography

Abstracts from literature relating to British (and European) plants are compiled by D. H. Kent for BSBI *Abstracts* (1971 on). Comprehensive annual lists of world vascular plant literature are provided by *The Kew Record of Taxonomic Literature* (1974 [for 1971] on. HMSO, London). The monthly *Royal*

Botanic Gardens, Kew. Library current awareness list covers literature about four years ahead of the *Kew Record.*

Geographical guide to the floras of the world. Part 2. *Misc. Publs U.S. Dept. Agric.* **797**: 1–797 pp.   (1961) **Blake, S. F.**

Index to European taxonomic literature for 1965. *Regnum Vegetabile* **45**: 1–166 (1970); *Regnum Vegetabile* **70**: 1–189 (for 1968). (1966–70) **Brummitt, R. K.**

Index to European taxonomic literature for 1966. *Regnum Vegetabile* **53**: 1–245 (1968); *Regnum Vegetabile* **61**: 1–202 (1969) (for 1967).

(1968–69) **Brummitt, R. K. and Ferguson, I. K.**

Guide to the standard floras of the world. Cambridge University Press, Cambridge. xx + 619 pp.   (1984) **Frodin, D. G.**

Progress in the study of the British flora since World War 2. *Webbia* **18**: 129–50.

(1963) **Kent, D. H.**

Progress in the study of the British flora, 1961–71. *Mem. Soc. Broteriana* **24**: 353–75.   (1975) **Kent, D. H.**

Index to European taxonomic literature for 1969. *Regnum Vegetabile* **80**: 1–160.

(1971) **Kent, D. H., Kovanda, M., and Brummitt, R. K.**

A bibliographic index of th British flora. Bournemouth. xix + 429 pp.

(1960) **Simpson, N. D.**

History

The history of the British flora, 2nd edn. Cambridge University Press, Cambridge. x + 541 pp.   (1975) **Godwin, H.**

The history of British vegetation, 2nd edn. English Universities Press, London. viii + 152 pp.   (1974) **Pennington, W.**

Glossary

Glossary of the British flora, 3rd edn. Cambridge University Press, Cambridge. xxiv + 96 pp.   (1964) **Gilbert-Carter, H.**

A glossary of botanic terms with their derivation and accent, 4th edn. London. x + 481 pp. Reprinted.   (1949) **Jackson, B. D.**

Botanical Latin, 2nd edn. David and Charles, Newton Abbot. xiv + 566 pp.

(1973) **Stearn, W. T.**

Vocabulary pp. 377–548.

Anatomy and morphology

Water plants. A study of aquatic angiosperms. Cambridge University Press, Cambridge. xvi + 436 pp, illustrated.   (1920) **Arber, A**

Monocotyledons. Cambridge University Press, Cambridge. xvi + 258 pp, illustrated.   (1925) **Arber, A.**

The natural philosophy of plant form. Cambridge University Press, Cambridge. xiv + 247 pp.   (1950) **Arber, A.**

Morphology of vascular plants. Macmillan, London and New York. xii + 560 pp, illustrated.   (1971) **Bierhorst, D. W.**

Morphology of vascular plants, 3rd edn. Harper and Row, New York. xv + 668 pp, illustrated.   (1973) **Bold, H.**

British timbers, their properties, uses and identification, 2nd edn. Black, London.
                                  (1946) **Boulton, E. H. B. and Jay, B. A.**

Comparative plant anatomy. Holt, Rinehart, and Winston, New York. xii + 146 pp.   (1961) **Carlquist, S.**

Juncales. *Anatomy of the monocotyledons.* Clarendon Press, Oxford. Vol. 4, 357 pp.
                                                    (1969) **Cutler, D. F.**

British hardwoods, their structure and identification. HMSO, London. *Forest Products Res. Bull.* **3**: vi + 53 pp.   (1929) **Chalk, L. and Rendle, B. J.**

Morphology of the angiosperms. McGraw-Hill, New York. ix + 518 pp, illustrated.
                                                    (1961) **Eames, A. J.**

Plant anatomy, 2nd edn. Pergamon Press, Oxford and New York. viii + 611 pp.
                                                        (1974) **Fahn, A.**

Holzanatomie der europäischen Laubhölzer und Sträucher, 2nd edn. Budapest.
                                                    (1959) **Greguss, P.**

Lebensgeschichte der Blütenpflanzen Mitteleuropas. E. Ulmer, Stuttgart. 4 vols.
                  (1904–38) **Kirchner, O. von, Loew, E., and Schröter, C.**

Gramineae. *Anatomy of the Monocotyledons.* Clarendon Press, Oxford. Vol. 1, lxii + 731 pp.   (1960) **Metcalfe, C. R.**

Cyperaceae. *Anatomy of the monocotyledons.* Clarendon Press, Oxford. Vol. 5, 597 pp.   (1971) **Metcalfe, C. R.**

Anatomy of the dicotyledons. Clarendon Press, Oxford. 2 vols.
                                      (1950) **Metcalfe, C. R. and Chalk, L.**

Textbook of wood technology. 1. Structure, identification, uses and properties of the commercial woods of the United States, 2nd edn. McGraw-Hill, New York. xiv + 643 pp.   (1964) **Panshn, A. J., Zeeuw, C. de, and Brown, H. P.**

Systematic anatomy of the dicotyledons. Clarendon Press, Oxford. 2 vols. (Translated into English by L. A. Brodle and G. E. Fritsch; revised by D. H. Scott).   (1908) **Solereder, H.**

Systematische Anatomie der Monokotyledonen. Berlin. Vols 1, 3, 4, and 6. (Never completed.)   (1928–33) **Solereder, H. and Meyer, F. J.**

Pollen and spore identification

An atlas of pollen of the trees and shrubs of Eastern Canada and the adjacent United States. University of Waterloo, Waterloo, Ontario. Parts 1–4 University of Waterloo Biology Series nos. 8–11.   (1972–79) **Adams, R. J. and Morton, J. K.** Covers many species widely cultivated in north-west Europe.

Leitfaden der Pollenbestimmung für Mitteleuropa und angrenzende Gebiete. G. Fischer, Stuttgart. xiv + 63 pp, 8 pls.   (1961) **Beug, H. J.**

An atlas of recent European moss spores. Akademiai Kiado, Budapest. 466 pp, 237 pls.   (1975) **Boros, Á. and Járai-Komlódi, M.**

Atlante dei principali pollini allergenici presenti in Italia. Università di Siena, Siena. 190 pp, illustrated.   (1981) **Ciampolini, F. and Cresti, M.**

An introduction to pollen analysis. Chronica Botanica, Waltham, Massachusetts. xv + 239 pp, illustrated. Revised printing.   (1954) **Erdtman, G.**

Pollen and spore morphology/plant taxonomy. Gymnospermae, Pteridophyta, Bryophyta. An introduction to palynology. Almqvist and Wiksell, Stockholm. Vols 2–3.
(1957–65) **Erdtman, G.**

Handbook of palynology. Morphology–taxonomy–ecology. An introduction to the study of pollen grains and spores. Munksgaard, Copenhagen. 486 pp, illustrated.
(1969) **Erdtman, G.**

Pollen morphology and plant taxonomy. An introduction to palynology, new edn. Hafner, New York and London. Vol. 1, xii + 553 pp, illustrated.
(1971) **Erdtman, G.**

An introduction to a Scandinavian pollen flora. Almqvist and Wiksell, Stockholm. 2 vols.   (1961–63) **Erdtman, G., Berglund, B., and Praglowski, J.**

Pollen and spore morphology/plant taxonomy, Pteridophyta (Text and additional illustration). An introduction to palynology. Almqvist and Wiksell, Stockholm. Vol. 4, 302 pp.   (1971) **Erdtman, G. and Sorsa, P.**

Textbook of pollen analysis, 3rd edn (by K. Faegri). Munksgaard, Copenhagen. 295 pp, illustrated.   (1975) **Faegri, K. and Iversen, J.**

The northwest European pollen flora. *Rev. Palaeobot. Palynology* **17**(3/4) on: *Suppls.*   (1974 on) **Jansen, C. R.** *et al.*

How to know pollen and spores. Wm. C. Brown Co, Dubuque, Iowa. 249 pp, illustrated.   (1969) **Kapp, R. O.**

The spores of the Pteridophytes. Hirokawa Publishing Company, Tokyo. 398 pp, illustrated.   (1972) **Kremp, G. O. W. and Kawasaki, T.**

An illustrated guide to pollen analysis. Hodder and Stoughton, London, Sydney, Auckland, and Toronto. viii + 133 pp, 48 pls.
(1978) **Moore, P. D. and Webb, J. A.**

World pollen and spore flora. Almqvist and Wiksell, Stockholm. Parts 1–14 Supplement to *Grana.* (1973 on) **Nilsson, S.**, ed.
Twelve angiosperm families treated.

Atlas of airborne pollen grains and spores in Northern Europe. Natur och Kultur, Stockholm. 159 pp, illustrated.
(1977) **Nilsson, S., Praglowski, J., and Nilsson, L.**

The northwest European pollen flora. Elsevier Scientific Publication, Amsterdam, Oxford, New York, and Tokyo. Includes:
Vol. 1, vi + 146 pp. (1976) **Punt, W.**, ed.
Vol. 2 (1980) viii + 266 pp; Vol. 3 (1981) vi + 138 pp; Vol. 4 (1984) ii + 370 pp. (1980, 1981, 1984) **Punt, W. and Clarke, G. C. S.**, eds.

BACTERIA

For the Cyanobacteria (Cyanophyta) see p. 279.

General

Bergey's manual of determinative bacteriology, 8th edn. Williams and Wilkins, Baltimore. xxvi + 1246 pp.
(1974) **Buchanan, R. E. and Gibbons, N. E.**, eds.

Bergey's manual of systematic bacteriology. Williams and Wilkins, Baltimore and London. Vol. 1, xxvii + 964 pp. (1984) **Kreig, N. R. and Holt, J. G.**, eds.

A guide to the identification of the genera of Bacteria, 2nd edn. Williams and Wilkins, Baltimore. xii + 303 pp. (1967) **Skerman, V. B. D.**

ACTINOMYCETES

Coryneform bacteria. Academic Press, London. 315 pp.
(1978) **Bousfield, I. J. and Callely, A. G.**, eds.

Taxonomy and classification of the Actinomycetes. In *Actinomycetales: characteristics and practical importance* (ed. G. Sykes and F. A. Skinner), pp. 11–12. Academic Press, London and New York.
(1973) **Cross, T. and Goodfellow, M.**

Classification. In *The biology of Actinomycetes* (ed. M. Goodfellow, M. Mordarski, and S. T. Williams), pp. 7–164. Academic Press, London.
(1984) **Goodfellow, M. and Cross, T.**

The biology of the actinomycetes. Academic Press, London. 544 pp.
(1984) **Goodfellow, M., Mordarski, M., and Williams, S. T.**, eds.

Biochemical, biological, and biomedical aspects of actinomycetes. Academic Press, Orlando, Florida. 643 pp.
(1984) **Ortiz-Ortiz, L., Bojalil, L. F., and Yakoleff, V.**, eds.

Medically important bacteria

Cowan and Steel's manual for the identification of medical bacteria. Cambridge University Press, London. xii + 238 pp. (1974) **Cowan, S. T.**

Plant-pathogenic bacteria

Isolation and preliminary study of bacteria from plants. *Rev. Pl. Path.* **49**: 213–18. (1970) **Bradbury, J. F.**

A guide to plant pathogenic bacteria. Commonwealth Mycological Institute, Kew. (In press). (1987) **Bradbury, J. F.**

Descriptions of pathogenic fungi and bacteria. Commonwealth Mycological Institute, Kew. Sets 2, 5, 13, 24, and 38. (1964 on)

Plant bacterial diseases. A diagnostic guide. Academic Press, Sydney. xvii + 393 pp.
(1983) **Fahy, P. C. and Persley, G. J.**, eds.

Manual of bacterial plant pathogens. Chronica Botanica, Waltham, Massachusetts. 186 pp. (1951) **Elliott, C.**

Methods for diagnosis of bacterial diseases of plants. Blackwell Scientific, Oxford. (In press). (1987) **Lelliott, R. A. and Stead, D. E.**

Laboratory guide for identification of plant pathogenic bacteria. American Phytopathological Society, St Paul, Minnesota. 72 pp.
(1980) **Schaad, N. W.**, ed.

VIRUSES

The viruses. Biochemical, biological and biophysical 'properties'. Academic Press, New York and London. 2 vols. (1959) **Burnet, F. M. and Stanley, W. M.**, eds.

Descriptive catalogue of viruses. *Comprehensive virology.* Plenum Press, New York. Vol. 1, x + 191 pp. (1974) **Frankel-Conrat, H.**

The viruses. Plenum Press, New York. 7 vols.
(1985 on) **Fraenkel-Conrat, H. and Wagner, R. R.** eds.
Animal viruses only, volumes on plant viruses *in prep.*

Classification and nomenclature of viruses. *Fourth Report of the International Committee on Taxonomy of Viruses.* Karger, Basle. 199 pp.
(1982) **Mathews, R. E. F.**

Animal viruses

Viruses of vertebrates, 4th edn. Baillière Tindall, London. 421 pp.
(1978) **Andrewes, C. H., Pereira, H. G., and Wildy, P.**

Plant viruses

An atlas of plant viruses. CRC Press, Buca Raton Florida. Vol. 1, 222 pp; Vol. 2,
284 pp.   (1985) **Francki, R. I. B., Milne, R. G., and Hatta, T.**

Handbook of plant virus infections. Comparative diagnosis. Elsevier/North-Holland,
Amsterdam. iv + 944 pp.   (1981) **Kurstak, E.**, ed.

Plant virus names. *Phytopath. Papers* **9**: i–ix + 1–204 and *Suppl.* **1**: 1–41 (1971).
(1968) **Martyn, E. B.**, ed.

Classification and nomenclature of viruses. *Fourth Report of the International
Committee on Taxonomy of Viruses*. Karger, Basle. 199 pp.
(1982) **Matthews, R. E. F.**
Reprinted from *Intervirology* **17**(1–3): 1982.

CMI/AAB descriptions of plant viruses. Commonwealth Mycological Institute
and Association of Applied Biologists, Kew. Nos. 1–295.
(1970–84) **Murant, A. F. and Harrison, B. D.**, eds.

AAB descriptions of plant viruses. Association of Applied Biologists, Wel-
lesbourne. Nos. 296 onwards.
(1985 on) **Murant, A. F. and Harrison, B. D.**, eds.
Standardized authoritative descriptions of upwards of 300 distinct plant viruses,
each prepared by an expert; a continuation of the preceding title.

A textbook of plant virus diseases, 3rd edn. Longman, London. x + 684 pp.
(1972) **Smith, K. M.**
Information on most viruses recorded to 1972.

Pflanzliche Virologie, 3rd edn. Akademie-Verlag, Berlin. Vol. 2, Die Virosen an
landwirtschaftlichen Kulturen, Sonderkulturen, und Sporenpflanzen in
Europa. x + 434 pp; Vol. 3, Die Virosen an Gemüsepflanzen, Obstgewach-
sen und Weinreben in Europa. viii + 389 pp; Vol. 4, Die Virosen an
Zierpflanzen, Geholzen und Wildpflanzen in Europa. viii + 528 pp.
(1977) **Klinkowski, M.**

Systematics Association Publications

1. Bibliography of key works for the identification of the British fauna and flora *3rd edition* (1967)
   Edited by G. J. Kerrich, R. D. Meikle and N. Tebble
2. Function and taxonomic importance (1959)
   Edited by A. J. Cain
3. The species concept in palaeontology (1956)
   Edited by P. C. Sylvester-Bradley
4. Taxonomy and geography (1962)
   Edited by D. Nichols
5. Speciation in the sea (1963)
   Edited by J. P. Harding and N. Tebble
6. Phenetic and phylogenetic classification (1964)
   Edited by V. H. Heywood and J. McNeil
7. Aspects of Tethyan biogeography (1967)
   Edited by C. G. Adams and D. V. Ager
8. The soil ecosystem (1969)
   Edited by H. Sheals
9. Organisms and continents through time (1973)†
   Edited by N. F. Hughes

Published by the Association (out of print)

Systematics Association Special Volumes

1. The new systematics (1940)
   Edited by Julian Huxley (Reprinted 1971)
2. Chemotaxonomy and serotaxonomy (1968)*
   Edited by J. G. Hawkes
3. Data processing in biology and geology (1971)*
   Edited by J. L. Cutbill
4. Scanning electron microscopy (1971)*
   Edited by V. H. Heywood
   Out of print

5. Taxonomy and ecology (1973)*
   Edited by V. H. Heywood
6. The changing flora and fauna of Britain (1974)*
   Edited by D. L. Hawksworth
7. Biological identification with computers (1975)*
   Edited by R. J. Pankhurst
8. Lichenology: progress and problems (1976)*
   Edited by D. H. Brown, D. L. Hawksworth and R. H. Bailey
9. Key works to the fauna and flora of the British Isles and northwestern Europe (1978)*
   Edited by G. J. Kerrich, D. L. Hawksworth and R. W. Sims
10. Modern approaches to the taxonomy of red and brown algae (1978)*
    Edited by D. E. G. Irvine and J. H. Price
11. Biology and systematics of colonial organisms (1979)*
    Edited by G. Larwood and B. R. Rosen
12. The origin of major invertebrate groups (1979)*
    Edited by M. R. House
13. Advances in Bryozoology (1979)*
    Edited by G. P. Larwood and M. B. Abbot
14. Bryophyte systematics (1979)*
    Edited by G. C. S. Clarke and J. G. Duckett
15. The terrestrial environment and the origin of land vertebrates (1980)*
    Edited by A. L. Panchen
16. Chemosystematics: principles and practice (1980)*
    Edited by F. A. Bisby, J. G. Vaughan and C. A. Wright
17. The shore environment: methods and ecosystems (2 Volumes) (1980)*
    Edited by J. H. Price, D. E. G. Irvine and W. F. Farnham
18. The Ammonoidea (1981)*
    Edited by M. R. House and J. R. Senior
19. Biosystematics of social insects (1981)*
    Edited by P. E. Howse and J.-L. Clément
20. Genome evolution (1982)*
    Edited by G. A. Dover and R. B. Flavell
21. Problems of phylogenetic reconstruction (1982)*
    Edited by K. A. Joysey and A. E. Friday
22. Concepts in nematode systematics (1983)*
    Edited by A. R. Stone, H. M. Platt and L. F. Khalil
23. Evolution, time and space: the emergence of the biosphere (1983)*
    Edited by R. W. Sims, J. H. Price and P. E. S. Whalley
24. Protein polymorphism: adaptive and taxonomic significance (1983)*
    Edited by G. S. Oxford and D. Rollinson
25. Current concepts in plant taxonomy (1983)*
    Edited by V. H. Heywood and D. M. Moore
26. Databases in systematics (1984)*
    Edited by R. Allkin and F. A. Bisby

27. Systematics of the green algae (1984)*
     Edited by D. E. G. Irvine and D. M. John
28. The origins and relationships of lower invertebrates (1985)‡
     Edited by S. Conway Morris, J. D. George, R. Gibson, and H. M. Platt
29. Infraspecific classification of wild and cultivated plants (1986)‡
     Edited by B. T. Styles
30. Biomineralization in lower plants and animals (1986)‡
     Edited by B. S. C. Leadbeater and R. Riding
31. Systematic and taxonomic approaches in palaeobotany (1986)‡
     Edited by R. A. Spicer and B. A. Thomas
32. Coevolution and systematics (1986)‡
     Edited by A. R. Stone and D. L. Hawksworth
33. Key works to the fauna and flora of the British Isles and northwestern
     Europe *5th edition* (1987)‡
     Edited by R. W. Sims, P. Freeman, and D. L. Hawksworth

*Published by Academic Press for the Systematics Association
†Published by the Palaeontological Association in conjunction with
the Systematics Association
‡Published by the Oxford University Press
for the Systematics Association